T0249559

Ecosystem Services

Ecosystem Services
Global Issues, Local Practices

Edited by

Sander Jacobs
Research Institute for Nature and Forest (INBO);
University of Antwerp. Department of Biology,
Ecosystem Management Research Group (ECOBE)

Nicolas Dendoncker
Department of Geography, University of Namur (UNamur).
Namur Research Centre on Sustainable Development (NAGRIDD).
Namur Centre for Complex Systems (naXys)

Hans Keune
Belgian Biodiversity Platform;
Research Institute for Nature and Forest (INBO);
Faculty of Applied Economics – University of Antwerp;
naXys, Namur Center for Complex Systems – University of Namur

AMSTERDAM • BOSTON • HEIDELBERG • LONDON
NEW YORK • OXFORD • PARIS • SAN DIEGO
SAN FRANCISCO • SINGAPORE • SYDNEY • TOKYO

ELSEVIER

Elsevier
525 B Street, Suite 1900, San Diego, CA 92101-4495, USA
225 Wyman Street, Waltham, MA 02451, USA

First edition 2014

Notice
No responsibility is assumed by the publisher for any injury and/or damage to persons
or property as a matter of products liability, negligence or otherwise, or from any use or
operation of any methods, products, instructions or ideas contained in the material herein.
Because of rapid advances in the medical sciences, in particular, independent verification
of diagnoses and drug dosages should be made.

Library of Congress Cataloging-in-Publication Data
A catalog record for this book is available from the Library of Congress

British Library Cataloguing in Publication Data
A catalogue record for this book is available from the British Library

For information on all Elsevier publications
visit our web site at store.elsevier.com

ISBN: 978-0-12-419964-4

Working together
to grow libraries in
developing countries

www.elsevier.com • www.bookaid.org

Contents

Part I

Ecosystem Service Basics

1. Inclusive Ecosystem Services Valuation

Nicolas Dendoncker, Hans Keune, Sander Jacobs and
Erik Gómez-Baggethun

2. Ecosystem Services and Their Monetary Value

Inge Liekens, Leo De Nocker, Steven Broekx,
Joris Aertsens and Anil Markandya

3. Biodiversity and Ecosystem Services

Sander Jacobs, Birgen Haest, Tom de Bie, Glenn Deliège,
Anik Schneiders and Francis Turkelboom

v

4. Ecosystem Service Indicators: Are We Measuring What We Want to Manage?

Wouter Van Reeth

5. Inquiring into the Governance of Ecosystem Services: An Introduction

Hans Keune, Tom Bauler and Heidi Wittmer

Part II
Ecosystem Services: Conceptual Reflections

6. Monetary Valuation of Ecosystem Services: Unresolvable Problems with the Standard Economic Model

John Gowdy and Philippe C. Baveye

7. Biodiversity and Ecosystem Services: Opposed Visions, Opposed Paradigms

Martin Sharman

8. Earth System Services—A Global Science Perspective on Ecosystem Services

Sarah Cornell

Part IV
Ecosystem Services: Tools & Practices

Part V

Ecosystem Service Reflections from Practice

Ecosystem services and their largely invisible values, so vital to recognize for sustainability, have at last begun to percolate into policies. *The Economics of Ecosystems and Biodiversity* (TEEB) reports have added much-needed awareness and societal debate about how much our well-being, economy, and even survival depend on biodiversity and ecosystems. At a local level, better informed practices are being implemented in a variety of socioeconomic and ecological contexts, while, internationally, institutions such as the Intergovernmental Science-Policy Platform on Biodiversity and Ecosystem Services (IPBES) have emerged.

However, our world is still rapidly approaching and even crossing planetary boundaries, including climate, biodiversity, nitrogen and phosphorous concentrations, ocean acidification and freshwater scarcity. Economies worldwide are still headed in the wrong direction, leading to resource exhaustion, social disparities, and persistent poverty. Increasing climate disruptions may cause price volatility and loss of arable land, and the poor will suffer the most from these disruptions. Urgent changes are needed to effect sustainable resource use.

This book originated from a lively community of practice on ecosystem services in a highly urbanized European region, representative for many developed countries. Practitioners, nongovernmental organizations (NGOs), policy makers at different levels, and scientists from many disciplines have united to implement ecosystem service approaches. This book—a product of this community of practice—is a "proof of concept" of a transdisciplinary approach involving a broad range of stakeholders to improve and link up knowledge and practice internationally.

From their experiences, a strong and clear plea emerges to reorient ecosystem service research and practice, bringing it back to its "sustainability" roots: to account for boundaries and fairness more than with just lip service or introductory texts; to document, communicate, and cope with uncertainties; to adopt an inclusive and transparent approach; and to evaluate the real impacts of various measures and instruments on materials flows.

The debates presented here are fundamentally important and have repercussions for any ecosystem service research or practice. As we need immediate changes if we want to steer clear of planetary boundaries and avoid large natural disasters, we have to share experiences, knowledge, and debates widely, within and across communities of practice, locally and globally.

The editors rightly argue that ecosystem service research and practice should urgently aim at a more limited and fairer resource use supported by transdisciplinary approaches and with real-life results. Their diverse contributions offer many practical lessons and tools to address the many challenges across a broad range of issues in assessing and managing ecosystem services.

In short, this book is a "must-read" for academicians conducting interdisciplinary ecosystem service research, practitioners and policy makers aiming to incorporate ecosystem services into their work, and students from the natural and social sciences. If ecosystem service practice is to live up to high expectations as well as urgent requirements to deliver sustainable resource use, the principles put forward in this book will have to be fully embraced.

Pavan Sukhdev

Pavan was Special Adviser and Head of UNEP's Green Economy Initiative, and lead author of their report "Towards a Green Economy." He was also appointed Study Leader for the G8+5 commissioned project on The Economics of Ecosystems and Biodiversity (TEEB).

Alain Peeters   RHEA Research Centre rue Warichet, Natural Resources, Human Environment and Agronomy (RHEA)

Anik Schneiders   Research Institute for Nature and Forest (INBO)

Anil Markandya   University of Bath, Department of Economics and International Development, United Kingdom and Scientific Director of Basque Center of Climate Change

Anne Teller   DG Environment, European Commission

Anne-hélène Prieur-Richard   DIVERSITAS, Reviewed by Francis Turkelboom (INBO); Belgian Biodiversity Platform; Research Institute for Nature and Forest (INBO); Faculty of Applied Economics – University of Antwerp; naXys, Namur Center for Complex Systems – University of Namur

Birgen Haest   The Flemish Institute for Technological Research (VITO)

Cédric Chevalier   Political advisor, Cabinet of the Walloon and the Wallonia-Brussels Federation Minister of Sustainable Development and Research

Conor Kretsch   COHAB

Corentin Fontaine   University of Namur (UNamur)

Dirk Van Gijseghem   Flemish Ministry of Agriculture, Division for Agricultural Policy Analysis

Dirk Vrebos   University of Antwerp, Ecosystem Management Research Group (ECOBE)

Dries Landuyt   University of Ghent, Laboratory of Environmental Toxicology and Aquatic Ecology-AECO; Flemish Institute for Technological Research (VITO)

Erik Gómez-Baggethun   Institute of Environmental Science and Technology, Universitat Autònoma de Barcelona; Social-Ecological Systems Laboratory, Department of Ecology, Universidad Autónoma de Madrid

Ferdinando Villa   Basque Centre for Climate Change (BC3); IKERBASQUE, Basque Foundation for Science

Francis Turkelboom   Research Institute for Nature and Forest (INBO)

Frederic Ghys   Sustainability analyst

Frédéric Huybrechs   Institute of Development Policy and Management (IOB), University of Antwerp

Gert Van Hecken Institute of Development Policy and Management (IOB), University of Antwerp

Glenn Deliège KU Leuven, Husserl-Archives, International Centre for Phenomenological Research

Guy Duke Environment Bank Ltd

Hans Keune Belgian Biodiversity Platform; Research Institute for Nature and Forest (INBO); Faculty of Applied Economics – University of Antwerp; naXys, Namur Center for Complex Systems – University of Namur

Heidi Wittmer Helmholtz Centre for Environmental Research—UFZ Division of Social Sciences, Department of Environmental Politics

Hilde Heyrman VLM

Ilse Simoens Research Institute for Nature and Forest (INBO)

Inge Liekens Flemish Research and Technology Organisation (VITO)

Jan Staes University of Antwerp, Ecosystem Management Research Group (ECOBE)

Jan Verboven Vlaamse Landmaatschappij regio West

Jeroen A.E. Panis Agency for Nature and Forests, Government of Flanders

Jim Casaer Research Institute for Nature and Forest (INBO)

Joachim H. Spangenberg Helmholtz Centre for Environmental Research—UFZ, Department Community Ecology

Johan Bastiaensen Institute of Development Policy and Management (IOB), University of Antwerp

John Gowdy Rensselaer Polytechnic Institute, Troy, New York

Joris Aertsens Flemish Research and Technology Organisation (VITO)

Jos Brils Senior adviser with Deltares in the area of sustainable management of natural resources; Co-Coordinator of the Dutch Community of Practice (CoP) on Ecosystem Services

Katrien Van der Biest University of Antwerp, Ecosystem Management Research Group (ECOBE)

Kelly Hertenweg DG Environment of Federal Public Service Health, Food Chain Safety and Environment

Kris Struyf Zwin Nature Centre, KUDDKE-HEIST

Kris Verheyen Forest and Nature Lab, Ghent University

Layla Saad Faun and Biotopes asbl

Leander Raes UG

Leen Franchois Research Department, Boerenbond

Leen Gorissen Unit Transition Energy and Environment, The Flemish Institute for Technological Research (VITO)

Leo De Nocker   Flemish Research and Technology Organisation (VITO)

Leo Declercq   Zwin Nature Centre, KUDDKE-HEIST

Leon C. Braat   Alterra, Wageningen University and Research, Wageningen

Lieve Janssens   Province of Antwerp, Department of Environment, Nature & Landscape Team

Linda Meiresonne   Research Institute for Nature and Forest (INBO)

Lucette Flandroy   DG Environment of Federal Public Service Health, Food Chain Safety and Environment

Maarten Stevens   Research Institute for Nature and Forest (INBO)

Machteld Gryseels   Brussels Environment, Direction Quality of the Environment and Nature Management Gulledelle

Marc Dufrêne   ULG-GxABT

Marije Schaafsma   University of East Anglia, Centre for Social and Economic Research on the Global Environment, England, United Kingdom

Marijke Thoonen   Research Institute for Nature and Forest (INBO)

Martin Hermy   KULeuven

Martin Sharman   Independent expert, Avenue des Orangers

Nathalie Pipart   Université Libre de Bruxelles, Institut de Gestion de l'Environnement et d'Aménagement du Territoire, Centre d'Etudes du Développement Durable

Nele Smeets   Flemish Research and Technology Organisation (VITO)

Nicolas Dendoncker   Department of Geography, University of Namur (UNamur). Namur Research Centre on Sustainable Development (NAGRIDD). Namur Centre for Complex Systems (naXys)

Olivier Beauchard   University of Antwerp, Ecosystem Management Research Group (ECOBE)

Patrick Meire   University of Antwerp, Ecosystem Management Research Group (ECOBE)

Paula Ulenaers   Vlaamse Landmaatschappij regio West

Perrine Raquez   University of Namur (UNamur)

Peter Goethals   University of Ghent, Laboratory of Environmental Toxicology and Aquatic Ecology-AECO

Philippe C. Baveye   Rensselaer Polytechnic Institute, Troy, New York

Pieter Vangansbeke   Unit Transition Energy and Environment, The Flemish Institute for Technological Research (VITO); Forest and Nature Lab, Ghent University

Pim Martens   ICIS, Maastricht University

Rik De Vreese   VUB

Rob D'Hondt   University of Ghent, Laboratory of Environmental Toxicology and Aquatic Ecology-AECO

Sabine Wallens   DG Environment of Federal Public Service Health, Food Chain Safety and Environment

Sander Jacobs   Research Institute for Nature and Forest (INBO); University of Antwerp. Department of Biology, Ecosystem Management Research Group (ECOBE)

Sarah Cornell   Stockholm Resilience Centre

Saskia Van Gaever   DG Environment of Federal Public Service Health, Food Chain Safety and Environment

Simon W. Moolenaar   Programme Manager SKB, Chair of the SNOWMAN Network, Strategic consultant Environment & Sustainability at Royal Haskoning DHV

Steven Broekx   Flemish Research and Technology Organisation (VITO)

Sylvie Danckaert   Flemish Ministry of Agriculture, Division for Agricultural Policy Analysis

Tanya Cerulus   Environment, Nature and Energy Department, Environmental, Nature and Energy Policy Division, Government of Flanders

Tom Bauler   Université Libre de Bruxelles, Institut de Gestion de l'Environnement et d'Aménagement du Territoire, Centre d'Etudes du Développement Durable

Tom de Bie   KU Leuven, Laboratory of Aquatic Ecology, Evolution and Conservation; Flemish Environment Agency (VMM), Catchment office Dijle-Zenne catchment

Wim Van Gils   Policy Director, Natuurpunt

Wouter Van Reeth   Research Institute for Nature and Forest (INBO)

Editorial for *Ecosystem Services — Global Issues, Local Practices*

No Root, No Fruit—Sustainability and Ecosystem Services

Sander Jacobs[1,2], Nicolas Dendoncker[3], and Hans Keune[4,5,6,7]

[1]*Research Institute for Nature and Forest (INBO),* [2]*University of Antwerp. Department of Biology, Ecosystem Management Research Group (ECOBE),* [3]*Department of Geography, University of Namur (UNamur). Namur Research Centre on Sustainable Development (NAGRIDD). Namur Centre for Complex Systems (naXys),* [4]*Belgian Biodiversity Platform,* [5]*Research Institute for Nature and Forest (INBO),* [6]*Faculty of Applied Economics – University of Antwerp,* [7]*naXys, Namur Center for Complex Systems – University of Namur*

The human species survives and prospers within well-defined planetary conditions. The concept of planetary boundaries [1] delimits this safe operating space with respect to the functioning of the Earth System. Since the 20th century, the Earth has entered a new epoch, the Anthropocene, where humankind constitutes the dominant driver of change to the Earth System as a global geological force in its own right [2], shifting the conditions beyond its own safe operating space ([3]; see also Chapter 7).

Among several planetary boundaries being crossed today [1], ever-accelerating biodiversity loss is particularly serious, given the vital importance of biodiversity for sustaining ecosystem functioning and preventing ecosystems from shifting into undesired states [4]. Biodiversity is essential for the Earth's functioning and our basic survival and well-being, which is not entirely correlated to consumption or monetary income [5], but relates to nature, social relationships, knowledge, and politics [6]. The societal implications of biodiversity loss were pointed out in the millennium ecosystem assessment, noting a 60% loss of ecosystem services in the last four decades of the 20th century alone [7, 8]. Globally, it is the poor who are facing the earliest and most severe impacts of this loss, but ultimately all societies and communities will suffer [9].

Sustainable development, defined as development that meets the needs of the present without compromising the ability of future generations to meet their own needs, by extending to all the opportunity to fulfill their aspirations for a better life [10], is the ultimate strategy to reverse this drift to human self-extinction (see also [11, 12]). The research field and concept of biodiversity, natural capital, and ecosystem services (ES) are indeed rooted in sustainability thinking (e.g., [13, 14, 15, 16, 17]). The explicit link between sustainability and ES assessments [18] stresses the importance of all three values of ES: ecological sustainability, social fairness, and economic efficiency. Conclusively, the final goal of ES valuation is to achieve a more sustainable resource use, contributing to the well-being of every individual, now and in the future [8] by providing an equitable, adequate, and reliable flow of essential ecosystem services to meet the needs of a burgeoning world population [19].

Until now, there has been a reluctance to fully embrace the message that by ignoring the dependence on our natural capital we are literally living at the expense of the poor and the future generations [20]. Still, the ecosystem service (ES) concept could be an effective lever to contribute to sustainable development with more than just lip service (Chapter 10). The ES concept has been picked up widely, percolated in many policy documents, and is being implemented in a variety of contexts. As the time left to effectively tackle sustainability challenges is running out, urgent refocusing of ES research and more importantly practice on its strong sustainability roots is essential. This conclusion directly arises from the methodological and conceptual challenges for ecosystem service valuations in this volume, echoes in many reflections from practice, and mirrors current scientific opinions on the topic (e.g., [19, 21, 22, 23] and others mentioned in this editorial).

There are no silver bullet solutions to be found in this volume. This is not surprising, as literature shows that "one size fits all" solutions are to be avoided [20] and complex problems will require diverse solutions [24]. Reading this volume will probably raise more questions, as awareness grows on the multitude of informed choices to be made on ever morecomplex matters. Once one is aware of this complexity, there seems to be no way back to easy solutions comfortably embedded within a single research or policy field. However, some key issues repeatedly and consistently surface throughout the variety of contributions, and they provide the focus and common ground for continued but innovative ecosystem service research and practice. Notwithstanding the diversity of insights found throughout this book, the key issues represent shared concerns among scientists, policy makers, and practitioners, and provide a frame of reference for practical implementation.

1. LIMITS

Limits (with related terms such as boundaries, resilience, carrying capacity, tipping points, and thresholds) refer to the ecological sustainability value of

Costanza and Folke [18]: the fact that ecosystem services, however efficiently used, depend on a limited amount of natural capital. Biophysical renewal of this capital occurs at a certain rate and involves complex and incompletely known ecological processes. Depletion by excessive use rates or negative effects on the supporting ecosystem will decrease the well-being of (future) beneficiaries. This relates to the concepts of the carrying capacity of nature [25], the planetary boundaries [1], and resilience [19]. Although this notion is intuitively evident and widely recognized throughout literature as a main challenge for implementation of the ES concept, it is hardly implemented in current practice. Many policy-related research papers and documents—beyond their introductions—almost exclusively emphasize efficiency (e.g., [26]) without questioning the indispensable (re)definition of the growth paradigm ([27]; Chapter 8) to stay within our ecological limits by defining boundaries to resource use.

Also in this volume, it is recognized that cost-benefit approaches can place underlying ecological assets at risk by overconcentration on changes in benefits (Chapter 2). Inclusion of underlying biodiversity for current and future service supply is still one of the main challenges in valuation (Chapter 11). This is why biodiversity management strategies should focus on maintaining and improving resilient systems, which increase long-term delivery of services (Chapter 3). Indicators could help in pointing out critical thresholds but rarely do so (Chapter 4; see also 20), while many economic indicators even implicitly but wrongly threaten the benefits of future generations through discounting rates (Chapters 6, 20). It stands out that ecological sustainability should be prominently included in the inclusive valuation framework (Chapter 1), despite the uncertainties and complexities involved (Chapters 14, 15). Deriving benefits from ecosystem services while respecting ecological boundaries requires shifts in current practices and often *painful choices* [10], as is illustrated for instance concerning agriculture (Chapter 22).

2. FAIRNESS

Although the first theme implicitly refers to intergenerational fairness and equity that can only be achieved by ensuring stable delivery of ES through time, sustainable equitable sharing also implies equity across regions and actors. Referred to variously as equity, solidarity, rights, common goods, benefit distribution, multiple values, option value, future values, and intangible values, it links to the social sustainability value of Costanza and Folke [18]. In the United Nations Rio declaration on environment and development from 1992 it is also known as the sustainable development principle, which states that *nations have the sovereign right to exploit their own resources, but without causing environmental damage beyond their borders*. Failure to respect this principle will (and already did) create environmental debts of some regions toward others (see, e.g., Chapter 1). International and global fairness are, however, often overlooked in the place-based ecosystem assessments, while global trade, for example, impacts highly on biodiversity and ES (Chapter 17). Sustainability has to *ensure that the poor get their fair share of the resources, and it requires that those who are more affluent adopt their lifestyles* [10]. Although a trade-off between global equity (fairness) and environmental goals (limits) is often presumed, recent analysis points out that significant synergies can be developed by adopting a unified approach [28], and ecosystem services approaches could be useful to address poverty alleviation [29]. Indeed, contributions from both science and practice point out that fairness and distribution issues among (local and global) individuals or groups of actors need to be better considered. Fairness should be at the core of valuation of ES. Moreover, monetary valuation cannot value everything (Chapters 1, 2), and when going beyond awareness raising, monetary valuation tools (e.g., Chapter19) should be used to complement and not substitute other legitimate reasoning to biodiversity conservation. This neatly responds to further critiques on strictly monetary approaches (Chapters 6, 9, 11) and the call for application of combined valuation approaches [20] within an integrated and multidimensional approach capturing the broader value of biodiversity and ecosystems (Chapter 3) and demonstrated in the human health chapter Chapter 16). The required inclusion of beneficiaries at the wider geographical scale (Chapter 11) is elaborated in the context of rural development (Chapter 21) and global trade (Chapter 17).

Therefore, a three-pillar valuation framework is proposed (Chapter 1), valuing for efficiency but in a democratic way, taking into account social fairness (i.e., distribution effects) and system boundaries. To help this process along, indicators capturing well-being and economic wealth in a more holistic way are needed (Chapter 4) and are to be used in deliberative methods (Chapter 15), which cope with stakeholder involvement and power aspects in ecosystem governance (Chapter 5; see also 20), to empowerment of regional policy makers and local stakeholders (Chapters 30, 41). Most practitioners point out that ethics and the need for a moral framework that integrates them are needed to implement ES (e.g., Chapters 31, 33, 35, 38).

3. COMPLEXITY

Ecosystem services research aims to analyze the relation between the natural environment and human society: the socioecological system. Socioecological systems are highly complex and poorly understood, as pointed out extensively in this book by referring to uncertainty, data gaps, risks, complexity, and the like. Generally, valuation applications disregard the uncertainty factor, and this is a real problem [20]. The gaps in our understanding of relating services to production of human well-being are serious problems for robust valuation of ecosystem services (Chapters 1, 4). From a critical complexity perspective, full and objective knowledge of complex issues is unattainable. Complex issues will always be characterized by uncertainties, unknowns, and ambiguity. Dealing with complex issues therefore requires choices, which by definition are normative, and by this we enter the domain of social debate (Chapter 13). Even the complexity of these issues is itself subject to negotiation (Chapter 15): The complexity to be taken into account and the approach for dealing with that complexity are part of context-specific negotiation among actors. Agreed-upon assumptions concerning the prevailing economic model, scaling aspects, and intangible values invoke decision risks to society and call for sound uncertainty quantification and honest communication (Chapter 14). Valuations should include a risk and uncertainty analysis to acknowledge the limitations of current knowledge (Chapter 11). Rather than focusing exclusively on minimizing uncertainties and ignoring complexity, herewith paralyzing or slowing down practice and discarding crucial issues, researchers and practitioners should adopt a more tolerant and pragmatic attitude toward complexity (Chapters 14, 15).

Concrete solutions to tackle complexity are sought in organizational aspects of ecosystem service assessment practices, such as analytical-deliberative approaches (Chapter 15). For instance, applying a consensus classification could decrease incompatibilities and promote comparability among assessments (Chapter 18). More importantly, inter- and transdisciplinarity are considered the only approaches to take into account nonuse values (Chapter 2), to perform a relative weighting of modeled supplied services in a bundle (Chapter 20), to realize an integrated valuation (Chapter 11), to reduce uncertainty and work inclusively (Chapter 1), to build understanding of socioeconomic systems (Chapter 11), to develop indicators (Chapter 4), and to increase policy uptake (Chapter 12). Conclusively, ecosystem services research and governance of complex socioecological systems have to incorporate the diversity of relevant stakeholders and issues (Chapter 5). This challenge is partly interdisciplinary: natural and social sciences have to be integrated; and partly transdisciplinary: apart from scientific knowledge, nonscientific knowledge, such as practical experience, is also relevant (see also [30, 31]). As research choices are partly normative, it is legitimate and reasonable to include stakeholders in the methodological decision-making process. Normative diversity and ambiguity can be reduced or at least can be made transparent by acknowledging moral and societal values and incorporating them in decision making (Chapter 1). This strategy can take into account complexity and organize critical mass (see also 19), although, simultaneously, pragmatic simplification and structuring of complexity is needed (Chapter 15).

4. REALITY CHECK

Ecosystem services, like biodiversity, is a mission-oriented concept (Chapter 3): Sustainable management of natural resources to increase human well-being has always been the final aim. But how does one check whether ES lives up to this expectation? Where best to judge the proof of concept? Throughout the book, contributors refer to "practice" to locate the concept's usefulness (Chapter 23). However, different, but partly overlapping, types of practices can be distinguished, such as scientific practice, discursive practice, policy practice, and on-the-ground practice. *Scientific* practices focus on methodological issues such as valuation approaches, evaluation and learning, performance indicators, and best practice methods. *Discursive* practices focus on meanings, for example, in public debate or in policy texts. *Policy* practices focus on governance and policy instruments. *On-the-ground* practices relate to concrete actions or developments, for example, in international trade, agriculture, biodiversity and nature management, land-use management, public health, financial business, and nature education.

With regard to practice in terms of knowledge production and policy making, the concept of ecosystem services risks remaining a paper concept without real-life implications. With regard to knowledge production, involvement of stakeholders should be aimed at ensuring the practicality of projects, as end users will have to use the results in their day-to-day practices (Chapter 27). This is essential to address obstacles for implementing ecosystem service concepts in practice. The other way around, theoretical reflections are crucial for real-life practice: Discussions in the book show how choices often remain implicit or not thoroughly discussed, substantially impacting the outcomes. Even though some practical tools already exist, practitioners feel that more research is needed on the concepts' definition, scope, tools, conditions of use, effects, and practice at political levels, in order to help adapt human activity, stop ecosystem and biodiversity degradation, and improve its status as a result (Chapter 31) A more integrated approach to environmental management seems to offer possibilities/avenues to free mainstream practice from blindly simplifying the difficulties at stake (Chapter 21).

Engaging in practice, however, presents challenges, as the position of both scientists (as objective technician) and decision makers (as taking an informed decision "for the benefit of all") is heavily challenged when implementing ES in practice (Chapter 14). Nevertheless, this seems the only proposed way to take the concept of ecosystem services where it belongs: real-life, sustainable practice.

A PRACTICAL GUIDE TO THE OUTER COMFORT ZONE

There is no clear roadmap for the challenge outlined here. Professionals have to navigate with only a few general common aims as a beacon. Rather than listing detailed methodological guidelines, which abound in the recent literature, we summarized the core ideas in a short checklist. Four points should be kept in mind and transparently addressed if ES research and practice are to contribute to a truly sustainable resource management.

The list presented here is not a guarantee of success. The four points surface as essential concerns throughout the contributions to this book, and failing to address even one of them could diminish the effectiveness of any ES approach.

The above-mentioned themes are strongly linked, as are the points in the checklist: Limits relate to fairness, as they consider intergenerational fairness by acknowledging risk of future decreases in service supplies; fairness considers the current distribution of benefits, as well as the transparency and legitimacy of complex research and decision-making processes; complexity again abounds in all three other themes, as does the need for critical reality checks. This said, different contexts of practice will pose specific challenges: For instance, pure awareness raising is a leap forward in a nonaware community, but in other contexts or later stages, focusing on it without on-the-ground realizations can be discarded as green washing. Thus, transparency on which points to prioritize throughout the process is required.

Point 1: Reduce Resource Use

Planetary boundaries are being crossed, risking a shift to atmospheric and climatologic conditions lethal for the human species. The resilience of many local systems is eroded, their thresholds are crossed, and future service supplies are jeopardized. Focusing research on the implementation of limits at different scales is essential. Because very few methods are available, the development of innovative multiple evidence-based approaches, expert judgments, and precautionary principles is needed. Aiming practice at reducing aggregated resource use is key, inasmuch as resource use efficiency, which is the prevailing focus, will not be sufficient.

Point 2: Redistribute Benefits Fairly

Sustainable development in a finite resource context inevitably requires redistribution to reduce poverty: We can no longer all have more. Focusing research on the assessment of values and benefits in a broad perspective is crucial, by representing people and groups that are not represented at all or hardly so: notably, indigenous people, third and fourth world populations, and so on. In practice, benefits have to shift to those global citizens who have the smallest share of benefits, at the cost of those who have an unequally large one.

Point 3: Accept and Acknowledge Complexity

Time has run out. We will have to act based on a limited understanding in a context too complex to completely grasp. Reducing uncertainty can no longer impede action. Acknowledging complexity and associated risks is essential to legitimize decisions, and this invokes transparent, multiple evidence-based approaches, including the values and beliefs of indigenous groups and multiple stakeholders. This means revising traditional science-policy-society interactions.

Point 4: Evaluate and Adapt Based on Reality Checks

Awareness does not necessarily produce behavioral change, neither personally nor institutionally. Planning and demonstration do not always result in the application of instruments, and implementation might produce unexpected effects. Critical evaluation of projects based on the physical reality of resource use reduction and the social reality of wel-being increase and equitable redistribution has to guide research and practice.

This short list represents the core of ES research and practice, if its aim is to deliver results beyond awareness-raising and value demonstration. ES—originally shaped as a partly strategic concept, linking biodiversity with human well-being and ethical considerations from an anthropocentric point of view—is used as a vehicle to mainstream the value of biodiversity within =prevailing economic decisions. Now that attention has been raised and the concept is on the verge of implementation at many different levels, more fundamental arguments have to move to the forefront to guide effective implementation in actually producing sustainable development and well-being. The role of intergovernmental organizations such as the Intergovernmental Panel on Biodiversity and Ecosystem Services (IPBES) in facilitating action research and establishing criteria of good practice will be essential. The role of independent NGOs and local communities of practice to instigate changes in the global legislative framework and the economic playfield will be vital. Ecosystem service research and practice requires innovative research, risky applications, new forms of cooperation, and critical self-evaluation.

The ecosystem service coin can now fall two ways. If the ES coin falls tails up, fundamental socioethical principles are ignored, focus remains on awareness and simplified valuation, and natural scientists are distracted into technical data-delivery for efficiency-only optimization assessments or green washing. Resource use effects continue to cross critical boundaries, macroclimatic and atmospheric conditions shift, and the human species dies out. If the ES coin falls heads up, the inescapable conclusions of ecosystem service, biodiversity, and sustainability science might be acknowledged throughout the economy. Resource use limitation and benefit redistribution are locally implemented in strong transdisciplinary cooperation and an economic model that takes into account the natural limits. Humankind might stand a chance to abound on this planet. It's time for the ES community to choose sides and revitalize the sustainability roots of current research and practice.

REFERENCES

1. Rockström, J., Steffen, W., and Noone, K. (2009, September). A safe operating space for humanity. Nature, 472–475.
2. Crutzen, P., 2002. Geology of mankind. Nature, 415, 23.
3. Steffen, W., Crutzen, J., and McNeill, J. R. (2007). The Anthropocene: Are humans now overwhelming the great forces of Nature? Ambio, 36(8), 614–621. Retrieved from http://www.ncb i.nlm.nih.gov/pubmed/18240674
4. Folke, C., Carpenter, S. R., Walker, B. H., Scheffer, M., Elmqvist, T., Gunderson, L. H., and Holling, C. S., 2004. Regime shifts, resilience and biodiversity in ecosystem management. Annual Review in Ecology, Evolution and Systematics, 35, 557–581.

5. Gowdy, J., Hall, C., Klitgaard, K., and Krall, L. (2010). What every conservation biologist should know about economic theory. Conservation Biology : The Journal of the Society for Conservation Biology, 24(6), 1440–1447. doi:10.1111/j.1523-1739.2010.01563.x

6. Wilkinson, Richard G., and Pickett, Kate. The Spirit Level: Why More Equal Societies Almost Always Do Better. London: Allen Lane; 2009.

7. MA (2003). (Millennium Ecosystem Assessment): A Conceptual Framework for Assessment. Washington, DC: Island Press; 2003.

8. MA (2005). Ecosystems and Human Well-being: Summary for Decision Makers. Washington, DC: Island Press; 2005.

9. CBD (2010). Secretariat of the Convention on Biological Diversity. (2010). Global Biodiversity Outlook 3. Montréal, 94 pages.

10. WCED. (1987). Our Common Future: Report of the World Commission on Environment and Development.

11. Griggs, D., Stafford-Smith, M., Gaffney, O., Rockström, J., Öhman, M. C., Shyamsundar, P., Steffen, W., et al. (2013). Sustainable development goals for people and planet. Nature, 495, 5–7.

12. Reyers, B., Polasky, S., Tallis, H., Mooney, H. A., and Larigauderie, A. (2012). Finding Common Ground for Biodiversity and Ecosystem Services. BioScience, 62(5), 503–507. doi:10.1525/bio.2012.62.5.12

13. Osborn, Fairfield. Our Plundered Planet. 1st ed. Boston: Little, Brown; 1948.

14. Vogt, W. (1949). Road to Survival. London: Victor Gollancz; 1949, p.335.

15. Leopold, A. A Sand County Almanac and Sketches from Here and There. 1949, p.226.

16. Ehrlich, P. R., and Ehrlich, A. (1970). Population, Resources, Environment: Issues in Human Ecology. Freeman, San Francisco, pp. 383.

17. Study of Critical Environmental Problems. (SCEP). Man's Impact on the Global Environment. 319 (1970).

18. Costanza, R., and Folke, C. Valuing ecosystem services with efficiency, fairness and sustainability as goals. Nature's Services: Societal Dependence on Natural Ecosystems (pp. 49–70). Washington, DC: Island Press; 1997.

19. Biggs, R., Schlüter, M., Biggs, D., Bohensky, E. L., BurnSilver, S., Cundill, G., Dakos, V., et al. (2012). Toward Principles for Enhancing the Resilience of Ecosystem Services. Annual Review of Environment and Resources, 37(1), 421–448. doi:10.1146/annurev-environ-051211-123836

20. Kumar, P., Brondizio, E., Gatzweiler, F., Gowdy, J., De Groot, D., Pascual, U., Reyers, B., et al. (2013). The economics of ecosystem services: From local analysis to national policies. Current Opinion in Environmental Sustainability, 5(1), 78–86. doi:10.1016/j.cosust.2013.02.001

21. Farley, J., and Costanza, R. (2010). Payments for ecosystem services: From local to global. Ecological Economics, 69(11), 2060–2068. doi:10.1016/j.ecolecon.2010.06.010

22. Sayer, J., Sunderland, T., Ghazoul, J., Pfund, J.-L., Sheil, D., Meijaard, E., Venter, M., et al. (2013). Ten principles for a landscape approach to reconciling agriculture, conservation, and other competing land uses. Proceedings of the National Academy of Sciences of the United States of America. doi:10.1073/pnas.1210595110

23. Morse-Jones, S., Luisetti, T., Turner, R. K., and Fisher, B. (2011). Ecosystem valuation: Some principles and a partial application. Environmetrics, 22(5), 675–685. doi:10.1002/env.1073

24. Ostrom, E. A general framework for analyzing sustainability of social–ecological systems. Science 2009, 325:419–422.

25. Arrow, K., B. Bolin, R. Costanza, P. Dasgupta, C. Folke, C. S. Holling, B. O. Jansson, S. Levin, K.-G. Maler, C. Perrings, and D. Pimentel. 1995. Economic growth, carrying capacity, and the environment. Science 268, 520–521.

26. European Commission: Our life insurance, our natural capital: An EU biodiversity strategy to 2020. COM(2011) 244; Brussels; 2011.

27. Daly, H. (2013). A further critique of growth economics. Ecological Economics, 88, 20–24. doi:10.1016/j.ecolecon.2013.01.007

28. Steffen, W., and Stafford Smith, M. (2013). Planetary boundaries, equity and global sustainability: Why wealthy countries could benefit from more equity. Current Opinion in Environmental Sustainability, 1–6. doi:10.1016/j.cosust.2013.04.007

29. Fisher, J. A., Patenaude, G., Meir, P., Nightingale, A. J., Rounsevell, M. D. A., Williams, M., and Woodhouse, I. H. (2013). Strengthening conceptual foundations: Analysing frameworks for ecosystem services and poverty alleviation research. Global Environmental Change. doi:10.1016/j.gloenvcha.2013.04.002

30. Nahlik, A. M., Kentula, M. E., Fennessy, M. S., and Landers, D. H. (2012). Where is the consensus? A proposed foundation for moving ecosystem service concepts into practice. Ecological Economics, 77, 27–35. doi:10.1016/j.ecolecon.2012.01.001

31. Reyers, B., Roux, D. J., Cowling, R. M., Ginsburg, A. E., Nel, J. L., and Farrell, P. O. (2010). Conservation Planning as a Transdisciplinary Process. Conservation Biology, 24(4), 957–965. doi:10.1111/j.1523-1739.2010.01497.x

Hendrik Segers[1], Dimitri Brosens[1], Hans Keune[2,3,4,5], Sander Jacobs[7,8], Nicolas Dendoncker[9], and Patrick Meire[6]

[1]*Belgian Biodiversity Platform, Belgian Science Policy Office (BELSPO),* [2]*Belgian Biodiversity Platform,* [3]*Research Institute for Nature and Forest (INBO),* [4]*Faculty of Applied Economics – University of Antwerp,* [5]*naXys, Namur Center for Complex Systems – University of Namur,* [6]*University of Antwerp, Ecosystem Management Research group (ECOBE),* [7]*Research Institute for Nature and Forest (INBO),* [8]*University of Antwerp. Department of Biology, Ecosystem Management Research Group (ECOBE),* [9]*Department of Geography, University of Namur (UNamur). Namur Research Centre on Sustainable Development (NAGRIDD). Namur Centre for Complex Systems (naXys)*

Ecosystem services connect biodiversity with human well-being. The potential of this perspective to inform and produce sustainable development is recognized by researchers, policy makers, and practitioners. However, being a relatively young concept, it invokes many debates and often crucial concerns that impact effective application and implementation. Scientific debates often seem distant from policy and practice, and real-life challenges often stay under the research radar. This book presents contributions from the "community of science and practice on ecosystem services" in Belgium, mostly co-authored and/or reviewed by international contributors. This introduction tells the story of how the book, and this community, came about.

The book provides a snapshot of ongoing debates globally and locally. It is the product of continuing exchange between the Belgian community of practice on ecosystem services and international networks. Initially, independent papers from local research projects evolved toward a larger number of collaboratively and interdisciplinary written chapters and contributions from practice. In this sense, the book is in itself proof of the concept of inter- and transdisciplinary cooperation: Challenges met in local practice relate to global debates and refocus them, while, equally, theoretical issues percolate to the practical level and foster common understanding of concepts and implementation.

Often, seemingly insurmountable differences between theoretical approaches and viewpoints dissolve when people cooperate in practice and semantic barriers are removed: For instance, when nonmonetary and monetary valuation experts reviewed each other's papers and agreed on terminology, they concluded that their main challenges were very similar after all and neatly mirrored the concerns of practitioners who contributed to the book.

A SHORT HISTORY: EMERGENCE OF THE BEES COMMUNITY OF PRACTICE

The first research projects related to ecosystem services (ES) in Belgium emerged in the early 1990s, and, following the mainstreaming effect of the Millennium Ecosystem Assessment, the number of ES research initiatives in Belgium grew exponentially (Figure 1), in line with the global trend.

While interdisciplinarity seemed common in these projects (with over half the projects involving both natural and social scientists), the focus was mostly on a single habitat and a few services. As ES analyses aim to serve sustainable resource use and land management decision support, this limitation was considered to be an important one, to be tackled in a network project.

An analysis of the BioBel database directly defined knowledge needs for this networking project among Belgian researchers: the BEES (Belgium Ecosystem Services) project. Funded by the Belgian federal science policy and coordinated by the University of Antwerp, it consisted of six thematic workshops and a final conference, organized by seven partners during the 2010 to 2012 period. The partnering research institutions aimed at broadening the audience by proposing a broad range of topics (classification, valuation, global trade, governance) and inviting participants from research institutions, administrations, and the private sector alike. Participation increased exponentially, and the project sparked great interest among involved research and practice groups in Belgium. Simply convening with people from other scientific disciplines and policy fields, and discussing challenges for ES research and practice together in workshops marked an urgently needed step to advance local expertise.

The active participation of policy makers, administrators, and NGOs in the inevitably ethical and societal debates characterizes the network, which started off as *inter*disciplinary but soon became truly *trans*disciplinary. Two conclusions stood out from the project: There was a need (1) to continue and broaden the network and link it closely to practice as well as international networks, and (2) to further develop the content of debates and make them available in a publication. Consequently, on April 26, 2012, a group of Belgian actors from science and policy decided to establish a community of practice on ecosystem services in Belgium. The BEES community (www.beescommunity.be), hosted by the Belgian Biodiversity Platform, aims to develop tools and practices that

- clarify ecosystem thresholds
- preserve the well-being of present and future generations
- halt or reverse ecosystem and biodiversity degradation
- integrate ecosystem services concepts in policy, management, business, and society
- facilitate the exchange of expertise and experiences

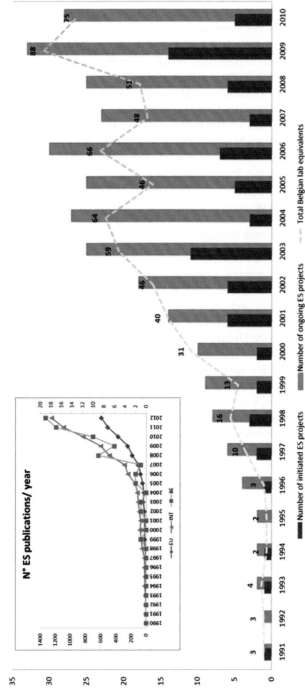

FIGURE 1 Ecosystem Service research activities in Belgium. Results from queries in the BioBel database (Belgian Biodiversity Platform, http://biobel.biodiversity.be, accessed in August 2010).

- transfer knowledge to and share needs from policy
- provide regular overviews of state of the art and best practices

BEES is now an open community in which all potentially interested organizations (policy, business, NGOs, science, consultancy, civil society) share interests, knowledge, and experiences, benefit from the knowledge and experience of others, and initiate practical collaboration. The BEES community is informal in its functioning, organization, and membership. It facilitates the engagement of local ecosystem services actors in international initiatives, such as the Intergovernmental Platform on Biodiversity and Ecosystem Services (IPBES), the EU Working Group on Mapping and Assessment of Ecosystems and their Services (MAES), The Economics of Ecosystems and Biodiversity (TEEB), and the Ecosystem Services Partnership (ESP).

READERS' GUIDE

This book has five main sections.

1. *Ecosystem Service Basics*, in which the main components of ES research are introduced and ongoing debates are illustrated with local examples.
2. *Ecosystem Service Global reflections*, in which a number of views on ecosystem service research and practice on a global scale provide the necessary overview of diverse viewpoints.
3. *Ecosystem Service Debates,* in which the issues presented in Section 1 are deepened and essential additions are made.
4. *Ecosystem Service Tools and Application,* in which specific examples of tools as well as applications of ES thinking to certain topics are gathered.
5. *Ecosystem Service Reflections from Practice,* which contains a broad overview of perspectives on ecosystem service practice, realizations, expectations, possible future developments, and needs, all written by policy makers, practitioners, and NGOs from different geographical scales.

We sincerely hope this book convinces you to leave your comfort zone of disciplines, daily practices, or traditional science-policy cooperation and truly engage in effective ecosystem service research and practice, and motivates you to take a personal step in that direction.

Part I

Ecosystem Service Basics

Shutterstock.com / © Kevin Eaves

Chapter 1

Inclusive Ecosystem Services Valuation

Nicolas Dendoncker[1], Hans Keune[2,3,4,5], Sander Jacobs[6,7] and Erik Gómez-Baggethun[8,9]

[1]Department of Geography, University of Namur (UNamur). Namur Research Centre on Sustainable Development (NAGRIDD). Namur Centre for Complex Systems (naXys), [2]Belgian Biodiversity Platform, [3]Research Institute for Nature and Forest (INBO), [4]Faculty of Applied Economics – University of Antwerp, [5]naXys, Namur Center for Complex Systems - University of Namur, [6]Research Institute for Nature and Forest (INBO), [7]University of Antwerp. Department of Biology, Ecosystem Management Research Group (ECOBE), [8]Institute of Environmental Science and Technology, Universitat Autònoma de Barcelona, [9]Social-Ecological Systems Laboratory, Department of Ecology, Universidad Autónoma de Madrid

1. INTRODUCTION: ON VALUE AND VALUATION

The word "value" comes from the Latin *valor*, which itself comes from the term *valere*, which can be translated as "being strong, having some kind of importance." De Groot et al. [1], for example, equate value with importance. This value or importance is not easy to determine. As Maris and Bechet [2] point out, values are contextual, relative to a certain place, a certain time, and a certain group of people facing a problem and engaged in collective action. According

Ecosystem Services. http://dx.doi.org/10.1016/B978-0-12-419964-4.00001-9

to Costanza [3], value ultimately originates in the set of goals to which a society aspires.

The word "valuation" can be defined as the act of assessing, appraising, or measuring value, as value attribution, or as framing valuation (how and what to value, who values). Valuation can thus refer to assessing a monetary value or a price but also an estimation or appreciation of worth or meaning. Thus semantics and etymology immediately reveal that by no means should valuation be equated with monetary valuation. Monetary valuation, or, as Liekens et al. (Chapter 2, this volume; [4]) call it, monetization, is only part of what ought to be envisaged when one is trying to value ecosystem services.

Ecological and social values should also be taken into account when valuing ecosystem services (ES) [5]. Daly [6] and later Costanza [3] recognized that, in order to conduct appropriate valuation of ecosystem services, one needs to consider a broad set of goals that include ecological sustainability and social fairness, along with the traditional economic goal of efficiency. However, papers dealing with economic valuation only, or briefly touching on other aspects of valuation but failing to detail the operational side, still largely dominate the literature on ES valuation. Likewise, case studies on ES valuation that explicitly deal with nonmonetary values (e.g., [7, 8, 9]) are still scarce compared to those dealing with monetary valuations. Over 20 years after Daly's paper [6], the first two goals of ecological sustainability and social fairness are still largely ignored.

In this chapter, we discuss the concept of valuation of ES. After clarifying the purpose of ES valuation, we propose to implement a three-pillar ecological-economic-social valuation process (adding to that proposed by 5) and briefly detail each pillar. We then suggest some promising methodologies to integrate these seemingly separate valuations into information that can be used for decision making. Finally, we conclude by framing valuation in the broader context of the transition path toward sustainability.

2. WHY DO WE VALUE?

Although this question may seem trivial, it is worth remembering that valuation of ES often applies to when we have to choose between different alternatives that lead to different outcomes in terms of ES provision (see also Liekens et al. this volume; [4, p.4]). As Costanza [3] puts it: "we humans have to make choices and trade-offs concerning ES and this implies and requires valuation, because any choice between competing alternatives implies that the one chosen was more highly valued." In other words, valuation is important if alternatives are competing. Valuation can be seen as the final step before decision making: For Daily [10], it translates the consequences of maintaining the status quo and opting for each alternative into comparable units of impact on human well-being, now and in the future. Valuation can also be useful in other policy contexts, including awareness raising, environmental accounting, design of incentives, and in informing litigation processes where economic compensations are claimed for environmental damage [11].

2.1. Valuing for Sustainability

Our society is following an unsustainable path, and we now fully realize the need to get on a sustainable one. The final goal of ES valuation is to contribute to a more sustainable and equitable resource use. It is from this perspective that Daily [10] mentions well-being as a unit for valuation. The ultimate goal of ES valuation is to improve the well-being of every individual, now and in the future. The Millennium Ecosystem Assessment says nothing else [12]. Of course, the indicators that best represent well-being are not easy to define and are context dependent [13].

We share the view of ecological economics, as represented by Costanza and Daly [14], that natural capital and manufactured capital are in a relation of complementarity rather than substitutability. In this strong sustainability view, economy is merely a subsystem of society, which is itself a subsystem of the environment, all being interdependent [15, 16, 17, 18]. Therefore, any valuation exercise should first assess the state of the ecological system (now and after expected changes), before dealing with the social and monetary aspects. In Section 3, we come back to these three pillars of valuation.

When valuing for sustainability, it is essential to respect the principles of sustainable development. The Brundtland Commission's report (1987) defined sustainable development as "development that meets the needs of current generations without compromising the ability of future generations to meet their own needs" [19, p.4]. The concept supports strong economic and social development, in particular for people with a low standard of living. At the same time it underlines the importance of protecting the natural resource base and the environment. Economic and social well-being cannot be improved with measures that destroy the environment. Intergenerational solidarity is crucial: All development has to take into account its impact on the opportunities for future generations.

Solidarity should not only be envisaged at the local scale but also at the global level. If project-based valuation is performed, the actors involved need to be aware of the global impacts of the choices made locally. For example, from an intragenerational equity perspective, it is pointless to implement sustainable agricultural systems locally (by, for example, reducing considerably the food provisioning services) if the local community then needs to rely on importing food that will have been produced elsewhere at the expense of biodiversity. This would contribute to increasing the already high ecological debt the global North owes to the global South (see, e.g., [20, 21]).

With regard to equity, Norgaard [22] mentions that sustainability is ultimately a distribution question, a matter of ethics or environmental justice within and between generations. This ethical principle has to be applied during the valuation process. For Vatn [23], choices made in the realm of the environment are fundamentally ethical in the sense that the preferences, valuations, and uses of goods by each of us influence what is left for others to value and use. The valuation process must therefore be a collective process engaging the relevant actors of society concerned with the issues at stake. Vatn [23] and Parks and Gowdy [24] argue further that fairness-based value would require that people vote their preferences as members of the community instead as individuals. This issue has important consequences for the valuation methodologies chosen.

2.2. Valuing Bundles

Often ES valuation and Payment for Ecosystem Services (PES) only relate to a single service. As Kosoy and Corbera [25] indicate, the valuation exercises of single ecosystem functions are often misleading because their search for marginal values may have no real meaning, particularly when the critical question is how to protect the resilience of ecosystems [26]. The authors warn that favoring the production of one ES may have detrimental effects on many others (see also [27]). Indeed, not valuing the whole variety of services has contributed heavily to today's biodiversity crisis. Hence, when valuing ES one should make sure to take all the relevant ES into account. In this respect, the approach followed by Van der Biest et al. [28], suggesting the need for a tool developed to capture the ecological complexity of delivery of ecosystem services bundles, is worth mentioning.

2.3. Valuing Change

Valuing bundles of ES does not mean valuing everything. Trying to quantify the present value of ecosystems as a whole makes little sense because without biodiversity human life could not exist (see also Jacobs et al., ch. 1.3 in [29]) The non-sense of quantitatively valuing—let alone pricing—biodiversity or ES as a whole is thus evident. Valuation of marginal or incremental change, however, is sometimes possible and potentially useful for ES management (see Liekens et al., this volume, [4]). Therefore, as already mentioned, we should value different alternatives, different scenarios of changes in management. This can be done using cost efficiency analysis to determine the cost (economic, ecological, and social) of different scenarios for achieving a desirable state. However, when systems are close to thresholds or tipping points, ecosystem service valuation will need to switch from choosing among resources or alternatives to valuing the avoidance of catastrophic ecosystem change [26].

3. VALUATION FOR SUSTAINABLE DEVELOPMENT—A THREE-PILLAR VALUATION FRAMEWORK

We argued earlier that when valuing for sustainability one needs to account for ecological, social, and economic aspects in an integrated way. For de Groot et al. [1], in order to make well-informed decisions about potential trade-offs between different management states, all costs and benefits should be taken into account, including ecological, sociocultural, and economic values and perceptions. However, as Spangenberg and Settele [30] indicate, ecological valuation based on biophysical accounting and economic valuation based on consumer preferences out of context or system characteristics currently prevail. For the authors, both tend to fall short of suitable methods in a sustainability framework, as sustainable development is based on the essential principles of prioritizing human needs, in particular those of the poor, and respecting environmental limits.

The challenge then is to reconcile ecological and monetary valuation with our goals as a society. In this respect an integrated three-pillar valuation framework seems appropriate. Very few valuation studies implement such a framework, with Fontaine et al. [9], Martín-López et al. [31], and the UK National Ecosystem Assessment [32] being important exceptions.

The National Ecosystem Assessment (NEA) indeed aimed for a comprehensive valuation framework by combining monetary and nonmonetary valuation of services that cannot be meaningfully assessed in monetary terms. In this sense, they defined three components of well-being: economic (monetary) value, health value, and shared social value.

3.1. Ecological Valuation

Ecological valuation (sometimes called biophysical valuation or ecosystem assessment) is the first necessary step of any valuation exercise. Social and monetary values depend on the actual production of services. For Daily [10], to establish sound policy, the production functions describing how ecosystems generate services need to be sufficiently characterized, and the interactions among these functions quantified as much as possible (see [33] and Van der Biest et al. this volume, [28]). Also, remaining or inherent uncertainty should be acknowledged (see Jacobs et al., this volume, [29]). According to de Groot et al. [1], the ecological value encompasses the health state of a system, with ecological indicators such as diversity and integrity [29]. These concepts are further developed in Jacobs et al., this volume [29].

3.2. Monetary Valuation

Monetary valuation, or monetization, of ES and the different methodologies available are discussed elsewhere in this book (see [4]).

Although most authors acknowledge the interest of monetary valuation, at least as an awareness-raising or pedagogical tool, or as a means of comparing the cost of alternatives to improve ES provision, some authors are more skeptical about its usefulness. For Spangenberg and Settele [30], the monetary valuation of ES fails to account for value in a broader sense and obliterates other social and ecological qualities embedded in these services, which are perceived by those who benefit from ES at different scales. According to Norgaard [22], current (monetary) valuation methods only help us see ES and their values from within our unsustainable economy: "We are trying to reach a sustainable economy by invoking the value of ES but doing so less effectively than needed because our point of view is the economy we have rather than the economy we are trying to attain." Its challenges and limitations are further discussed in Liekens et al., this volume [34].

Independently from these critiques, it is important to stress again that not all value dimensions are commensurable and even less so if money is the unit of measure. For Vatn [35], the price of even the simplest commodity only captures a subset of the dimensions of its importance, worth and meaning to humans [36]. The real issue is how monetization or economic valuation fits in within the broader framework of valuation for societal transition to sustainability.

3.3. Social Valuation of ES

Evaluating ES implies handling both normative (what should be) and cognitive (what is) complexities [23] and uncertainties. Some authors argue that the uncertainties related to the concepts of ES and ES valuation are so large (e.g., uncertainties in understanding ecosystem processes, different values for different actors, changing values through time, etc.) that there is a need for a change in scientific posture when studying ES [37]. For Funtowicz and Ravetz [38], when societal and scientific uncertainties are strong, scientists need to give up their role of experts and rather launch a dialogue among researchers, decision makers, and citizens. In such a "postnormal" posture, the key point is the quality of the interaction leading to decision making.

Barnaud et al. [37] argue that the point of views of all stakeholders should be confronted from the start of a project to build a collective consensus on what ES should be prioritized and what potential trade-offs and synergies, as well as distributional issues, should be discussed, and assessed collectively. Collective decisions through deliberation allow actors to decide who gives up what as part of a negotiation and collective learning process. This issue relates to Sen's view [39] that value formation should occur through public discussion. De Groot et al. [1] add that for proper valuation, and trade-off analysis, all scales and associated stakeholders should be taken into account. The question then is who are the stakeholders? If social valuation has to be done at the level of the community [3], it is important to define the community.

Deliberative methods and multicriteria analysis have been proposed as potentially relevant techniques for collective valuation of ES [23, 40, 41, 42]. Plural values including monetary considerations might be addressed through some forms of small group deliberative monetary valuation [41]. For Vatn [23], deliberative methods may involve people as citizens or stakeholders, communicating in groups over the relevant arguments to find a common solution in the form either of a consensus or a compromise. Deliberative valuation automatically escapes the potential problem of aggregation of values that may have different units (from monetary to qualitative) of measure: In the case of deliberative valuation, the focus is on agreeing rather than aggregating.

Deliberative methods can also be used to best define payment for ecosystem services. For Kosoy and Corbera [25], pluralism involves the development of consensus-building processes, so as to gather existing knowledge, views, and diverse values, and to define the most appropriate combination of monetary and nonmonetary incentives. Aiming for consensus building is, however, not the only option for participatory approaches: Consensus building may turn a blind eye to relevant or important diversity of knowledge or opinions. Differences of opinion among scientists or stakeholders, for example, can be very informative for decision makers as being an important part of reality [43].

The valuation process needs to be as transparent as possible if it has to be used for decision making. Society can make better choices about ecosystems if the valuation issue is made as explicit as possible [3]. The challenge for scientists is to develop tools for communicating valuation results that are scientifically sound, explicit about cognitive uncertainties and ambiguities (see Jacobs et al. this volume, [29]), and yet easily apprehended by decision makers. As regards normative uncertainties and ambiguity, acknowledging societal and moral values and incorporating them in a process of deliberative decision making allows reducing them or at least making them transparent [2].

4. IS VALUATION OF ES ENOUGH FOR PROPER ENVIRONMENTAL DECISION MAKING?

As Daily argues [10], valuation is a way of organizing information to help guide decisions, but it is not a solution or an end in itself. Rather, it is one tool in the much larger politics of decision making. Wielded together with financial instruments and institutional arrangements that allow individuals to capture the value of ecosystem assets, however, the process of valuation can have favorable effects.

When valuing ES, the questions of politics and governance cannot be ignored (see Keune, this volume, [15]). Valuation exercises always take place in a given institutional setting [44]. Because environmental resources are common and complex goods, this institutional setting should ideally favor social rationality and communicative action, warranting that a societal perspective is taken and that the procedure must be able to treat weakly comparable or incommensurable value dimensions [44, 45].

At the global level, we need new institutions and far more resources devoted to environmental governance (Norgaard, [22]). Norgaard also argues that property rights need to be reallocated. He cautions that setting the boundaries of an ES analysis as a project and doing a project-by-project type of analysis may not be relevant: "The major changes need to be accomplished at the level of national and global politics, not project analysis."

Nevertheless, if changes do need to occur at top levels, we believe that they may start at the bottom, and this justifies bottom-up project-based actions and ES valuations. These changes should be envisaged and implemented with sustainability as the final objective, keeping in mind its social, ecological, and economic principles. These projects should therefore be collectively constructed, engaging scientists and stakeholders (including citizens), and rely on ecological and economic valuation to foster social transformation. We believe with Sunstein ([46], cited in [23]) that it is our responsibility to create a future that will provide both present and coming generations the opportunity to live good lives.

REFERENCES

1. De Groot, R.S., Alkemade, R., Braat L., Hein, L., and Willemen, L. 2010. Challenges in integrating the concept of ecosystem services and values in landscape planning, management and decision making. *Ecological Complexity* 7, 260–272.
2. Maris, V., and Bechet, A. 2010. From adaptive management to adjustive management: A pragmatic account of biodiversity values. *Conservation Biology* 24(4), 966–973.
3. Costanza, R. 2000. Social goals and the valuation of ecosystem services. *Ecosystems* 3, 4–10.
4. Liekens, I., De Nocker, L., Broekx, S., Aertsens, J., and Markandya, A. Ecosystem services and their monetary value.
5. De Groot, R.S., Wilson, M., and Boumans, R.M.J. 2002. A typology for the classification, description and valuation of ecosystem functions, goods and services. *Ecological Economics* 41, 393–408.
6. Daly, H.E. 1992. Allocation, distribution, and scale: towards an economics that is efficient, just, and sustainable. *Ecological Economics* 6, 185–193.
7. Calvet-Mir, L., Gomez-Baggethun, E.G., and Reyes-Garcia, V. 2012. Beyond food production: Ecosystem services provided by home gardens. A case study in Vall Fosca, Catalan Pyrenees, Northeastern Spain. *Ecological Economics* 74, 153–160.
8. Martin-Lopez, B. Iniesta-Arandia, I., Garcia-Llorente, M., Palomo, I., Casado-Arzuaga, I., Garcia Del Amo, D., Gomez-Baggethun, E., Oteros-Rozas, E., Palacios-Agundez, I., Willaarts, B., Gonzalez, J.A., Santos-Martin, F., Onaindia, M., Lopez-Santiago, C., and Montes, C. 2012. Uncovering Ecosystem Service Bundles through Social Preferences. *PLOS1* 7(6), 1–11.
9. Fontaine, C., Dendoncker, N., de Vreese, R., Jacquemin, I., Marek, A., Van Herzele, A., Devilllet, G., Mortelmans, D., and François, L. 2013. Towards participatory valuation of ecosystem services under land use change. *Journal of Land Use Science* (in press). DOI:10.1080/17474 23X.2013.786150
10. Daily, G. 2000. The value of nature and the nature of value. *Science* 289(5478), 395–396.
11. Gómez-Baggethun, E., and Barton, D.N. 2013. Classifying and valuing ecosystem services for urban planning. *Ecological Economics* 86, 235–245.
12. MEA. 2005. Millennium Ecosystem Assessment: Ecosystems and Human Well-being: Synthesis. Washington, DC: Island Press.
13. Zorondo-Rodrigez, F., Gomez-Baggethun, E., Demps, K., Ariza-Montobbio, P., Garcia, C., and Reyes-Garcia, V. 2012. What defines quality of life? The gap between public policies and locally defined indicators among residents of Kodagy, Karnataka (India). Social Indicators Research. DOI 10.1007/s11205-012-9993-z
14. Costanza, R., and Daly, H. 1992. Natural capital and sustainable development. *Conservation Biology* 6(1), 37–46.

15. Keune, H. Governance of ecosystem services.
16. Gómez-Baggethun, E., de Groot, R., Lomas, P.L., and Montes, C. 2010. The history of ecosystem services in theory and practice: From early notions to markets and payment schemes. *Ecological Economics* 69(6), 1209–1218.
17. Daly, H.E. 1977. Steady state economics. San Francisco: W.H. Freeman.
18. Norgaard, R.B. 1994. Development betrayed: The end of progress and a coevolutionary revisioning of the future. New York: Routledge.
19. Our Common Future, Report of the World Commission on Environment and Development, World Commission on Environment and Development, 1987. Published as Annex to General Assembly document A/42/427, Development and International Co-operation: Environment August 2, 1987.
20. Simms, A. 2005. *Ecological Debt: The Health of the Planet and the Wealth of Nations*. Pluto Books. Melbourne: Australia
21. Srinivasan, U.T., Carey, S.P., Hallstein, E., Higgins, P.A.T., Kerr, A.C., Kateen, L.E., Smith, A.B., Watson, R., Harte, J., and Norgaard, R.B. 2008. The debt of nations and the distribution of ecological impacts from human activities. *PNAS* 105(5), 1768–1773.
22. Norgaard, R.B. 2010. Ecosystem services: From eye-opening metaphor to complexity blinder. *Ecological Economics* 69(6), 1219–1227.
23. Vatn, A. 2008. An institutional analysis of methods for environmental appraisal. *Ecological Economics* 68, 2207–2215.
24. Parks, S., and Gowdy, J. 2013. What have economists learned about valuing nature? A review essay. *Ecosystem Services* 3, 1–10.
25. Kosoy, N., and Corbera, E. 2010. Payments for ecosystem services as commodity festishism. *Ecological Economics* 69(6), 1228–1236.
26. Limburg, K.R., O'Neill, R., Costanza, R., and Farber, S. 2002. Complex systems and valuation. *Ecological Economics* 41, 409–420.
27. Raudsepp-Hearne, C., Peterson, G.D., and Bennett, E.M. 2010. Ecosystem services bundles for analyzing tradeoffs in diverse landscapes. *PNAS* 107(11), 5242–5247.
28. Van der Biest, K. Jacobs, S., D'Hondt, R., Landuyt, D., Staes, J. Meire, P. and Goethals, P. EBI: An index for delivery of ecosystem services bundles.
29. Jacobs, S., Haest B., de Bie, T., and Turkelboom, F. The biology of ecosystem services.
30. Spangenberg, J., and Settele, J. 2010. Precisely incorrect? Monetising the value of ecosystem services. *Ecological Complexity* 7, 327–337.
31. Martín-López, B., Gómez-Baggethun, E., García-Llorente, M., Montes, C. 2013. Trade-offs across value-domains in ecosystem service assessment. *Ecological Indicators*, http://dx.doi.org/10.1016/j.ecolind.2013.03.003.
32. NEA 2011. *The UK National Ecosystem Assessment: Technical Report*. Cambridge: UNEP-WCMC.
33. Kremen, C. 2005. Managing ecosystem services: What do we need to know about their ecology? *Ecology Letters* 8, 468–479.
34. Liekens, I., and De Nocker, L. Valuation of ecosystem services: challenges and policy use.
35. Vatn, A. 2000. The environment as a commodity. *Environmental Values* 9, 493–509.
36. Gómez-Baggethun, E., and Ruiz-Pérez, M. 2011. Economic valuation and the commodification of ecosystem services. *Progress in Physical Geography* 35, 613–628.
37. Barnaud, C., Antona, M., and Marzin, J. 2011. Vers une mise en débat des incertitudes associées à la notion de service écosystémique. VertigO 11(1), 24 p.
38. Funtowicz, S.O., and Ravetz, J.R. 1994. The worth of a songbird: Ecological economics as a postnormal science. *Ecological Economics* 10(3), 197–207.

39. Sen, A. 1995. Rationality and social choice. *American Economics Review* 85, 1–24.
40. Howarth, R., and Wilson, M. 2006. A theoretical approach to deliberative valuation. *Land Economics* 83, 1–16.
41. Spash, C.L. 2008. How much is that ecosystem in the window? The one with the bio-diverse trail. *Environmental Values* 17, 259–284.
42. Keune, H., and Dendoncker, N. Negotiated complexity in ecosystem services science and policy making.
43. van de Kerkhof, M. 2006. A dialogue approach to enhance learning for sustainability—A Dutch experiment with two participatory methods in the field of climate change. The *Integrated Assessment Journal* 6(4), 7–34
44. Vatn, A. 2005. Institutions and the Environment. Chentelham: Edgar Elgar.
45. Martínez-Alier, J., Munda, J., and O'Neill, J. 1998. Weak comparability of values as a foundation for ecological economics. *Ecological Economics* 26, 277–286.
46. Sunstein, C.R. 1993. Endogenous preferences, environmental law. *Journal of Legal Studies* 22(2), 217–254.

Chapter 2

Ecosystem Services and Their Monetary Value

Inge Liekens[1], Leo De Nocker[1], Steven Broekx[1], Joris Aertsens[1] and Anil Markandya[2]

[1]*Flemish Research and Technology Organisation (VITO),* [2]*University of Bath, Department of Economics and International Development, United Kingdom and Scientific Director of Basque Center of Climate Change*

1. WHY SHOULD WE MONETIZE ES?

Many ecosystems and the services they provide are at risk. One reason is the institutional framework we use to guide decision making. Markets, policy processes, or other social mechanisms often do not take ecosystems services or their impact on human well-being into account. Ecosystem services are not fully captured in commercial markets or adequately quantified in terms that are comparable with economic services such as euros; therefore they are often given too little weight in policy decisions [1]. By ensuring that projects and policy appraisals fully consider the costs and benefits of the natural environment and by highlighting much more clearly the implications for human well-being, ecosystem service valuation can provide policy with new insights [2].

The monetary valuation of ecosystem services offers a promising approach to highlight the relevance of ES to society and the economy, to serve as an element in the development of cost-effective policy instruments for nature restoration and management, and to use in impact assessments in cost-benefit analysis. Monetary valuation may also be useful in developing payments for ecosystem services [3].

Ecosystem Services. http://dx.doi.org/10.1016/B978-0-12-419964-4.00002-0

Monetary valuation can help provide feedback on the consequences of actions to our society. Our society, like all societies, derives resources from the environment, but we have distanced ourselves from that same environment, and as a result do not see the consequences of our behavior. For example, many people don't know where agricultural products come from or how they are grown. Furthermore, monetary valuation can help communicate the value of nature to different people using a language easily understood by dominant economic and political views around the world [4]. Making values explicit can help build support for new instruments—for example, using Payment for Ecosystem Services (PES) to change the decision equation facing landowners, investors, and other users of natural resources. Monetary valuation is also increasingly being used to inform impact assessments of proposed legislation and policies (e.g., the Water Framework Directive).

Monetary valuation can be particularly effective in enabling informed trade-offs in cost-benefit analysis, where the focus lies on assessing the marginal change in the provision of an ecosystem service relative to the provision of the same service in an alternative scenario.

2. WHAT IS MONETARY VALUATION?

The goal of monetary valuation is to value the so-called total economic value (TEV) of an ecosystem. The TEV consists of use value and nonuse value (see Figure 2-1). By definition, use values are derived from the actual use of the environment. They are sometimes further divided into two categories: (1) *Direct use value*, related to the benefits obtained from direct use of ecosystem services.

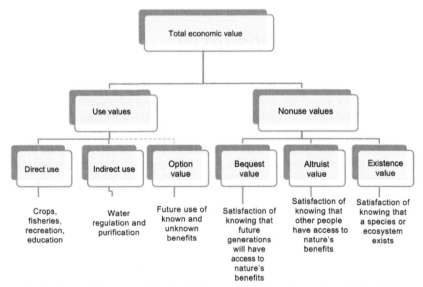

FIGURE 2-1   Value types within the TEV approach. *Source: Adapted from TEEB, 2012.*

Such use may be extractive, which entails consumption (for instance, of food and raw materials), or nonextractive (e.g., aesthetic benefits from landscapes). (2) *Indirect use values* are usually associated with regulating services, such as air quality regulation or erosion prevention, which can be seen as public services that are generally not reflected in market transactions. The option value is defined as the value of future use of known and unknown ecosystem services. Nonuse values on the other hand are noninstrumental. They reflect satisfaction that individuals derive from the knowledge that biodiversity and ecosystem services are maintained and that other people have or will have access to them [5]. In the first case, nonuse values are usually referred to as *existence values*, while in the latter they are associated with *altruist values* (in relation to intra-generational equity concerns) or *bequest values* (when concerned with inter-generational equity) [4].

Economic theory is based on the premise that individuals have preferences for different market and nonmarket goods. These preferences have a degree of substitutability; if the quantity of one good is reduced, the quantity of a different good can be increased to leave the person no worse off. The trade-offs made during this substitution reveal something about the values held for each good. Measurements of these values are expressed as either willingness to pay (WTP), the maximum amount a person would be willing to pay for an increment of a good, or willingness to accept the minimum amount a person would require as compensation for the loss of an increment of a good [6].

In economics value is always associated with scarcity and trade-offs, that is, something only has (economic) value if we are required and willing to give up something to get or enjoy it. This concept of valuation is thus anthropocentric in nature. Morse-Jones et al. [7] indicate that "monetary valuation cannot value everything—that is, not all benefits provided by ecosystem services are fully translatable into economic terms" (p. 676)—for example, some ecological values such as the value of one species to the survival of another species [8]. Therefore, it should be used to complement rather than substitute other legitimate rasoning for biodiversity conservation.

3. WHAT ARE WE VALUING?

It is important to know exactly what is being valued. Something can be valuable in two ways: instrumentally or intrinsically. Instrumental (or utilitarian) means that something has value because it is useful for something else. Intrinsic means that something has value in and of itself, not because something else deems it valuable [9]. As virtually no anthropogenic activity is possible without our ecosystems, they are worth in that sense an almost inifinite amount. This is not very interesting and useful to measure. Of greater interst is the effort to value the benefits to society associated with improving ecosystems or with preventing their degradation, but this is much more difficult [3]. What the precise benefits are to society is not always easy to define. A multitude of definitions and classification schemes for ecosystem services exist [4, 10, 11, 12, 13]. One of the most widely cited is the Millennium Ecosystem Assessment (MA) definition, which describes ecosystem services as "the benefits that people obtain from ecosystems" [11]. As stated before, it clarifies the anthropocentric focus of the ecosystem service concept. It classifies services into *supporting, regulating, provisioning,* and *cultural*. This framework provides an excellent platform for moving toward a more operational classification system (see Figure 2-2) which explicitly links changes in ecosystem services to changes in human welfare [14].

Haines-Young and Potschin [14] classify MA services into intermediate, final ecosystem services, and benefits. It is very context dependent when a service is intermediate or final. For example, clean water provision is a final service to a person requiring drinking water, but it is an intermediate service to a recreational angler. Importantly, a final service is often but not always the same as a benefit. For example, recreation is a benefit to the recreational angler, but the final ecosystem service is the provision of the fish population. For the purpose of monetary valuation we are only interested in the benefits, so the issue of potential double-counting is minimized. For example, we only value the benefit of recreation instead of the provision of a fish population, habitat, and so on.

As already stated, monetary valuation is about the benefits of improving ecosystems or preventing them from deterioration. So it is about small or marginal changes in the conditions of natural assets. However, Morse-Jones et al. [7] conclude that "a review of existing literature suggests that in fact very few studies do look at changes in that sense [15, 16]. Instead the majority focus on

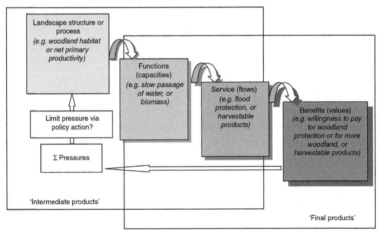

FIGURE 2-2 The relationship between biodiversity, ecosystem function, and human well being [14].

'point estimates' of the value of services from a given ecosystem" (p. 679). Marginal changes are relatively small, incremental changes rather than large state-changing impacts. It is the change by an additional unit, and it depends largely on the situation: It can be a tree or a whole forest, but it is not the total of forests in a country.

Ecosystem assessments should be set within the context of contrasting scenarios, recognizing that both the values of ecosystem services and the costs of actions can be measured as a function of changes between alternative options.

Bateman et al. [17] show that "an obvious concern in adopting such an approach is that an over-concentration on changes in benefits may place underlying ecological assets at risk [13, 18], thereby risking over-exploitation and system change or collapse e.g. focusing on the fish only and neglect functions as water quality, breeding areas etc. leads to the overexploitation as we know today. This has to be guarded against by imposing the constraint that ecosystem assets are not run down to unsustainable levels" and by valuing bundles of ecosystem services rather than a single service. See also Chapter 11.

Therefore, the monetary valuation process should include the following steps:

1. Identify the policy context and purpose of the assessment.
2. Identify and make explicit all relevant ecosystem services, users, and stakeholders of the ecosystem and delineate relevant scale(s).
3. Quantify in biophysical terms the changes in ecosystem services flows in response to a given intervention in order to provide a solid ecological underpinning to the monetary valuation.
4. Value. What is this change worth and to whom?

4. THE ECONOMIST TOOLBOX

4.1. Monetary Valuation Techniques

Only a small proportion of the benefits ecosystems deliver has any sort of presence in economic markets. Great efforts have been made to somehow create a value for nonmarketed goods and services. A variety of approaches can be used to estimate the value of ecosystem services. These approaches fall into two main categories: monetary valuation and nonmonetary valuation techniques. This chapter describes only the monetary valuation techniques. These consist of techniques that estimate economic values—valuation approaches—and techniques that produce estimates equivalent to prices—pricing approaches. It is important to know that the price of a good or service and its economic value are distinct and can differ greatly: pricing approaches are not able to capture the consumer surplus element of value.

Valuation approaches are devided into revealed and stated preferences. Pricing methods are adjusted market prices and avoided (damage) costs.

The appropriate tools depend on the characteristics of the goods or services ([19] and overviews made in, e.g., [6, 20, 21]).

For provisioning services (such as the production of fuel, fiber, food, and medicinal plants, which are so called direct use values), estimating economic values would seem to be fairly straightforward, as these services are largely traded on markets and so have a **market price.** This is somewhat deceptive inasmuch as a number of limitations to market prices exist. Markets are often distorted (monopolies, subsidies, with not all costs accounted for, such as for pollution). If possible, we need to take market distortions into account and correct the existing market prices [22].

Methods used to value regulating and cultural services that are not sold on the market often require a number of assumptions as well as copious amounts of data and intensive statistical analysis. Probably the most serious problems facing robust valuation of ecosystem services are the gaps in our understanding of the underpinning science relating those services to the production of human well-being.

Regulating services is mostly valued through avoided (damage) costs (costs that we would have incurred if the service was absent or costs of replacing a service with human-made systems)—for example, avoided damage costs for flooding or avoided investment costs in wastewater treatment to estimate the value of water quality. The major underlying assumptions of these approaches are that the nature and extent of physical damage expected is predictable (an accurate damage function is available) and that the costs to replace or restore damaged assets can be estimated with a reasonable degree of accuracy.

Cultural services such as amenity values and recreation values are mostly valued through revealed preferences or stated preferences techniques. The stated preferences technique is the only method that takes nonuse values into account.

An overview of the existing methods is given in Table 2-1.

The revealed preferences methods are the hedonic pricing method and the travel cost method.

Hedonic pricing is based on the fact that the prices paid for goods or services that have environmental attributes differ depending on those attributes. Thus, a house in a clean environment will sell for more than an otherwise identical house in a polluted neighborhood. Hedonic price analysis compares the prices of similar goods to extract the implicit value ("shadow price") that buyers place on the environmental attributes. This method assumes that markets are transparent and work reasonably well, and it would not be applicable where markets are distorted by policy or market failures. Moreover, this method requires a very large number of observations; it is very data intensive and statistically complex to analyze. Its applicability is limited to environmental attributes. The advantage of the method is that it is a well-established technique and is based on actual observed behavior.

The **travel cost** method enables the economic value of recreational use (an element of direct use value) for a specific site to be estimated. The method

TABLE 2-1 Various Valuation Methods Applied to Ecosystem Services

Valuation Method	Value Types	Overview of Method	Common Types of Applications	Examples of Ecosystem Services Valued	Example Studies
Adjusted market prices	Use	Market prices adjusted for distortions such as taxes, subsidies and noncompetitive practices	Food, forest products, R&D benefits	Crops, livestock, multipurpose woodland, etc.	Bateman et al. (2003), Godoy et al. (1993)
Production function methods	Use	Estimation of production functions to isolate the effect of ecosystem services as inputs to the production process	Environmental impacts on economic activities and livelihoods, including damage costs avoided, due to ecological regulatory and habitat functions	Maintenance of beneficial species; maintenance of arable land and agricultural productivity; support for aquaculture; prevention of damage from erosion and siltation; groundwater recharge; drainage and natural irrigation; storm protection; flood mitigation	Ellis and Fisher (1987), Barbier (2007)
Damage cost avoided	Use	Calculates the costs that are avoided by not allowing ecosystem services to degrade	Storm damage; supplies of clean water; climate change	Drainage and natural irrigation; storm protection; flood mitigation	Badola and Hussain (2005), Kim and Dixon (1986)
Averting behaviour	Use	Examination of expenditures to avoid damage	Environmental impacts on human health	Pollution control and detoxification	Rosado et al. (2000)
Revealed preference methods	Use	Examine the expenditure made on ecosystem related goods (e.g., travel costs; property prices in low pollution areas)	Recreation; environmental impacts on residential property and human health	Maintenance of beneficial species, productive ecosystems and biodiversity; storm protection; flood mitigation; air quality, peace and quiet, workplace risk	See Bockstael and McConnell (2006) for the travel cost method and Day et al. (2007) for hedonic pricing.
Stated preference methods	Use and nonuse	Uses surveys to ask individuals to make choices between different levels of environmental goods at different prices to reveal their willingness to pay for those goods	Recreation; environmental quality, impacts on human health, conservation benefits	Water quality, species conservation, flood prevention, air quality, peace and quiet	See Carson et al. (2003) for contingent valuation and Adamowicz et al. (1994) for discrete choice experiment approach

Source: From [17]: adapted from [22], [23], [24], [25], and [26].

requires that the costs incurred by individuals traveling to recreation sites—in terms of both travel expenses (fuel, fares, etc.) and time (e.g., foregone earnings)—is collected. The basic assumption is that these travel costs serve as a proxy for the recreational value of visiting a particular site. The advantage of the method is that it is a well-established technique and is based on actual observed behavior. Disadvantages are that it is only applicable to recreational sites, and it is difficult to account for the possible benefits derived from travel and multi-purpose trips. It is very resource intensive and statistically complex to analyze.

Contingent valuation (CV) is an example of a stated preference technique. It is carried out by asking consumers directly about their WTP to obtain an environmental service (or, in some circumstances, their willingness to accept). A detailed description of the service and how it will be delivered is provided. The valuation can be obtained in a number of ways, such as asking respondents to name a figure (classical CV), asking them whether they would pay a specific amount (dichotomous or polychotomous choice), or having them choose from several options (choice modeling). By phrasing the question appropriately, CV can be used to value any environmental benefit. Moreover, since it is not limited to deducing preferences from available data, it can be targeted to address specific changes in benefits that a particular change in ecosystem condition might cause.

Because of the need to describe the service being valued in detail, interviews in CV surveys are time consuming. In designing CV surveys, it is important to identify the relevant population to ensure that the sample is representative and to pretest the questionnaire to avoid bias. A potentially important limitation when applying these methods to ecosystem services is that respondents cannot typically make informed choices if they have a limited understanding of the issue in question. Choosing the right approach to improve the sample group's understanding of biological complexity and the question at hand without biasing respondents is a challenge for stated preference methods.

Choice modeling consists of asking respondents to choose their preferred option from a set of alternatives where the alternatives are defined by attributes (including price). The alternatives are designed so that the respondent's choice reveals the marginal rate of substitution between the attributes and the item that is traded off (for example, money). Choice modeling has several advantages. First, control of the stimuli is in the experimenter's hand, as opposed to the low level of control generated by real market data. Second, control of the design yields greater statistical efficiency. Third, the attribute range can be wider than that found in market data. The method also minimizes some of the technical problems (such as the strategic behavior of respondents) that are associated with CV. The disadvantages associated with the technique are that the responses are hypothetical and therefore suffer from problems of hypothetical bias (similar to CV) and that the choices can be complex when there are many attributes and alternatives. The econometric analysis of the data generated by choice modeling is also relatively complex.

All of these tools have their strengths as well as their shortcomings. They are affected by uncertainty, stemming from incomplete knowledge about ecosystem dynamics, human preferences, and technical issues in the valuation process. When deciding which valuation tools to use, one should consider these shortcomings. An extensive review of these issues is provided in [27] and [28]. A combination of valuation techniques is required to comprehensibly value ecosystem goods and services.

4.2. Benefits Transfer

Benefits transfer (BT) refers to applying the results of previous environmental valuation studies to new decision-making contexts. Benefits transfer is commonly defined as the transposition of monetary environmental values estimated at one site (study site) to another site (policy site). The study site refers to the site where the original study took place, whereas the policy site is a new site where information is needed about the monetary value of similar benefits. BT methods can be divided into four categories: (1) unit BT, (2) adjusted unit BT, (3) value function transfer, and (4) meta-analytic function transfer [19].

Unit BT involves estimating the value of an ecosystem service at a policy site by multiplying a mean unit value estimated at a study site by the quantity of that ecosystem service at the policy site. Unit values are generally either expressed as values per household or as values per unit of area. Adjusted unit transfer involves making simple adjustments to the transferred unit values to reflect differences in site characteristics. The most common adjustments are for differences in (household) income between study and policy sites and for differences in price levels over time or between sites.

Value or demand function transfer methods use functions estimated through valuation applications (travel cost, hedonic pricing, contingent valuation, or choice modeling) for a study site, together with information on parameter values for the policy site to transfer values. Parameter values of the policy site are plugged into the value function to calculate a transferred value that better reflects the characteristics of the policy site (e.g., size of the area, sociodemographic factors of the beneficiaries).

Lastly, meta-analytic function transfer uses a value function estimated from multiple study results, together with information on parameter values for the policy site, to estimate values. The value function therefore does not come from a single study but from a collection of studies. This allows the value function to include greater variation in both site characteristics (e.g., socioeconomic and physical attributes) and study characteristics (e.g., valuation method) that cannot be generated from a single primary valuation study.

Recently, guidelines have been established to improve benefits transfer (e.g., European Aquamoney project [29, 30]). Despite the uncertainties and difficulties of transferring valuation approaches and results between projects and regions (see also Chapter 11), benefit transfer can be a practical, swift, and cheap way to get an estimate of the value of local ecosystems, particularly when the aim is to study a large number of diverse ecosystems.

5. MONETARY VALUATION OF ES IN BELGIUM

During the last decade, the number of available valuation examples of external costs of pollution and valuation studies of ESS has gradually increased in Belgium. For Flanders, the state of play in valuation methods and examples in practice is summarized in the LNE manual "Milieubaten of milieus-chadekosten—waarderingsstudies in Vlaanderen (Environmental benefits or environmental damage costs—valuation studies in Flanders)" [31]. Examples include the valuation of avoided health costs due to air pollution and cost-benefit analyses related to nature development, including the value of ecosystem services. In the Walloon Region valuation exercises are also becoming increasingly available. Similar exercises on health costs related to air pollution and, recently, valuation of benefits related to the Water Framework Directive are the best known examples.

In a large number of studies, the link between the biophysical processes of the ecosystem and the ecosystem services is not taken into account when quantifying the biophysical change in the ecosystem service. In most cases, results of a single study are transferred to the study site. This is partly because, too often, ecological and economic studies are carried out separately; as a result, the most reliable ecological and economic information cannot be brought together.

Another reason is that relatively few original valuation studies exist in Belgium. We summarize them in Table 2-2.

Spatial issues such as distance decay and substitutes, especially in the stated preference studies, are hardly taken into account, except for the ones where it is indicated. Nevertheless, these studies show that there is a willingness to pay for quality improvements of rivers and nature areas.

As there is a large demand for information about costs and benefits of ecosystems, the Flemish department of environment, Nature and Energy, ordered a study to provide guidance on which figures to use and how to use them in a cost-benefit analysis. A manual was published in 2010 [44] and a web-based tool was developed (see also Chapter 19 in this book).

6. CONCLUSION

Monetary valuation can provide useful information about changes to welfare that will result from ecosystem management actions. Despite the substantial progress made, no valuation method is perfect. Each method has advantages and disadvantages, and should be carefully chosen based on the specific goals of the study.

"Valuation techniques have limitations that are as yet unresolved (see also Chapter 11 in this book). Valuation practitioners should present their results as

TABLE 2-2 Original Valuation Studies in Belgium

Study	Ecosystem Services	Method	Comments	Source
Economische waardering van bossen: een case studie van Heverleebos-Meerdaal woud	Recreation value and nonuse value	Transport cost Contingent valuation method	Other services are also looked at making use of benefit transfer	[33]
Economische waarderingsstudie van natuur en landschap voor kosten-batenanalyses	Amenity and nonuse value Value regulating services	Choice experiment Avoided costs	Biophysical change is quantified by specific functions based on literature Choice experiment specifically designed for BT, including distance decay	[34]
Natuurherstel in de Hemmepolder	Amenity value Effect on shrimp population (nursery function)	Contingent ranking method. Market value	Improvement of waterquality is also looked at. Benefit transfer is used	[35]
Ontwikkelen van socio-economische beoordelingscriteria die het mogelijk maken de kostprijs van een degradatie van het mariene milieu objectief te bepalen	Nonuse value	Contingent valuation method	Other impacts of an oil disaster on the marine ecosystem through replacement cost method	[36]
Economische waardering van parken				[37]
'Economische waardering van natuurgebieden. Case-study: Meldertbos'	Recreational value Nonuse value	Travel cost Choice experiment		[38]
MKBA voor de actualisatie van het Sigmaplan	Recreational value Nonuse value Flood protection value	Contingent valuation method Avoided damage costs	Other benefits of water ecosystems are valued through benefit transfer. Quantification of the biophysical changes was done by a location specific model	[39]

Continued

TABLE 2-2 Original Valuation Studies in Belgium—cont'd

Study	Ecosystem Services	Method	Comments	Source
Economic valuation of the non-market benefits of the European Water Framework Directive: an international river basin application of CVM	Improvement ecological status rivers	Contingent valuation method	Distance decay included	[40]
Aquamoney: case study of the Scheldt river	Improvement ecological status rivers	Contingent valuation method Choice experiment	Distance decay and substitutes included	[41]
Evaluation économique des bénéfices environnementaux non-marchands et de la valeur de non-usage relaissés suite à la mise en oeuvre des plans de gestion de l'eau et l'atteinte des objectifs environnementaux de la Directive Cadre Eau pour les eaux de surface en Région wallonne	Improvement ecological status rivers	Contingent valuation method	Scale issues concerning payment for local river versus all rivers in the Walloon region	[42]
La fonction récréative des massifs forestiers wallons: analyse et evaluation dans le cadre d'une politique forestière intégrée (thèse de doctorat)	Recreation value of forests		Spatial issues; substitutes	[43]
VOTES (Valuation of Terrestrial Ecosystem Services in a Multifunctional peri-urban space	Recreation value Biodiversity/quality of life	Travel cost Deliberative valuation	Spatial issues captured	[44]

such, and policy makers should interpret and use valuation data accordingly" [4; Key message, Chapter 5]. We must be careful to use monetary valuation in appropriate ways. Most people in Western societies are not concerned about basic survival, but many environmentalists use the survival of the human species as an argument for conservation. People are concerned, however, about quality of life. It is in this area that environmental arguments and monetary valuations can be most effective [46]. This means that we only use monetary valuation in valuing marginal changes and in comparing scenarios, and we only value the benefits (final services) in order to avoid double-counting.

Despite difficulties, limitations, and issues surrounding ecosystem service valuation, there does seem to be a general consensus that the value of ecosystem services often outweighs economic use and that protecting ecosystem services is, or should be, one of the most important responsibilities of today's politicians, resource managers, and society in general [12, 15, 46].

As this consensus exists, and one uses monitization in the correct way, we need to go one step further than performing valuation studies for scientific research and start applying ecosystem valuation in different policy fields. This requires good guidance on how to use estimated values, taking into account uncertainties and the limitations of monetary valuation. It is also important to realize that ecosystem service valuation can be a useful tool but cannot alone provide all the information needed to solve a problem. These economic arguments are best used along with and to support political and social considerations [47].

REFERENCES

1. Costanza, R., dArge, R., deGroot, R., Farber, S., Grasso, M., Hannon, B., Limburg, K., Naeem, S., ONeill, R. V., Paruelo, J., Raskin, R. G, Sutton, P., and van den Belt, M. 1997. The value of the world's ecosystem services and natural capital. *Nature* 387(6630), 253–260.
2. Defra. 2007. *An Introductory Guide to Valuing Ecosystem Service.* London: Department for Environment, Food and Rural Affairs.
3. Markandya, A. 2011. Challenges in the Economic Valuation of Ecosystem Services. BEES Workshop IV: Ecosystem Services and Economic Valuation, Brussels, May 18, 2011.
4. TEEB. 2012. The Economics of Ecosystems and Biodiversity: Ecological and Economic Foundations (P. Kumar, ed.). Abingdon and NewYork: Routledge.
5. Kolstad, C. 2000. Energy and Depletable Resources: Economics and Policy, 1973–1998. *Journal of Environmental Economics and Management*, 39(3), 282–305. Available at: http://dx.doi.org/10.1006/jeem.1999.1115
6. Freeman, A.M., III. 2003. The Measurement of Environmental and Resource Values: Theory and Methods. Washington, DC: Resources for the Future.
7. Morse-Jones, S., Luisetti, T., Turner, R.K., and Fisher, B. 2010. Ecosystem Valuation: Some principles and a partial application, *Environmetrics* 22(5), 675–685. http://onlinelibrary.wiley.com/doi/10.1002/env.1073/abstract
8. Farber, S. 2002. Economic and ecological concepts for valuing ecosystem services. *Ecological Economics* 41(3), 375–392. Available at: http://dx.doi.org/10.1016/S0921-8009(02)00088-5
9. Meffe, G.K., and Carroll, C.R. 1997. Principles of Conservation Biology. Sunderland, MA: Sinauer Associates.

10. Potschin, M., and Haines-Young, R. 2011. Ecosystem Services: Exploring a geographical perspective. *Progress in Physical Geography* 35(5), 575–594.

11. Millennium Ecosystem Assessment. 2005. Ecosystems and Human Well-being: Synthesis. Washington, DC: Island Press.

12. Daily, G.C. 1997. Valuing and safeguarding Earth's life support systems. *In* Nature's Services: Societal Dependence on Natural Ecosystems (G.C. Daily, ed.), pp. 365–374. Washington, DC: Island Press.

13. Turner, R.K. 1999 The place of economic values in environmental valuation. *In* Valuing Environmental Preferences (I.J. Bateman and K.G. Willis, eds.). Oxford: Oxford University Press.

14. Haines-Young, R. and M. Potschin (2010): The links between biodiversity, ecosystem services and human well-being. In: Raffaelli, D. & C. Frid (eds.): Ecosystem Ecology: a new synthesis. Cambridge University Press, BES, p.110–139.

15. Balmford, A., Bruner, A., Cooper, P., Costanza, R., Farber, S., Green, R.E., Jenkins, M., Jefferiss, P., Jessamy, V., Madden, J., Munro, K., Myers, N., Naeem, S., Paavola, J., Rayment, M., Rosendo, S., Roughgarden, J., Trumper, K., and Turner, R.K. 2002. Economic reasons for conserving wild nature. *Science* 297, 950–953.

16. Turner, R. K., Paavola, J., Cooper, P., Stephen Farber, Jessamy, V., and Georgiou, S. 2003. Valuing nature: lessons learned and future research directions. *Ecological Economics* 46, 492–510.

17. Bateman, I.J., Mace, G.M., Fezzi, C., Atkinson, G., and Turner, R.K. 2011. Economic Analysis for Ecosystem Service Assessments, Special Issue on Conservation and Human Welfare: Economic Analysis of Ecosystem Services (Guest Editors: Brendan Fisher, Steve Polasky, and Thomas Sterner). *Environmental and Resource Economics* 48, 177–218, doi: 10.1007/s10640-010-9418-x.

18. Gren, I-M., Folke, C., Turner, R.K., and Bateman, I.J. 1994. Primary and secondary values of wetland ecosystems. *Environmental and Resource Economics* 4, 55–74.

19. Brouwer R. (2000). Environmental value transfer: state of the art and future prospects, *Ecological Economics* 32(1), 137–152.

20. Champ, P.A., K.J. Boyle, and T.C. Brown. 2003. *A Primer on Nonmarket Valuation*. Dordrecht: Kluwer.

21. Hanley, Nick, and Edward B. Barbier. 2009. *Pricing Nature: Cost-Benefit Analysis and Environmental Policy-making*. London: Edward Elgar.

22. De Groot, R.S., Wilson, M.A., and Boumans, R.M.J. 2002. A typology for the classification, description, and valuation of ecosystem functions, goods, and services. *Ecological Economics* 41, 393–408.

23. Heal, Geoffrey M., Edward B. Barbier, Kevin J. Boyle, Alan P. Covich, Stephen P. Gloss, Carl H. Hershner, John P. Hoehn, Catherine M. Pringle, Stephen Polasky, Kathleen Segerson, and Kirstin Shrader-Frechette. 2005. *Valuing ecosystem services: Toward better environmental decision making*. Washington, DC: National Academies Press.

24. Barbier, E.B. 2009. Ecosystems as Natural Assets. *Foundations and Trends in Microeconomics* 4(8), 611–681.20

25. Bateman, I.J. 2009. Bringing the real world into economic analyses of land use value: Incorporating spatial complexity, *Land Use Policy*, 26, 30–42, doi:10.1016/j.landusepol. 2009.09.01029

26. Kaval, P. 2010. A summary of ecosystem service economic valuation methods and recommendations for future studies. Working Paper Series #10, Department of Economics, University of Waikato.

27. Kontoleon, A., and Pascual, U. 2007. Incorporating Biodiversity into Integrated Assessments of Trade Policy in the Agricultural Sector. Volume II: Reference Manual. Chapter 7. Economics and Trade Branch, United Nations Environment Programme. Geneva. Available at: http://www.unep.ch/etb/pdf/UNEP%20T+B%20Manual.Vol%20II.Draft%20June07.pdf

28. Hadley, D., D'Hernoncourt, J., Franzén, F., Kinell, G., Söderqvist, T., Soutukorva, Å, and Brouwer, R. 2011. Monetary and nonmonetary methods for ecosystem services valuation—Specification sheet and supporting material, Spicosa Project Report, University of East Anglia, Norwich13.

29. Brouwer, R., Barton, D.N., Bateman, I.J., Brander, L., Georgiou, S., Martín-Ortega, J., Pulido-Velazquez, M., Schaafsma, M., and Wagtendonk, A. 2009. *Economic Valuation of Environmental and Resource Costs and Benefits in the Water Framework Directive: Technical Guidelines for Practitioners. Institute for Environmental Studies.* Amsterdam, the Netherlands: VU University.

30. Bateman, I., R. Brouwer, S. Ferrini, M. Schaafsma, D. Barton, A. Dubgaard, B. Hasler, S. Hime, I. Liekens, S. Navrud, L. De Nocker, R. Ščeponavičiūtė, and D. Semėnienė. 2011. Making Benefit Transfers Work: Deriving and Testing Principles for Value Transfers for Similar and Dissimilar Sites Using a Case Study of the Non-Market Benefits of Water Quality Improvements Across Europe. *Environmental and Resource Economics* 50, 365–387.

31. LNE. 2008. Milieubaten of milieuschadekosten–waarderingsstudies in Vlaanderen. D\2007\3241\314

32. Moons, E., Eggermont, K., Hermy, M., and Proost, S. 2000. *Economische waardering van bossen—een case-study van Heverleebos-Meerdaalwoud.* 356 p. Leuven: Garant.

33. Liekens, I., Schaafsma, M, De Nocker, L., Broekx, S., Staes, J., Aertsens, J., and Brouwers, R. 2013. Developing a value function for nature development and land use policy in Flanders, Belgium. *Land Use Policy* 30(1), 549–559.

34. Liekens I., Bogaert, S., De Nocker, L., Maes, J., Libbrecht, D., De Smet, L., and Nunes, P. 2006. Maatschappelijke kosten-batenanalyse van het natuurherstelproject Hemmepolder, p. 105.

35. Van Biervliet, K., Bogaert, G., Deconinck, M., Le Roy, D., and Bogaert, S. 2003. Ontwikkelen van socio-economische beoordelingscriteria die het mogelijk maken de kostprijs van een degradatie van het mariene milieu objectief te bepalen, 113 p. Opgenomen in: Federaal wetenschapsbeleid: PODO I, Programma: "Duurzaam beheer van de Noordzee," Eindrapport: Beoordeling van de mariene degradatie in de Noordzee en voorstellen voor een duurzaam beheer—MARE-DASM Te bestellen via http://www.belspo.be/belspo/home/publ/rappMNDD2_nl.stm

36. Bogaert, S., Van Hoof, V., and Le Roy, D. 2004. Economische waardering van parken, Ecolas, Ministerie van de Vlaamse Gemeenschap—Afdeling Bos en Groen, p. 151.

37. Lambrechts, W. 2006. *Economische waardering van natuurgebieden. Case-study: Meldertbos.* Brussels: Vrije Universiteit Brussel, p. 116.

38. Broekx, S., Smets, S., Liekens, I., Bulckaen, D, Smets, S., and De Nocker, L. 2011. Designing a long-term flood risk management plan for the Scheldt estuary using a risk based approach, *Natural Hazards* 57, 245–266.

39. Brouwer, R., Beckers, A., Courtecuisse, A., Vanden Driessche, L., and Dutrieux, S. 2007. Economic valuation of the non-market benefits of the European Water Framework Directive: An international river basin application of the contingent valuation method, report SCALDIT p. 37.

40. Liekens, I., De Nocker, L., Schaafsma, M., Wagtendonk, A., Gilbert, A., Broekx, S., and Brouwer, R. 2009. AQUAMONEY Case Study Report International Scheldt Basin, Vrije universiteit Amsterdam (IVM), October 2009.

41. ACTeon, Espace Environnement, CEESE—ULB. 2009. Evaluation économique des bénéfices environnementaux non-marchands et de la valeur de non-usage réalisés suite à la mise en oeuvre des plans de gestion de l'eau et l'atteinte des objectifs environnementaux de la Directive Cadre Eau pour les eaux de surface en Région wallonne.

42. Colson, V. 2009. La fonction récréative des massifs forestiers wallons: analyse et evaluation dans le cadre d'une politique forestière intégrée (Thèse de doctorat). 277 p. Gembloux: Faculté universitaire des Sciences agronomiques.

43. Fontaine, C.M., De Vreese, R., Jacquemin, I., Marek, A., Mortelmans, D., Dendoncker, N., Devillet, G., François, L., and Van Herzele, A. 2012. Valuation of Terrestrial Ecosystem Services in a Multifunctional Peri-Urban Space (The VOTES project). Final Report. Brussels: Belgian Science Policy.

44. Liekens, I., Schaafsma, M., Staes, J., De Nocker, L., Brouwer, R., and Meire, P. 2010. Economische waarderingsstudie van ecosysteemdiensten voor MKBA. Een handleiding. In opdracht van LNE, Brussels.

45. Power, T.M. 2001. The contribution of economics to ecosystem preservation: Far beyond monetary valuation. *In* Managing Human-Dominated Ecosystems: Proceedings of the Symposium at the Missouri Botanical Garden, St. Louis, Missouri, March 26–29, 1998.

46. Salzman, J., Thompson, B.H., Jr., and Daily, G.C. 2001. Protecting ecosystem services: Science, economics, and law. *Stanford Environmental Law Journal* 20, 309–332.

47. Toman, M.A. 1997. Ecosystem Valuation: An Overview of Issues and Uncertainties. *In* Ecosystem Function and Human Activities: Reconciling Economics and Ecology. (R.D. Simpson and N.L. Christensen Jr., eds.), pp. 25–44. New York: Chapman and Hall.

Chapter 3

Biodiversity and Ecosystem Services

Sander Jacobs[1,2], Birgen Haest[4], Tom de Bie[5,6], Glenn Deliège[7], Anik Schneiders[3] and Francis Turkelboom[3]

[1]Research Institute for Nature and Forest (INBO), [2]University of Antwerp. Department of Biology, Ecosystem Management Research Group (ECOBE), [3]Research Institute for Nature and Forest (INBO), [4]The Flemish Institute for Technological Research (VITO), [5]KU Leuven, Laboratory of Aquatic Ecology, Evolution and Conservation, [6]Flemish Environment Agency (VMM), Catchment office Dijle-Zenne catchment, [7]KU Leuven, Husserl-Archives, International Centre for Phenomenological Research

Chapter Outline

1. INTRODUCTION

Fully understanding the concept of ecosystem services (ES) requires thorough knowledge of the ecological principles that characterize interactions between organisms and their environment. The scales of these interactions vary from microbe to landscape and global level, and from milliseconds to millions of years, further complicated by multiscale energy and material fluxes. This results in an enormously complex biological interacting system that has filled the agenda of many researchers across the world, in a search to understand, determine, and characterize the entanglement of organisms, biological processes, and their environment, and—perhaps the most essential aspect—the importance of natural ecosystems for human well-being and survival.

The link between both concepts—biodiversity and ecosystem services—is obvious. But due to the complexity of both terms [1, 2, 3, 4], discussions

Ecosystem Services. http://dx.doi.org/10.1016/B978-0-12-419964-4.00003-2

are often narrowed to specific components, provoking many debates. Because ecosystem service assessments are intended to provide guidance for ecosystem management, the confusion over how to treat biodiversity is potentially a serious problem [4]. A clarification of the biodiversity concept in relation to ecosystem services is needed.

2. BIODIVERSITY

From its inception, the term *biodiversity* has had a dual purpose [5, 6, 7]. On the one hand it had to serve a scientific goal in analyzing and remediating the perceived extinction-crisis; on the other, it had to thematize the extinction crisis as a pressing concern that should be high on the political and societal agenda. Serving two masters, the term *biodiversity* has thus a claim to both objectivity and normativity. Keeping these two components apart is crucial for the viability of biodiversity as a scientific concept, although in practice both components are often muddled together, leading to a concept that has only limited use for either science or policy [6].

The term *biodiversity* was launched at the 1986 National Forum on Bio-Diversity [8], a conference conceived within the circles of the then recently founded Society for Conservation Biology with the express aim of introducing the term into the scientific and especially the political community, as well as into society as a whole [5]. The Society for Conservation Biology itself had been founded out of a growing concern among a group of mainly North American biologists for the rate and scale at which natural habitats were being transformed by humans and the loss of species that such transformations entailed [9]. Moreover, there was a growing discontent among certain biologists about the limits of contemporary nature conservation legislation in dealing with the extinction crisis and the failure of the science of biology to come up with practical solutions for countering the crisis. This created the need for both a term expressing these concerns and a scientific affiliation researching conservation issues [7].

The term *biodiversity*, in its normative component, has the express purpose of changing the way in which one thinks about nature [5]. Whereas traditional nature conservation in the United States had focused on nature as wilderness (which found its consecration in the Wilderness Act of 1964) and on single endangered species (Endangered Species Act of 1973), such focus was found wanting in the thematization of the extinction crisis. What the members of the Society for Biological Conservation saw as the primary problem was not the loss of wilderness or individual species, but a threat to life as a whole. Hence, the term *biodiversity* was proposed as giving expression to this threat to life as a whole [5], as of course nothing could be a more pressing concern than the fact that life as a whole is being jeopardized [6]. As such, biodiversity came to stand for all of biological life and especially the way in which biological life is structured in a complex web of interacting, hierarchically structured levels of organization.

Seen in this way, biodiversity is a glitzy shorthand for the intricate biological organization of life on Earth, useful for policy contexts and as a rhetorical device in societal debate. However, when used as a scientific term, it becomes too broad and vague to be used in any kind of analysis of the causes, consequences, and solutions for the extinction of biological life. For this reason, when used in scientific studies or in concrete conservation schemes, it needs to be operationalized by giving it a much more exacting definition. Yet, because its general scope is so broad, the whole of biological life, operationalization of the term leads to the use of a plethora of sometimes conflicting operational definitions, usually depending on the goal of the study in question or the background and interests of the researchers [6, 7]. In this sense, the normative aspect of biodiversity clashes with its scientific pretentions.

Although concern about the decline of biodiversity is accepted worldwide, the biodiversity framework is too broad and vague to be applied in daily routines [2]. Nature managers, farmers, foresters, or politicians, each of them reduces the meaning and values of biodiversity, pronouncing the aspects linked to their own practice. It is important to describe an overall picture of biodiversity and to clarify the most important characteristics, before engaging in any biodiversity–ecosystem service debate.

Biodiversity encompasses three important attributes—composition, structure, and function—all nested in a range of organizational levels—genetic, species-population, community-ecosystem, and landscape [2] (see Figure 3-1). The "diversity" part of the concept strictly refers to the degree of variation. In statistics, diversity reflects not only the number of different biological types but also how evenly they are distributed within a certain area. Diversity can be the number and evenness of species in an ecosystem or a landscape, the amount of gene heterozygosis within a population, or the amount and complexity of organization levels within a community. Typical diversity indexes for multiple-scale biodiversity are alpha (α), beta (β), and gamma (γ) diversity, reflecting the local species diversity, the differences in diversity between local sites, and the total diversity within the landscape ($\alpha * \beta = \gamma$) [10].

Not all these components of biodiversity are equally implemented in a policy or management context. The most widely used policy definition is formulated within the convention on biological diversity organized by the United Nations in 1992: "Biodiversity or biological diversity means the variability among living organisms from all sources including, inter alia, terrestrial, marine and other aquatic ecosystems and the ecological complexes of which they are part: this includes diversity within species, between species and of ecosystems."

Although this definition roughly includes all attributes, it pronounces the compositional part. As a result, a huge amount of indicators—such as red lists, umbrella species, or endangered habitats—is developed. Nature conservation policy—such as the European Habitat and Bird Directive—also focuses on the conservation and restoration of these selected species and habitats. The logical outcome is a narrowing of the debate toward a limited part of the biodiversity.

In science as well as in the policy and management context, the need for more integrated indicators—including the structural and functional aspects—increased. More elaborated indexes for diversity depend on distribution of species, functional groups, traits, or combinations of these typologies.

3. BIODIVERSITY WITHIN THE FRAMEWORK OF ECOSYSTEM SERVICES

The integrated approach mentioned above—especially the part focusing on the functional aspects—has gained increasing attention in ecological and environmental sciences over the last two decades and has created an ecological foundation for the framework of ecosystem services [11, 12, 13, 14]. This debate, now revived as "stabilizing ecosystem functions," is interpreted as stabilizing the basis of the ES supply.

It is to be expected that some conservation goals (as number and abundance of protected species) will not always coincide with goals for optimizing ecosystem service delivery. However, the spatial estimation of ecosystem services and biodiversity across all types of ecosystems on a regional scale often remains quite crude and difficult, because relationships are often blurred by missing or unreliable data and many studies often include a limited number of species as well as ecosystem services [15].

Within ecosystem services discourses and frameworks, the position of the term *biodiversity* has shifted during the last decade. In early schemes, biodiversity was sometimes listed as "'just another" service. Although the importance of (broadly defined) biodiversity underpinning all ecosystem services is central in, for example, the Millennium Ecosystem Assessment [16], the positioning within its classification framework has caused confusion and discussion, sometimes leading to skepticism toward the ecosystem service concept altogether. Within later frameworks, such as the Economy of Ecosystems and Biodiversity [17], biodiversity is more implicitly mentioned as "habitat and supporting services" and is depicted as a category overarching (supporting) all other services. Still, this gives the impression that these services could be quantified and valued along with the others. However, in the UK National Ecosystem Assessment, these supporting services are regarded as "intangible," and no exact quantification or pricing attempt is made [18]. Generally, it is accepted that pricing biodiversity or the entire support system in practice comes down to determining the price of virtually everything, which is senseless (Chapter 1 in this volume). Notwithstanding this evolution toward a nonmonetary, noncommodity consensus on biodiversity versus ES, discussion still smolders as (priceable) benefits from biodiversity-dependent services are often communicated as *thé value of biodiversity*. Also, the adverse effects of ES optimization on (single) species conservation goals are often wrongly perceived as proven negative impacts on biodiversity or the ecosystem as a whole. The reason for ongoing and inhibiting confusion is the use of the single term *biodiversity* for different concepts, as well as the genuinely different roles it plays in ecosystem processes and services [4].

The fact that biodiversity valuation indicators are often limited to "the amount of species" or "endangered species" creates a lot of confusion and puts the "biodiversity—ES debate" in a wrong perspective. Reducing the ecosystem service concept to monetary pricing (e.g., Chapter 9 in this volume) and neglecting the potential to include multiple values sharpen this misconception.

4. BIODIVERSITY AND ECOSYSTEM FUNCTIONING

Mace et al. [4] do a very good job of disentangling different layers of meaning of biodiversity within the ecosystem service concept, and guide the way out of many fruitless debates. First, *biodiversity is a regulator of ecosystem processes*. This aspect will receive the largest attention in our contribution. However, the debate is confused by the use of *biodiversity as final ecosystem service*, linked to direct provision of goods and services (e.g., functional biodiversity linked to crop production), and *biodiversity as a good*, where biodiversity in itself is the object valued by humans (e.g., biodiversity conservation attaches a range of cultural values to biodiversity) [4].

Different relationships between biodiversity in terms of species richness and its corresponding magnitude of ecosystem functions are amply described in the literature [12, 19, 20, 21, 22, 23]. In most theoretical as well as experimental graphs, the magnitude of a function levels off with the increase of species diversity. A further increase can make the system more resilient but will not increase the magnitude, and additional species "matter less." This relationship is illustrated by the grassland experiments of Tilman [24] describing the increase of biomass production with increasing species diversity. In a linear relation, every species matters equally, and functioning increases with every single species. Rare species as well as dominant species contribute to the ecological functions. A concave curve describes a relationship where the ecological function depends on one or only a few specific species, and losing only a small part of the biodiversity could mean a total loss of the corresponding function.

In reality, more unpredictable relationships are described [21], and the gradient between a high and low level of biodiversity in relation to the functional magnitude is often not well known. However, all relations generally point to increased functioning with higher biodiversity.

These relationships can also be found on a landscape scale, with site patches as biodiversity units. In highly fragmented landscapes with a low connectivity, losing small patches will often result in a gradual decline of the related ecosystem functions, showing a linear relationship. In highly connected networks, local losses tend to be repaired by inputs from linked patches, until a critical level is attained and the system collapses [22].

4.1. Biodiversity as Insurance of Ecosystem Service Delivery

Generally, ecosystem functions are thus stabilized and diversified by a higher biodiversity [25]. This implies that ecosystem service delivery by these functions is insured. To use the ecosystem service perspective as a tool for the management of natural resources, understanding the inherent dependence on biodiversity through this insurance function is essential. In other words: which functions are important for maintenance of the ecosystems' functioning?

Several recent studies have empirically shown links between the functioning of ecosystems and most of the known aspects of biodiversity, for example, genetic diversity [26], trophic diversity [27], and resource diversity [28]. Functional structure of communities, that is, the composition and diversity of functional traits, appears to be a good predictor of ecological processes (e.g., [29]).

The current challenge is to include different facets of biodiversity simultaneously, which strongly increases the complexity of the studies. By focusing on the relationships among genes, traits, phylogeny, and biotic and abiotic factors, more insight into the mechanisms that underpin the functioning of ecosystems will be gained.

Because ecosystem functions are the biophysical foundation of the services an ecosystem delivers, a variety of empirically established links between ecosystem services and biodiversity have been reported. For example, Hajjar et al. [30] looked at the potential benefits of crop genetic diversity in enhancing the provision of services. Increasing crop diversity has shown to be directly or indirectly useful in pest and disease management and enhanced pollination services and soil processes. Potvin and Gotelli [31] showed that more diverse plantings increased yields in tropical tree plantations, due to the increased growth of individual trees. Ostfeld and Keesing [32] investigated the risk of Lyme disease and argued that increased species diversity within host communities may reduce the risk of exposure to Lyme disease. Most of this research on the relationship between aspects of biodiversity and several ecosystem services has been done on terrestrial grasslands, plants, and insects. Information on aquatic systems [33, 34] and microbial communities is lacking, but new technologies offer promising shifts in this research field (e.g., [35]).

In summary, recent reviews and meta-analyses gather evidence that biological diversity may boost the functioning of ecosystems and their ability to provide society with the goods and services needed to prosper ([36, 37]; Figure 3-1).

When a single service in a certain context is considered, two important mechanisms occur: First, more diverse communities are more likely to contain and become dominated by strong performers (i.e., the sampling effect; [25]). This is very relevant for habitats that are strongly fragmented because these strong performers may not be able to reach the habitats. Second, niche differentiation and facilitation may increase the performance of communities above that expected from the performance of individual species alone (i.e., the complementarity effect; [14]). The two mechanisms do not exclude each other.

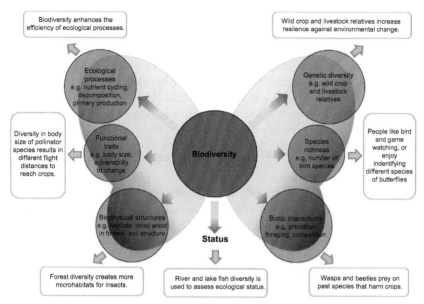

FIGURE 3-1   The multifaceted role of biodiversity to support the delivery of ecosystem services. *Source Figure 1: European Union (2013) Mapping and Assessment of Ecosystems and their Services: An analytical framework for ecosystem assessments under Action 5 of the EU Biodiversity Strategy to 2020. Discussion paper* (http://ec.europa.eu).

The majority of studies focus on only one or a few functions or services. Caliman et al. [38] indicate in their review paper that, although scientific attention to the topic has been growing, there is a strong bias toward short-term experimental studies in which biodiversity was experimentally manipulated either in the laboratory or in the field. They furthermore conclude and suggest that the research emphasis should now be more directed toward the study of integrative and larger-scale real-world observational studies.

In reality, ecosystems indeed deliver multiple ecosystems services, which are supplied by a large number of functions that in turn all depend on different physical circumstances, species, and ecological preferences. Only recently have studies that characterize biodiversity and multiple ecosystem functions and services emerged. These studies suggest that the importance of biodiversity increases even more when one considers multiple functions [39] because different species often influence different functions, and studies focusing on individual processes in isolation will underestimate levels of biodiversity required to maintain multifunctional ecosystems.

The ability of ecosystems to maintain functioning under varying circumstances, stresses or disturbances, or the resilience of the ecosystem, also ensures the delivery of services. This insurance function of biodiversity is likely due to "partial" redundancy, which acts as a buffer: Species may have similar functions but operate optimally under different environmental conditions, in this way replacing each other to maintain basic ecosystem processes [40]. The overlap in functionality combined with differences in ecological preferences enhances resilience or the system's capacity to recover from external pressure. This simply occurs by replacement of a disappearing species by a species with similar functioning but different preferences and reactions to the cause of extinction.

However, the less species and functional redundancy, the smaller this buffer. This principle is illustrated by the rivet metaphor: When one rivet pops from a plane's wing, the other rivets will easily take the extra pressure. In case of functional redundancy between species, the initial loss of specific species caused by external pressures may not significantly impact ecosystem functioning because the functions are compensated by species with similar characteristics. However, with every rivet removed, the chance that the next one will pop increases exponentially, and at a certain point the plane suddenly crashes. As ecosystem diversity decreases further and a certain threshold is crossed, the loss of species may result in a sudden and drastic decrease of ecosystem functioning and cause cascading ecosystem changes and regime shifts [41, 42, 22].

The loss of resilience caused by the loss of species has been shown to be a crucial factor in triggering regime shifts [43]. Regime shifts are large, abrupt, persistent changes in the structure and function of a system, and they can substantially affect the flow of ecosystem services on which societies rely [44]. Loss of species across trophic levels may alter food web interactions, and decreasing key mediators of ecosystem functioning [45] and loss of organisms with specific functional traits may cause direct reductions in ecological processes relevant to ecosystem services [46]. In terms of ecosystem services, the effects of such regime shifts can be huge, but predicting shifts is difficult.

The importance of biodiversity increases further when we consider the multifunctionality of ecosystems at increasing spatial and temporal variability (e.g., [27, 48, 49]): Species that control ecological functions in a particular year or location often differ from those that control processes in other years or locations and under other environmental conditions.

Conclusively, a larger suite of species, structures, and processes will exhibit differential responses to a given environmental perturbation and therefore, when considered together, create a stabilizing function that preserves the integrity of a service [50]. Consequently, to ensure a stable supply of ecosystem functions and services in a changing world, biodiversity conservation and restoration through determining ecological or biodiversity value is essential [4].

4.2. Food for Thought

It is generally accepted that biodiversity plays an important role in the extent and stability of the services provided by ecosystems [21]. Moreover, loss of biodiversity puts at risk essential ecosystem service delivery such as provision of food and clean water, or climate regulation [36, 37]. Recently, both Schneiders et al. [51] and Maes et al. [52] showed, respectively, at the Flemish regional and European continental spatial scale, a clear link between a decline in both biodiversity and vital ecosystem services with an increase in land-use intensity.

Ascertaining our own survival's dependence on biodiversity, one conclusion stands out: Species extinction and habitat degradation rates, which are currently at least 100 times higher than the average over the past history of life on Earth [52], are a direct and imminent threat to our own well-being and survival.

The debate on biodiversity loss and ecosystem services should focus on the societal consequences of biodiversity decrease, or, more constructively, on opportunities for nature conservation and restoration to maintain and increase ES delivery and human well-being.

An ecosystems approach is not an alternative paradigm to conservation, but conservation is simply part of a bigger societal picture [54]. The evidence base of using selected indicator groups for protection of our natural capital is weak [55], and the conservation community can be more explicit about the societal values of biodiversity (from utilitarian to cultural) [4]. Provided a broad multitude of values is taken into account, ecosystem services are an opportunity rather than a threat to biodiversity conservation.

Combined research involving natural and social sciences as well as societal stakeholders can focus on teasing out the supporting ecosystem functions that produce and maintain service delivery, and on establishing straightforward ways of incorporating physical limits concerning biodiversity in natural resource management policies. Strategies for sustainable ecosystem management can maintain resilience and transform degraded ecosystems into more resilient configurations that increase long-term delivery of bundles of services.

REFERENCES

1. Norgaard, R.B. 2010. Ecosystem services: From eye-opening metaphor to complexity blinder. *Ecological Economics* 68, 1219–1227.
2. Noss, R.F. 1990. Indicators for monitoring biodiversity: A hierarchical approach. *Conservation Biology.* Vol. 4, pp. 355–364.
3. Reyers, B., Polasky, S., Tallis, A., Mooney, H.A., and Larigauderie, A. 2012. Finding common ground for biodiversity and ecosystem services. *BioScience* 62, 503–507.
4. Mace, G.M., Norris, K., and Fitter, A.H. 2012. Biodiversity and ecosystem services: A multilayered relationship. *Trends in Ecology and Evolution* 27(1), 19–26. doi:10.1016/j.tree.2011.08.006
5. Takacs, D. 1996. *The Idea of Biodiversity.* Johns Hopkins University Press, Baltimore, MD.
6. Haila, Y. 1994. Making the Biodiversity Crisis Tractable: A Process Perspective. In: Oksanen, M., Pietarinen, J. *Philosophy and Biodiversity.* Cambridge University Press, pp. 54–84.
7. Sarkar, S. 2005. *Biodiversity and Environmental Philosophy.* Cambridge Washington DC: Cambridge University Press.
8. Wilson, E. (eds.). 1988. *BioDiversity.* Washington, DC: National Academy Press.
9. Soulé, M. 1985. What Is Conservation Biology?. *BioScience* 35(11), 727–734.
10. Whittaker, R.H. 1960. Vegetation of the Siskiyou Mountains, Oregon, and California. *Ecological Monographs* 30, 279–338.

11. Balmford, A., Bruner, A., Cooper, P., Costanza, R., Farber, S., Green, R.E., Jenkins, M., Jefferiss, P., Jessamy, V., Madden, J., Munro, K., Myers, N., Naeem, S., Paavola, J., Rayment, M., Rosendo, S., Roughgarden, J., Trumper, K., and Turner, K. 2002. Economic reasons for conserving wild nature. *Science* 297, 950–953.

12. Balvanera, P., Pfisterer, A.B., Buchmann, N., He, J-S., Nakashizuka, T., Raffaelli, D., and Schmid, B. 2006. Quantifying the evidence for biodiversity effects on ecosystem functioning and services. *Ecology Letters* 9, 1146–1156.

13. Hooper, D.U., Chapin, F.S., Ewel, J.J., Hector, A., Inchausti, P., Lavorel, S., Lawton, J.H., Lodge, D.M., Loreau, M., Naeem, S., Schmid, B., Setälä, H., Symstad, A.J., Vandermeer, J., and Wardle, D.A. 2005. Effects of biodiversity on ecosystem functioning: A consensus of current knowledge. *Ecological Monographs* 75, 3–35.

14. Loreau, M., and Hector, A. 2001. Partitioning selection and complementarity in biodiversity experiments. *Nature* 412, 72–76.

15. Naidoo, R., A. Balmford, R. Costanza, B. Fisher, R.E. Green, B. Lehner, T.R. Malcolm, and T.H. Ricketts. 2008. Global mapping of ecosystem services and conservation priorities. *Proceedings of the National Academy of Sciences* 105(28), 9495–9500.

16. MA. 2005. *Ecosystems and Human Well-being: Summary for Decision Makers*. Washington, DC: Island Press.

17. TEEB. 2010. The Economics of Ecosystems Biodiversity. *The Economics of Ecosystems and Biodiversity: Mainstreaming the Economics of Nature: A Synthesis of the Approach Conclusions and Recommendations of TEEB*. Malta: Progress Press.

18. UK National Ecosystem Assessment. 2011. The UK National Ecosystem Assessment, Synthesis of the Key Findings. Cambridge: UNEP-WCMC.

19. Haines-Young, R., and M. Potschin. 2010. The links between biodiversity, ecosystem services and human well-being. Chapter 7 in: Raffaelli, D. and C. Frid (eds.). *Ecosystem Ecology: A New Synthesis*. Cambridge: BES Ecological Reviews series, pp. 110–139.

20. Kremen, C. 2005. Managing ecosystem services: What do we need to know about their ecology? *Ecological Letters* 8, 468–479.

21. Naeem, S., Bunker, D.E., Hector, A., Loreau, M., and Perrings, C. 2009. *Biodiversity, Ecosystem Functioning and Human Wellbeing*. New York: Oxford University Press.

22. Scheffer, M., Carpenter, S.R., Lenton, T.M., Bascompte, J., Brock, W., Dakos, V., van de Koppel, J., van de Leemput, I.A., Levin, S.A., van Nes, E.H., Pascual, M., and Vandermeer, J. 2013. Anticipating critical transitions. *Science* 338, 344–349.

23. M.W. Schwartz, M.W., Brigham C.A., Hoeksema, J.D., Lyons, K.G.,·Mills, M.H., and van Mantgem, P.J. 2000. Linking biodiversity to ecosystem function: Implications for conservation ecology. *Oecologia* 122, 297–305.

24. Tilman, D., C.L. Lehman, and C.E. Bristow. 1998. Diversity-stability relationships: Statistical inevitability or ecological consequence? *The American Naturalist* 151, 277–282.

25. Cardinale, B.J., Srivastava, D.S., Duffy, J.E., Wright, J.P., Downing, A.L., Sankaran, M., and Jouseau, C. 2006. Effects of biodiversity on the functioning of trophic groups and ecosystems. *Nature* 443, 989–992.

26. Cook-Patton, S.C., McArt, S.H., Parachnowitsch, A.L., Thaler, J.S., and Agrawal, A.A. 2011. A direct comparison of the consequences of plant genotypic and species diversity on communities and ecosystem function. *Ecology* 92, 915–923.

27. Duffy, J.E., Cardinale, B.J., France, K.E., McIntyre, P.B., Thébault, E., and Loreau, M. 2007. The functional role of biodiversity in ecosystems: Incorporating trophic complexity. *Ecology Letters* 10, 522–538.

28. Langenheder, S., Bulling, M.T., and James, M.S. 2010. Bacterial Biodiversity-Ecosystem Functioning Relations Are Modified by Environmental Complexity. *PloS ONE 5*, e10834.
29. Mouillot, D., Villéger, S., Scherer-Lorenzen, M., and Mason, N.W.H. 2011. Functional structure of biological communities predicts ecosystem multifunctionality. *PloS ONE 6*, e17476.
30. Hajjar, R., Jarvis, D.I., and Germill-Herren, B. 2008. The utility of crop genetic diversity in maintaining ecosystem services. *Agriculture, Ecosystems and Environment* 123, 261–270.
31. Potvin, C., and Gotelli, N.J. 2008. Biodiversity enhances individual performance but does not affect survivorship in tropical trees. *Ecology Letters* 11, 217–223.
32. Ostfeld, R.S., and Keesing, F. 2000. Biodiversity and Disease Risk: The Case of Lyme Disease. *Conservation Biology* 14(3), 722–728. doi:10.1046/j.1523–1739.2000.99014.x
33. Worm, B., Barbier, E.B., Beaumont, N., Duffy, E., Folke, C., Halpern, B.S., Jackson, J.B.C., Lotze, H.K., Micheli, F., Palumbi, S.R., Sala, E., Selkoe, K.A., Stachowicz, J.C., and Worm, R.W. 2006. Impacts of Biodiversity Loss on Ocean Ecosystem Services. *Science* 314, 787–790.
34. Lecerf, A., and Richardson, J.S. (2010). Biodiversity–ecosystem function research: Insights gained from streams. *River Research and Applications* 26, 45–54.
35. Roux, X. le, Recous, S., and Attard, E. 2011. Soil microbial diversity in grasslands and its importance for grassland functioning and services. In: Lemaire, G., Hodgson, J., and Chabbi, A. (eds.). *Grassland Productivity and Ecosystem Services*, Wallingford, UK: CAB International, pp. 158–165.
36. Cardinale, B.J., Duffy, E., Gonzalez, A., Hooper, D.U., Perrings, C., Venail, P., Narwani, A., Mace, G.M., Tilman, D., Wardle, D.A., Kinzig, A.P., Daily, G.C., Loreau, M., Grace, J.B., Larigauderie, A., Srivastava, D.S., and Naeem, S. 2012. Biodiversity loss and its impact on humanity. *Nature* 486, 59–67.
37. Naeem, S., Duffy, J.E., and Zavaleta, E. 2012. The functions of biological diversity in an age of extinction. *Science* 336, 1401–1406.
38. Caliman, A., Pires, A. F., Esteves, F. A., Bozelli, R. L., and Farjalla, V. F. 2009. The prominence of and biases in biodiversity and ecosystem functioning research. *Biodiversity and Conservation* 19(3): 651–664. doi:10.1007/s10531-009-9725-0
39. Hector, A., and Bagchi, R. 2007. Biodiversity and ecosystem multifunctionality. *Nature* 448, 188–190.
40. Yachi S., and Loreau, M. 1999. Biodiversity and ecosystem productivity in a fluctuating environment: The insurance hypothesis. *Proceedings of the National Academy of Sciences USA* 96, 1463–1468.
41. Leadley, P., Pereira, H.M., Alkemade, R., Fernandez-Manjarrés, J.F., Proença, V., Scharlemann, J.P.W., and Walpole, M.J. 2010. Biodiversity Scenarios: Projections of 21st century change in biodiversity and associated ecosystem services. Secretariat of the Convention on Biological Diversity, *Montreal. Technical Series no.* 50, 132 pages.
42. Scheffer, M., Carpenter, S., Foley, J.A., Folke C., and Walker B. 2001. Catastrophic shifts in ecosystems. *Nature* 413, 591–596.
43. Folke, C., Carpenter, S. Walker, B., Scheffer, M., Elmqvist, T., Gunderson, L., Holling, C.S. 2004. Regime shifts, resilience, and biodiversity in ecosystem management. *Annual Review of Ecology, Evolution, and Systematics* 35, 557–581.
44. Biggs, R., et al. 2009. Turning back from the brink: Detecting an impending regime shift in time to avert it. *Proceedings of the National Academy of Sciences* 106, 826–831
45. Estes, J.A., Terborgh, J., Brashares, J.S., Power, M.E., Berger, J., Bond, W.J., Carpenter, S.R., Essington, T.E., Holt, R.D., Jackson, J.B.C., Marquis, R.J., Oksanen, L., Oksanen, T., Paine, R.T., Pikitch, E.K., et al. 2011. Trophic Downgrading of Planet Earth. *Science* 333, 301–306.

46. Diaz, S., Lavorel, S., de Bello, F., Quétier, F., Grigulis, K., and Robson, T.M. 2007. Incorporating plant functional diversity effects in ecosystem service assessments. *Proceedings of the Natlional Academy of Sciences USA* 104:20684–20689.

47. Stachowicz, J.J., Graham, M., Bracken, M.E.S., and Szoboszlai, A.I. 2008. Diversity enhances cover and stability of seaweed assemblages: The role of heterogeneity and time. *Ecology* 89, 3008–3019.

48. Isbell, F., Calcagno, V., Hector, A., Connolly, J., Harpole, W.S., Reich, P.B., Scherer-Lorenzen, M., Schmid, B., Tilman, D., van Ruijven, J., Weigelt, A., Wilsey, B.J. 2011. High plant diversity is needed to maintain ecosystem services. *Nature.* 477:199–202.

49. Zavaleta, E.S., Pasari, J.R., Hulvey, K.B., and Tilman, G.D. 2010. Sustaining multiple ecosystem functions in grassland communities requires higher biodiversity. *PNAS* 107, 443–1446.

50. Elmqvist, T., C. Folke, M. Nyström, G. Peterson, J. Bengtsson, B. Walker, et al., 2003. Response diversity, ecosystem change and resilience. *Frontiers in Ecology and the Environment* 1, 488–494.

51. Schneiders, A., Van Daele, T., Van Landuyt, W., and Van Reeth, W. 2011. Biodiversity and ecosystem services: Complementary approaches for ecosystem. *Ecological Indicators* 21, 123–133.

52. Maes, J., Paracchini, M.L., Zulian, G., Dunbara, M.B., and Alkemade, A. 2012. Synergies and trade-offs between ecosystem service supply, biodiversity, and habitat conservation status in Europe. *Biological Conservation* 155, 1–12.

53. Smith, F.D.M., May, R.M., Pellew, R., Johnson, T.H., and Walter, K.R. 1993. How much do we know about the current extinction rate? *TREEl* 8, 375–378.

54. Rands, M.R.W., et al. 2010. Biodiversity conservation: Challenges beyond 2010. *Science* 329, 1298–1303.

55. Lindenmayer, D.B., and Likens, G.E. 2011. Direct measurement versus surrogate indicator species for evaluating environmental change and biodiversity loss. *Ecosystems* 14: 47–59.

Chapter 4

Ecosystem Service Indicators: Are We Measuring What We Want to Manage?

Wouter Van Reeth

Research Institute for Nature and Forest (INBO)

Not everything that can be counted counts, and not everything that counts can be counted.
—William Bruce Cameron, 1963

Chapter Outline

1. INTRODUCTION

1.1. A New Discourse and Strategy in Biodiversity Policy

Since the 1990s, the importance of biodiversity and ecosystems in supporting economic welfare and human well-being has become a central line of argumentation for halting biodiversity loss. Policy makers and other stakeholders are embracing the idea that ecosystems are capital assets that yield a valuable flow of services, providing they are properly managed [1]. Ecosystem conservation and restoration with the purpose of safeguarding or optimizing flows of ecosystem services has become a clearly visible part of policy objectives at an

Ecosystem Services. http://dx.doi.org/10.1016/B978-0-12-419964-4.00004-4

international and European level [2, 3]. From there the concept of ecosystem services is being linked with legislation and policy programs that were already in place in national and regional legislation in countries around the world. As a result, international study projects such as the Millennium Ecosystem Assessment (MA—www.maweb.org) and The Economics of Ecosystems and Biodiversity (TEEB—www.teebweb.org) argue for the development of indicators to support and inform this policy cycle. The first sets of indicators have already been proposed [4].

This chapter investigates the extent to which ecosystem service indicators have become part of the knowledge base from which reports on the status and trends of biodiversity and ecosystems draw, with the Flanders region as a case study. It intends to gauge the capacity of policy agencies to inform decision makers on the stock and trend of natural capital and on the magnitude of resulting shifts in flows of ecosystem services and socioeconomic benefits. As such, it allows assessment of how well we are currently able to measure what we (want to) manage under the new global and European biodiversity strategies.

1.2. On Indicators

The MA suggests the selection and development of appropriate indicators of ecosystem conditions, ecosystem services, human well-being, and drivers of ecosystem change [5]. These indicators are to be highly relevant to policy makers, easily understood, and effectively convey the key findings about the impact of ecosystem change on human well-being. From the idea that "you cannot manage what you do not measure," TEEB for national and international policy makers recommends the development of a compact set of headline biodiversity and ecosystem service indicators [6]. These indicators should fit into a coherent framework for analysis "that addresses functional relationships between nature and human well-being." They should also be linked to "smart"[1] policy targets in order to serve as a management dashboard and to monitor policy progress. The streamlined set of headline indicators should be supported by more detailed and elaborate sets of indicators for measurement and monitoring, or for evaluation of specific policy instruments. Ecosystem service indicators should also be used to complement traditional macroeconomic indicators.

In public administration and policy science literature, indicators have been defined in various ways [7]. These definitions are usually based on their content (what they measure and communicate), their format (how that content is expressed and presented), or their functionality (how they can be used). In terms of their functionality, they can facilitate the communication between scientists, policy makers, and stakeholders. They can be used to clarify, to some extent, the complex relationships between ecosystems, the services they generate, their impact on stakeholders and society at large, and vice versa. They can also help

---

1. SMART: specific, measurable, accepted, realistic, time-specific.

to monitor, evaluate, and report progress on policy implementation and the distance to policy targets. Along with these definitions there is often a normative undertone, referring to why and how they should or should not be used.

Indicators are usually selective representations of a phenomenon. It is this selectivity that gives indicators their focus and policy relevance. But at the same time this selectivity introduces values, policy targets, political choices, and value judgments—in short, a certain bias. Explicit acknowledgment of this selectivity can help avoid misinterpretations or political abuse of the conveyed information.

Indicators can be expressed on a quantitative or qualitative scale, based on a data collection that can be repeated (in principle) across time and space. Typical representations are line, bar, or pie graphs and maps. A reference to a policy target (e.g., distance to target), historical observations (e.g., time series), or observations on other locations (e.g., experiments, benchmarks) can be included in the presentation to assist in or suggest interpretation. As such, indicators also imply the use of a certain type of data, knowledge, and expertise. Knowledge that is not easily harnessed into indicator formats may thus have a more difficult time in reaching stakeholders, policy makers, and politicians in indicator-based communications.

2. A SYSTEMS APPROACH FOR THE DEVELOPMENT, INTERPRETATION, AND ASSESSMENT OF INDICATORS

Ecosystems are a form of renewable natural capital [1, 8–10]. Capital is a stock that yields an output, which may be a flow of tangible goods (e.g., biomass, freshwater) of biophysical services (buffering floods, erosion control) or even relational services (identity and social relations depending on landscape features) (see Figure 4-1). If the goods produced by an ecosystem are left

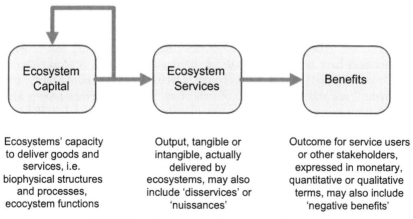

FIGURE 4-1   Components of ecosystem performance.

untouched, they become part of the capital stock and renew or transform it, to support other or future flows of goods and services (see the feedback loop in Figure 4-1). Timber from a forest or fish from the sea are typical examples. On the other hand, if these goods are harvested and removed from the ecosystem, they can be directly consumed by humans. Often, this consumption will imply use as a resource in a production process, together with input from manufactured, human, and social capital.

Intensifying the harvest or extraction of ecosystem goods can reduce the capital stock's ability to provide these same goods in the future, or to perform other services. If more wood is removed from a forest than it can regenerate, both the stock or future flow of goods (e.g., timber) and the capacity to supply other services (e.g., purify air, filter water, and prevent landslides) will decrease. The future performance of the ecosystem is thus undermined, and the exploitation of the ecosystem may become unsustainable if the capital stock decreases below some minimum level. A sustainable flow of goods and services that does not undermine ecosystem health and performance (also called the maximum sustainable yield) can be regarded as a form of natural income that increases economic prosperity and human well-being. However, in our economic accounts and information systems, any harvest and consumption of ecosystem goods and services is regarded as income, even when it exceeds sustainability limits.

When ecosystems are viewed as functioning entities with a natural capital base that delivers output (goods and services) and generates outcome (benefits), then ecosystem performance indicators need at least to capture the capital base, output, and outcomes of ecosystems. However, existing indicator assessments or reviews tend to cover the components of ecosystem performance only partially, or they fail to distinguish between stocks, output, and outcome. Some have focused on the flow of services (e.g. [11]). Others have distinguished between stocks and flows (e.g. [12, 13] but mix supply and demand. In other cases, quantitative measures of service volume are juxtaposed with monetary benefit measures, thus mixing output indicators with outcome indicators [6, 11].

Our understanding of the functional relationships between human activities, natural capital, ecosystem services, and human well-being is still very incomplete [12, 14]. Also, the interdependencies between biodiversity and ecosystem services have not been cleared out thoroughly, although general patterns have been proposed [15, 16]. In this complex web of causality, nonlinearity, and uncertainty, use and interpretation of indicators are tricky. Review reports and articles have suggested embedding sets of policy indicators in a DPSIR[2]-like framework or model [5, 6, 11, 12, 16–21]. Such a framework may serve as a structure to organize data and present policy-relevant information. Positioning the indicators along a logical structure can also assist in their interpretation. Furthermore, it can be used as a framework for action, highlighting areas of interest on which policy measures and programs should focus. We propose an approach built on frameworks that have been proposed and used by the World Resources

---

2. DPSIR: driving forces, pressure, state, impact, response.

Institute (WRI), TEEB, and UNEP's World Conservation Monitoring Centre to structure and analyze ecosystem service indicators (see Figure 4-2). WRI's Ecosystem Service Indicators Database (www.esindicators.org) is an example of how this could be set up in practice.

In Figure 4-2, *biodiversity* refers to the condition and biological integrity of ecosystems and the status and trends of species from an intrinsic value point of view, regardless of their functional contribution to ecosystem services and human well-being. Indicators of intrinsic value are not designed to assess, and do not necessarily represent, ecosystems' service-providing capacity, or vice versa (see Chapter 3 for a more extensive treatment of the biodiversity–ecosystem service relationship). *Ecosystem capital* represents an anthropocentric perspective on ecosystem structures (e.g., species populations, habitats), processes, and functions that generate flows of goods and services that may be appropriated

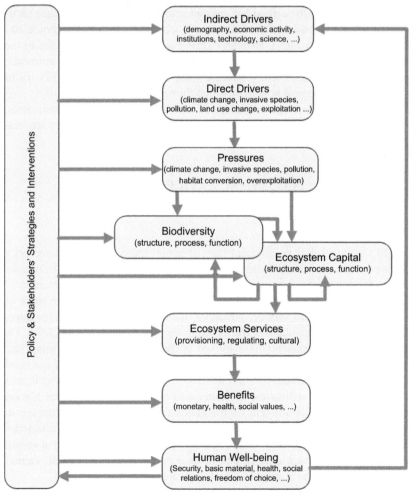

FIGURE 4-2   A systems approach to ecosystem service indicators (adapted from [10, 12, 13, 28]).

for human use. (e.g., service-providing units [22] and ecosystem service pro-viders [23]). *Ecosystem services* are the output delivered by the ecosystem for human use. This output includes what is sometimes referred to as disservices or nuisances [24–26]. For instance, farmers may suffer reduced crop yield and a decrease in personal income because of damage from wild herbivores residing in nearby forests. Green infrastructure in built areas may cause material damage or may require inspection, maintenance, and replacement in order to avoid damage costs. The *benefits* constitute the outcome of this use of ecosystem services. They can refer to levels of health, safety, or other individual or collective experiences affected by (changes in) ecosystem services. They can be expressed in terms of quantity, utility, happiness, euros, or any other form of appreciation expressed by humans. *Human well-being* is often put on the opposite of a continuum with poverty or "ill-being." It is not restricted to just material or economic wealth, although those are certainly part of it. It refers to the extent to which people's basic material needs for a good life are met, and whether they experience secu-rity, good health, good social relations, and generally freedom of choice and action [5, 27]. *Policy and stakeholder strategies and interventions* refer to the choices and behavior (production, consumption) through which governments, companies, individuals, NGOs, interest groups, and other stakeholders try to shape the socioecological system of which they are a part.

Table 4-1 presents some examples of Flemish ecosystem service indicators, structured according to this systems approach. The Flemish indicators are ana-lyzed in the next section of this chapter.

3. CASE STUDY: ECOSYSTEM SERVICE INDICATORS IN FLANDERS

3.1. Approach

Post-MA research has pointed out that, although many indicators have been proposed, "there is no consensus on a manageably small set that can be consis-tently applied and that services the needs of decision makers and researchers" [29, p.258]. Several reports and articles and reports have tried to take stock of the existing ecosystem service indicators, identify data and knowledge gaps, and provide recommendations for indicator development (e.g. [6, 11–13, 18, 30–33]). These assessments have looked at the scope of indicator sets, their scientific qualities, as well as their communicative abilities. A widely accepted practice in economics and accounting is to use "different costs for different purposes." Also for indicators, the criteria to define what constitutes a good indicator are highly contingent on the context in which the indicators are to be used (e.g., scale, policy domain, purpose of use) and on the characteristics of the user (e.g., knowledge with regard to the subject matter). As a result, assessments of indicator quality, and of their communicative abilities, cannot be walled off from the context in which they are used.

TABLE 4-1 Illustration of 16 Indicators of Ecosystem Performance.

Indicator Class	Indicator	Data Unit	Format	Source*
Ecosystem Capital				
Ecosystem Processes & Functions				
Water cycling	Infiltration potential of soil in Flanders	qualitative scale	regional map	Staes et al., 2010:28
	Water retention potential in Flanders	qualitative scale	regional map	Staes et al., 2010:30
Capital Stocks				
Water cycling	Ground water level	concept	concept	Staes et al., 2010:89
Carbon stocks	Carbon stocks in soil & biomass of forests	$kg\ C/m^2$	regional map	Lettens et al., 2010:66
Primary production	Standing wood stock (volume, species, age class)	m^3 and %	graph	Van der Aa, 2010:125
Air quality	Average annual PM10-concentrations in Flanders	$\mu g/m^3$	regional map	De Meulenaer et al., 2010:223
Ecosystem Services				
Provisioning Services				
Nutrition	Crop production	ton / year	number	Turkelboom, 2010:112
Materials	Wood sales from public forests & forest groups	m^3 / year	graph	Van der Aa, 2010:126

Continued

TABLE 4-1 Illustration of 16 Indicators of Ecosystem Performance—cont'd

Indicator Class	Indicator	Data Unit	Format	Source*
Regulating Services				
Regulation of wastes, pollution & nutrients	Nutrient removal (N, P) in the Scheldt estuary	Kg N, P/ha*year	number	Vandevenne, 2010:53
Climate regulation	Carbon sequestration by farmal and forests	ton C/ha*year	number	Lettens et al., 2010:70–71
Cultural Services				
Recreation services	Vicinity of green space for built areas	km	regional map	Jacobs et al., 2010:145
Benefits from Ecosystem Services				
Benefits from Provisioning Services				
From nutrition	Production from Flemish agriculture and horiculture	€/year	graph	Jacobs et al., 2010:113
From materials	Wood harvest	€/year	number	Jacobs et al., 2010:127
Benefits from Regulating Services				
From regulation of wastes, pollution & nutrients	N-removal by grass buffer strips	€/ha*year	number	Jacobs et al., 2010:262
From climate regulation	Benefits from carbon sequestration	€/ton C	number	Lettens et al., 2010:68–69
Benefits from Cultural Services				
From recreation services	Urban residents with access to green space	%	graph	Simoens, 2010:146

*The names in the source column refer to the lead author of the chapter of (Jacobs et al., 2010) [34] from which the indicators were compiled.

At the time of our analysis (the second half of 2011), the Flemish government had not approved any explicit policy strategy with regard to ecosystem services. We therefore focus our assessment on the scope of a set of indicators that had then been proposed to the Flemish biodiversity policy community. The indicators were compiled from the texts of two reports presented in 2010. One was the Flemish Biodiversity Indicators report, which is compiled on an annual basis by the Research Institute for Nature and Forest (INBO), based on data from various sources [34]. The second report was an exploratory study by Ecosystem Management Research Group (Ecosystem Management) (ECOBE) (University of Antwerp) and INBO, commissioned by the Flemish Agency for Nature and Forest (ANB) [35]. This study was the first in Flanders to collect policy-relevant knowledge and information ona wide range of ecosystem services on a regional scale.

The conceptual framework presented in Section 2 of this chapter was used as a reference to assess the scope. The division of the classes into subclasses is based on the CICES-classification, adapted for Belgium (see Chapter 18 and [36, 37]). The classification for drivers and pressures is based on that of the MA [5].

3.2. Results

The data collection resulted in a total of 152 indicators (see [38] for a complete list). About two-thirds of the indicators are supported by empirical data in the form of graphs, maps, or numbers. One-third is presented as a concept, with no data available. The set includes 29 European headline indicators (Streamlining European Biodiversity Indicators—SEBI) that were used to monitor the progress of Flanders under the European 2010 Biodiversity Strategy [34]. Eighty-nine other indicators or concepts also refer to the Flemish regional level; an additional 34 indicators or concepts refer to the subregional or local level. The Flemish regional indicators are not yet part of recurrent policy reporting. Thirteen indicators were presented as a graph, 10 as a map, and 35 simply as a number in a text. An additional 31 concepts or proposals for regional indicators are mentioned in words, and in some instances are illustrated or documented with data from a national or global scale.

The SEBI indicators for which data were available (graph, map, or number) reflect a classical pressure/state/response-discourse. The SEBI-indicators do not really cover drivers and ecosystem performance. The other regional and local indicators, however, collected from Jacobs et al., 2010 [34] focus much more on the components of ecosystem performance (ecosystem capital, ecosystem services, and benefits). However, in this study no indicators are suggested with regard to drivers, except for production and consumption trends in agriculture and forestry. Also, no indicators were suggested for human well-being.

About half of the indicators refer to aspects of ecosystem performance, with an equal distribution between ecosystem capital, service output, and benefits (see Table 4-2). The capital indicators focus on capital stocks rather than on ecosystem processes and functions. Output and benefit indicators refer mostly to regulating services, with some attention to provisioning and cultural services.

TABLE 4-2 Flemish Indicators of Ecosystem Performance. [34, 35]

			SEBI	Regional	Local	Concept
Ecosystem capital & integrity	Ecosystem processes & functions	Nutrient cycling		3	4	1
		Water flow regulation				1
	Natural capital stocks	Carbon stocks in ecosystems		1		1
		Ground water levels				1
		Soil formation				1
		Nutrient stocks in ecosystems		1		1
		Primary production		4		1
		Protective natural structures		1		1
		Air quality		1		1
Ecosystem services	Provisioning services	Nutrition		2		1
		Water supply				
		Materials		4		1
		Energy				1
	Regulation & maintenance services	Regulation of wastes, pollution & nutrients	1	3	3	10
		Water & mass flow regulation		4	4	2
		Regulation of climate		1		
		Regulation of biotic environment				
	Cultural services	Recreation services		1		
		Experiential services		1		1
		Intellectual services		2		2

Benefits					
Benefits from provisioning services	From nutrition		3		
	From water supply		2		
	From materials		2		1
	From energy				
Benefits from regulation & maintenance services	From regulation of wastes, pollution & nutrients		3	4	2
	From water & mass flow regulation			4	3
	From regulation of climate		1		
	From regulation of biotic environment		1		1
Benefits from cultural services	From recreation services	1	3	3	
	From experiential services				
	From intellectual services				
Total		2	34	22	28

While the 34 regional ecosystem performance indicators are evenly distributed over ecosystem capital, service output, and benefits, the local indicators focus more on benefits than on natural capital and service output. A number of 34 regional indicators of ecosystem performance may seem a lot, and certainly more than there are SEBI indicators. However, these 34 cover only 5 out of 11 classes for service output and 6 out of 9 classes for benefits. If we connect the indicators to the 56 service types of the Belgian CICES-classification, then only 4 out of 22 provisioning service types have indicators; for the regulating services, this is 7 out of 22. For the 12 cultural service classes, this analysis was not possible because most of the indicators referred to more than one cultural service (see Section 3.3, Discussion). Also, for many of the benefit indicators there was not a direct 1:1-link with the 56 different service types.

A more detailed analysis of the results is available in [38].

3.3. Discussion

The analysis shows the limited capacity of Flemish biodiversity agencies to collect and present indicators to monitor progress toward the 2020-target under the new biodiversity strategy [2, 3].

Methodological Limitations

The method and data used for this case study pose some limitations. By limiting the data collection to the two aforementioned reports, the set of collected indicators does not cover all that was available in Flanders. Other explorative reports from government agencies as well as recent research projects contain indicators that cover aspects of ecosystem performance. Because of the limited data collection, there is also some selection bias. The higher count of indicators for regulating services compared to provisioning services may reflect the focus and composition of the research teams (e.g., the functioning of hydrological systems, nutrient cycling, carbon sequestration, erosion control, cultural services) and the absence of some fields of expertise of provisioning services (e.g., energy, water provision, hunting, and wild food). The results presented above should therefore be interpreted with some caution.

All indicators collected represent single-service indicators. Indicators that synthesize information on trade-offs were not yet included. Nevertheless, informing decisions on the impact of alternative choices in land use or spatial planning (ex ante) or providing accountability for the consequences of these choices (ex post) is one area in which ecosystem services is deemed to be most promising [18]. Although more recent examples exist in Flanders where maps for different services have been juxtaposed, or where ecosystem services and benefits were synthesized in spider graphs, the indicators are still very experimental and evolving (e.g. [39, 40]. These examples usually refer to the local or landscape level. Upscaling this knowledge or finding alternative approaches to inform decision makers at the regional level remains a major challenge.

Besides spatial or sectoral trade-offs, choices between services and benefits for the current versus the next generation are part of land-use and spatial planning choices. Thus far, benefit indicators based on monetary valuations have been based on the standard 4% discounting standard [41, 42]. Alternative perspectives using declining or negative discount rates have to date received little or no attention in monetary indicators in Flanders.

From Capital Stocks to Service Flows

In some cases, conceptual inconsistencies in the underlying classifications create ambiguity on how to position the indicators in a systems approach or conceptual framework. The CICES-classification approaches some of the regulating and cultural services as stocks rather than as flows. For instance, the cultural ecosystem service "physical and intellectual interactions with ecosystems and landscapes' is operaas "area for outdoor activities."[43, p.6]" Similarly, the regulating service "protection against peak events—natural flood protection & sediment regulation" is operationalized as "natural flood plains & wetlands" [43, p.5]. A similar approach was observed for indicators in ecosystem assessments [31]. In some cases it may be impractical or too expensive to monitor the actual service output flow. For instance, monitoring "recreation output" would then require counting and surveying vacationers or other people conducting outdoor recreational activities across Flanders and converting these data into "recreation units." In the case of peak protection, measuring the service would require waiting for an extreme event to assess whether and where the service was actually provided. In these cases, a capital stock indicator may be a pragmatic and sensible solution. If it can be assumed or modeled that a higher or different level of natural capital stock (e.g., interpreting the extent of a recreation area or natural protective structures) is a proxy for an increased service capacity, potential service use, and/or for increased potential benefits, then measuring and reporting capital stock can offer credible and useful information on the expected service output and benefits. Other examples of such "stock for service indicators" can be found in Table 4-1.

There are also instances where the assumed causal relationship between natural capital stock and service output or between service output and benefit cannot be taken for granted. For instance, the same landscapes typically support recreational, social, and mental cultural services (see classification in [43]. Proxy measures based on capital stocks cannot distinguish between these services and implicitly assume that changes in capital stock affect these services evenly. In reality, changes in the capital stock are more likely to trigger uneven changes for different services and service users. If, say, an increase in the recreation area makes it more popular for vacationers during the weekend, it may negatively impact local residents, who appreciate the area for its identity, sense of place, or silence. A stock-based indicator may in this case result in a rhetoric of "more is better," while for some services or some stakeholders the outcome is just the opposite. The disconnect between natural capital stocks and ecosystem services is likely to be stronger where different service users attach different values to the same capital stock.

From Service Flows to Benefits

The causal relationship between service output and outcome (benefit) is an elusive one. Mainstream ecosystem services literature and manuals that define ecosystem services as "the benefits that people derive from nature" implicitly posit this relationship as evidence. In reality, there may be a certain service output level beyond which the benefit level does not increase further, or even decreases. An ongoing increase in the number of visitors in a natural landscape may lead to a congestion of service users, a decrease in the benefit per service user, and ultimately in a decrease in the total benefit. Moreover, if the service use exceeds the ecological carrying capacity of the area, then the capital basis may erode and negatively affect service output and benefits in the future. This knowledge is not new; in fact, much of the work in natural resource economics is focused on the relationship between natural capital stocks, levels of provisioning output, and the resulting economic benefits. Indicators such as maximum sustainable yield or maximum sustainable harvest remind us that stock, output, and outcome indicators are only interchangeable within certain ecological, economic, and social boundaries.

From Benefits to Human Well-being

None of the 152 collected indicators directly referred to human well-being. The benefit indicators could be argued to cover at least partly the contribution to human well-being (e.g., monetary value of health benefits). But the constituents of well-being, as described in the MA, are not yet part of the Flemish biodiversity indicators or indicator concepts. Part of the explanation for this gap lies in the conceptual challenges in defining clear and unambiguous indicators. Perhaps the fact that the indicators still largely reflect the knowledge base of the natural sciences, and to some extent that of environmental economics, can explain their absence. Several international organizations and forums compile country indexes to capture well-being and economic wealth in a more holistic way than standard macroeconomic indicators do. Examples include the net national product (NNP [44]), the index of sustainable economic welfare (ISEW [45]), the genuine progress indicator (GPI [46]), the human development index (HDI—http://hdr.undp.org/en/statistics/hdi/), and the happy planet index (HPI—http://www.happyplanetindex.org/).

Challenging the idea of linking well-being trends to (changes in) ecosystems is the notion that well-being is shaped by many factors other than ecosystem changes. Similarly, the benefits associated with ecosystem services do not result from ecosystems alone but from a mix of natural capital, manufactured capital, human capital, and social capital. Most people conducting outdoor activities of leisure in a natural environment prefer at least some basic infrastructure that facilitates their walks, bike rides, or horseback trips in the open. Such an infrastructure may suffer from vandalism and may require expensive repairs if there is not enough social capital to treat it well. Addressing these knowledge gaps again requires interdisciplinary research on the part of academics and intersectoral cooperation on the part of policy workers and stakeholders.

From Measurement to Management

Performance-oriented reforms in public administration and public policy since the early 1980s have introduced a more managerial discourse in the public and not-for-profit sector [47]. Developing "smart" results-oriented policy objectives and targets, monitoring progress through indicators, and including them in policy plans, budgets, management contracts, or evaluation reports has become a fairly common practice in many policy programs and policy domains, including environmental and biodiversity policy. Also, the idea that measures and indicators increase managerial capacity and improve decision making (cf. "you cannot manage what you do not measure") is still advocated as common managerial knowledge. The fact, however, that performance measurement and indicators may also pose barriers to improving performance is often not perceived [48]. Gaming strategies may be set up in which individuals or organizations react strategically to the publication or use of measures and indicators [49]. This may mean that policy is implemented selectively in such a way that performance scores improve, without necessarily improving policy performance. Alternatively, gaming may also mean that policy targets and indicator definitions are adjusted to what can successfully be reported via performance measures. This results in symbolic or rhetorical performance management practices that "hit the target but miss the point" [50, 51]. More measurement of ecosystem performance, then, is not necessarily conducive to better management of ecosystem performance.

From Management to Policy, Politics, and Power

Management has been defined as "taking responsibility for the performance of a system" [52]. Responsibility entails a top-down perspective of control, as well as a bottom-up perspective of accountability. In a public-sector environment, this might involve control over a government agency or policy program, or accountability toward a parliament, to the constituency during an election round, or just as well to the financial markets, as recent history has taught us.

Indicators are developed and used as part of a managerial language of efficiency, control, and accountability. This language imposes a moral order and reveals itself as "a political activity concerned with the creation, maintenance and manipulation of power" [48]. It affects how performance standards and indicators are set and what rewards or sanctions are attached to them. It determines what kind of information is needed and how it should or should not be used. This in turn influences and determines the measurement policy and the meanings derived from these measures. These meanings may be used, abused, or misused. This is the stage on which performance measures are performed. It is not uncommon for managerial textbooks and manuals to portray measures and indicators as mere technical tools that allow registering and monitoring reality. In reality however, measures and indicators are rather the opposite. They are specifically designed to have an impact on that reality. They are used to motivate an organization or stakeholders to pursue and achieve specific organizational objectives and policy targets. They are designed to render accountability and to help establish legitimacy with regard to a certain course of action taken. There is a fine line between the use of indicators to inform decision makers in a political debate, and using indicators to prove a case. There is a fine line indeed between, on the one hand, developing and presenting measures that provide accountability about a certain course of action or, on the other hand, presenting them in a way to rationalize and justify it.

Implications for the Development and Assessment of Indicators

Assessments of existing sets of indicators therefore need not only address the question of whether their design meets scientific standards of validity, accuracy, and reliability and whether they convey their message in a clear and unambiguous way (e.g. [11, 33]. They should also pay attention to their use in a managerial/political context (for what purposes are they used?); to the organizational or policy outcomes of this use (do they help to reach policy targets, or do they have dysfunctional effects?); and to the power implications of their use (who wins or loses from the use of indicators?). Criteria may be developed to support indicator assessments in a participatory process. That assessment, however, should not only focus on the indicators themselves. It should also include an evaluation of how the indicators were used, by whom, and with what consequences, as part of a continuing search for valid, meaningful, functional, and legitimate indicators.

4. CONCLUSIONS AND RECOMMENDATIONS

Are we measuring what we want to manage? If the managerial aspiration is one of a society that is able to sustain or restore the planet's ecosystem performance (or that of a smaller region), then Flanders is currently only able to measure this to a limited extent. Our analysis of 152 Flemish indicators indicated that they cover only a minority of our natural capital stocks, ecosystem services, and related benefits. Most indicators lack empirical data or are still in a conceptual stage. The available indicators in 2011 also refer to single services. Indicators on synergies and trade-offs between ecosystem services, to support decision making with regard to service bundles (see e.g. Chapter 20), were not yet available for policy makers, at least not on a regional level. Also, the links between ecosystem services and benefits and their impact on human well-being are not yet covered by the available indicators.

Current indicators often mix natural capital stocks with service flows (output) and benefits (outcome). This can reduce construct validity or may increase the ambiguity of the indicators. An indicator or a measure that captures ecosystem health does not reveal the benefits actually enjoyed. An indicator that shows the amount or distribution of benefits does not show whether this is sustainable and so on [12, 13]. Since ecosystem performance constitutes a multidimensional and multifaceted construct, indicators need to distinguish stocks and flows, ecosystem quantity and quality, supply and demand of services, volumes and values, and their temporal and spatial distribution between regions or stakeholders.

Policy and governance information systems should emulate this diversity and complexity. Sets of indicators will need to refer to these different dimensions of ecosystem performance, and they will have to be based on different types of measurement and data analysis, through interdisciplinary and transdisciplinary approaches. Clearly, a tension exists between the requirement of completeness and thoroughness of understanding, and there is a need to limit sets of indicators to prevent information overload for users or exorbitant costs of data collection.

The mix-up of constructs in indicator lists sometimes relates to fundamental challenges in capturing complex relationships through indicators as well as to practical measurement problems, especially for the nontangible or immaterial cultural services. Pragmatic solutions have been proposed, although these may result in reduced construct validity or in strategic use of the indicators.

The available indicators still reflect the research focus of particular centers of expertise, leaving several sections of the ecosystem framework blank. In the coming years, policy agencies and research centers will have to broaden the research and knowledge base in order to cover a wider range of ecosystem services as well as their link with human well-being. The current collection of indicators needs to be expanded and improved before it can fully support the use of ecosystem service concepts in policy making. With a significant proportion of the ecosystem services still lacking indicators, the aggregation of information across different services still represents a major challenge. Without such aggregations, indicator sets risk facing the choice between becoming too selective and incomplete, or becoming too exhaustive and overpopulated.

Feld et al. (2009) [30] observed that since the MA, international and national policies have not yet adequately stimulated the development of comprehensive indicator systems suited to detect and measure the state and trends in biodiversity and their implication for ecosystem service provision. This observation also applies to Flanders. An ecosystem service strategy that transcends the separate policy domains may help provide guidance on priority indicators. In addition to a governmentwide approach, more sectoral or thematic policy programs like common agricultural policy, integrated water policy, climate change, protected areas, sustainable urban development, and tax and subsidy reform can help prioritize the development of indicators. Capturing the low-hanging fruit by selectively including currently available indicators in policy documents and processes, can help to spark interest and build awareness among policy makers and stakeholders. Reporting processes can serve as a vehicle to convert and improve existing sets of indicators. Outlines a stepwise approach for such a process. It would appear wise to heed the advice formulated in the early stages of the MA: "The more involved decision-makers and stakeholders are in the selection of indicators, the greater will be their acceptance of results of the assessments" (MA, 2003) [5].

At the same time, ecosystem performance cannot be entirely captured or wholly explained, let alone be managed, by means of indicators alone. Indicators should therefore be part of a broader knowledge basis that includes other knowledge types and formats, including (local) stakeholder knowledge, to help explain and understand the observed indicator trends. Indicators, however, can help point out critical thresholds, can provide early warnings for irreversible changes, or can suggest general trends that can be used to inform debates in policy and society [5]. They can also help single out issues that merit more attention, or consultation with policy implementers or local stakeholders.

Political and strategic use of indicators should probably be regarded as standard practice rather than as a deviation. On the one hand, this requires some tolerance for ambiguity and subjectivity on the part of the producers of indicators, an idea that may fit uneasily in natural science communities. On the other hand, indicator-based communications will have to be submitted periodically to more thorough assessments and evaluations. These reviews should not only address the indicators' validity and accuracy or their communicative abilities but also determine whether their application has led to functional or dysfunctional decisions and policy outcomes.

LIST OF ABBREVIATIONS

ANB   Agentschap voor Natuur en Bos—Agency for Nature and Forest
COP   Conference of the Parties
DPSIR Drivers, Pressure, State, Impact, Response
ECOBE Onderzoeksgroep Ecosysteembeheer (Ecosystem management research Group), University of Antwerp
INBO   Instituut voor Natuur en Bosonderzoek—Research Institute for Nature and Forest
LARA   Landbouwrapport—Agricultural Report
LNE   Leefmilieu, Natuur en Energie—Environment, Nature and Energy
MIRA   Milieurapport—Environmental Report
NARA   Natuurrapport—Nature Report
SEBI   Streamlining European Biodiversity Indicators
VRIND Vlaamse Regionale Indicatoren—Flemish Regional Indicators

REFERENCES

1. Daily, G.C. 2000. Management objectives for the protection of ecosystem services. *Environmental Science and Policy* 3(6), 333–339.
2. Convention on Biological Diversity. 2000. *Strategic Plan for Biodiversity 2011–2020 and the Aichi Targets: Living in Harmony with Nature.* Montreal, Canada.
3. EC. 2011. Our life insurance, our natural capital: An EU biodiversity strategy to 2020. Communication from the Commission to the European Parliament, the Council, the Economic and Social Committee and the Committee of the Regions.

4. UNEP. 2011. Report of the ad hoc technical expert group on indicators for the strategic plan for biodiversity 2011–2020. United Nations Environmental Programme, Convention on Biological Diversity, p. 55.

5. MA. 2003. *Ecosystems and Human Well-Being: A Framework for Assessment.* Washington, DC.

6. ten Brink, P., et al. 2011. Strengthening indicators and accounting systems for natural capital, in the Economics of Ecosystems and Biodiversity. TEEB for National and International Policy Makers (P. ten Brink, ed.), pp. 79–128. Bonn: TEEB.

7. Halachmi, A., and G. Bouckaert. 1996. *Organizational Performance and Measurement in the Public Sector. Toward Service, Effort and Accomplishment Reporting.* Westport, CT: Quorum.

8. Costanza, R., and H.E. Daly. 1992. Natural capital and sustainable development. *Conservation Biology* 6(1), 37–46.

9. MA. 2005. *Ecosystems and Human Well-being: Synthesis.* Washington, DC.

10. de Groot, R.S., et al. 2010. Integrating the ecological and economic dimensions in biodiversity and ecosystem service valuation' In *The Economics of Ecosystems and* Biodiversity: Ecological and Economic Foundations (P. Kumar, ed.), pp. 9–40. London: Earthscan.

11. Layke, C. 2009. *Measuring Nature's Benefits: A Preliminary Roadmap for Improving Ecosystem Service Indicators.* Working paper, Washington, DC: World Resources Institute, 36 pp.

12. de Groot, R.S., et al. 2010. Challenges in integrating the concept of ecosystem services and values in landscape planning, management and decision making. *Ecological Complexity* 7(3), 260–272.

13. UNEP-WCMC. 2011. *Developing Ecosystem Service Indicators: Experiences and Lessons Learned from Sub-global Assessments and Other Initiatives.* Montreal, Canada.

14. Daily, G.C. 1997. *Nature's Services: Societal Dependence on Natural Ecosystems.* Washington, DC: Island Press.

15. Elmqvist, T., et al. 2010. Biodiversity, ecosystems and ecosystem services. In *The Economics of Ecosystems and Biodiversity: Ecological and Economic Foundations* (P. Kumar, ed.). pp. 41–111. London: Earthscan.

16. Braat, L., & P. ten Brink (eds.). 2008. *The Cost of Policy Inaction: The Case of Not Meeting the 2010 Biodiversity Target* Alterra report 1718, Wageningen, Netherlands: Alterra, 312 pp.

17. Müller, F., and B. Burkhard. 2012. The indicator side of ecosystem services. *Ecosystem Services* 1(1), 26–30.

18. van Oudenhoven, A.P.E., et al. 2012. Framework for systematic indicator selection to assess effects of land management on ecosystem services. *Ecological Indicators* 21(0), 110–122.

19. Mace, G.M., et al. 2011. Conceptual framework and methodology. In The UK National Ecosystem Assessment Technical Report. 2011, UK National Ecosystem Assessment, pp. 11–25. Cambridge: UNEP-WCMC.

20. EC. 2013. The Economic Benefits of the Natura 2000 Network. Luxembourg: Publications Office of the European Union, p. 74.

21. Kandziora, M., B. Burkhard, and F. Müller. 2013. Interactions of ecosystem properties, ecosystem integrity and ecosystem service indicators—A theoretical matrix exercise. *Ecological Indicators* 28(May), 54–78.

22. Luck, G.W., et al. 2009. Quantifying the contribution of organisms to the provision of wcosystem aervices. *Bioscience* 59(3), 223–235.

23. Kremen, C. 2005. Managing ecosystem services: What do we need to know about their ecology? *Ecology Letters* 8, 468–479.

24. Zhang, W., et al. 2007. Ecosystem services and dis-services to agriculture. *Ecological Economics*, 64, 253–260.

25. Dunn, R.R. 2010. Global mapping of ecosystem disservices: The unspoken reality that nature sometimes kills us. *Biotropica* 42(5), 555–557.

26. Lyytimäki, J., and M. Sipilä. 2009. Hopping on one leg—The challenge of ecosystem disservices for urban green management. *Urban Forrestry and Urban Greening* 8, 309–315.

27. McMichael, A., et al. 2005. Linking ecosystem services and human well-being. In *Ecosystems and Human Well-being*. Volume 4: *Multiscale Assessments* (MA, ed.), pp. 43–60. Washington, DC: Island Press.

28. Institute, W.R. 2009. Ecosystem Service Indicators Database.

29. Carpenter, S.R., et al. 2006. Millennium Ecosystem Assessment: Research needs. *Science* 314, 257–258.

30. Feld, C.K., et al. 2009. Indicators of biodiversity and ecosystem services: A synthesis across ecosystems and spatial scales. *Oikos* 118(12), 1862–1871.

31. Layke, C., et al. 2012. Indicators from the global and sub-global Millennium Ecosystem Assessments: An analysis and next steps. *Ecological Indicators* 17(0), 77–87.

32. Mace, G.M., and J.E.M. Baillie. 2007. The 2010 Biodiversity Indicators: Challenges for science and policy. *Conservation Biology* 21(6), 1406–1413.

33. Reyers, B. 2011. Measuring biophysical quantities and the use of indicators. In *The Economics of Ecosystems and Biodiversity: Ecological and Economic Foundations* (P. Kumar, ed.), pp. 113–147. London: Earthscan.

34. Van Daele, T., et al. 2010. *Biodiversity Indicators 2010. The State of Nature in Flanders (Belgium)*. Brussels.

35. Jacobs, S., et al. 2010. Ecosysteemdiensten in Vlaanderen. Een verkennende inventarisatie van ecosysteemdiensten en potentiële ecosysteemwinsten.

36. Haines-Young, R. and Potschin, M. 2011. Common International Classification of Ecosystem Services (CICES): 2011 update: Paper prepared for discussion at the expert meeting on ecosystem accounts organised by the UNSD, the EEA, and the World Bank, London, December 2011. Centre for Environmental Management, University of Nottingham, UK.

37. UNCEEA. 2012. Proposed Common International Classification of Ecosystem Services. 7th meeting of the UN Committee of Experts on Environmental-Economic Accounting, Rio de Janeiro, June 11–13, 2012.

38. Van Reeth, W. 2013. *Ecosystem Service Indicators: Are We Measuring What We Want to Manage?* Brussels: Research Institute for Nature and Forest.

39. Jacobs, S., et al. 2011. *Ecosysteemdiensten in de Zwinstreek. Verkennende studie in het kader van het project "Recreatie & Ecotoerisme in de Zwinstreek" (REECZ)*. Antwerp: Universiteit Antwerpen, Ecosystem Management Research Group, p. 136.

40. Van Der Biest, K., and S. Jacobs. 2011. *Wat biedt de Nete aan ecosysteemdiensten en kunnen we die beheren?*. Universiteit Antwerpen, Ecosystems Management Research Group. presentation held at the annual meeting of ANKONA, the Antwerp Community for Nature Studies on 12 February 2011.

41. Ochelen, S., and B. Putzeijs. 2007. *Milieubeleidskosten. Begrippen en berekeningsmethoden*. Brussel: Flemish Government, Department of Environment, Nature and Energy.

42. Liekens, I., et al. 2010. *Economische waardering van ecosysteemdiensten, een handleiding*. 2010: Brussels: Flemish Government, Department of Environment, Nature and Energy.

43. Turkelboom, F., P. Raquez, S. Jacobs et al. 2012. *Adapted CICES classification for Belgium v.4*. INBO, FUNDP, UA-ECOBE, UGent, DEMNA, VUB, VITO, ANB, VLM, 8 pp.

44. Dasgupta, P., and K.-G. Mäler. 2000. Net national product, wealth and social well-being. *Environment and Development Economics* 5(1&2), 69–93.

45. Daly, H.E., and J. Cobb. 1989. *For the Common Good. Redirecting the Economy toward Community, the Environment and a Sustainable Future*. Boston: Beacon Press.
46. Lawn, P.A. 2003. A theoretical foundation to support the Index of Sustainable Economic Welfare (ISEW), Genuine Progress Indicator (GPI) and other related indexes. *Ecological Economics* 44, 105–118.
47. Bouckaert, G., and J. Halligan. 2008. *Managing Performance: International Comparisons*. New York: Routledge.
48. Bouckaert, G. 1995. Improving performance measurement. In *The Enduring Challenges in Public Management. Surviving and Excelling in a Changing World* (A. Halachmi and G. Bouckaert, eds.), pp. 379–412. San Francisco: Jossey-Bass.
49. Van Dooren, W., et al. 2006. *Issues in Output Measurement for "Government at a Glance."* (P.G. Committee, ed.). Paris: OECD.
50. Carwardine, J., et al. 2009. Hitting the target and missing the point: Target-based conservation planning in context. *Conservation Letters* 2, 3–10.
51. Hood, C. 2006, July–August. Gaming in targetworld: The targets approach to managing British public services. *Public Administration Review*, 515–521.
52. Metcalfe, L. 1993. Public management: From imitation to innovation. In *Modern Governance. New Government-Society Interactions* (J. Kooiman, ed.), pp. 173–189. London: Sage.

Chapter 5

Inquiring into the Governance of Ecosystem Services: An Introduction

Hans Keune[1,2,3,4], Tom Bauler[5] and Heidi Wittmer[6]

[1]*Belgian Biodiversity Platform,* [2]*Research Institute for Nature and Forest (INBO),* [3]*Faculty of Applied Economics – University of Antwerp,* [4]*naXys, Namur Center for Complex Systems – University of Namur,* [5]*Université Libre de Bruxelles Institut de Gestion de l'Environnement et d'Aménagement du Territoire Centre d'Etudes du Développement Durable,* [6]*Helmholtz Centre for Environmental Research—UFZ Division of Social Sciences, Department of Environmental Politics*

1. INTRODUCTION

The management of ecosystems is the object of long-standing, cross-disciplinary debates, both in practice and in academia, and originated in defining the objectives needed to locally harness the conservation of species and preservation of ecosystems. It resulted in extensive inquiry into the institutional arrangements that define the relationship between nature and humans. The much more recent shift in conceptualizing this relationship in terms of *ecosystem services* has quite logically revived parts of these centennial debates and the domain of inquiry, at a time when a quite profound debate is developing in the study of socio-institutional arrangements itself. In the present chapter, we propose to present a brief, introductory account of some of the more stringent perspectives and proposals on how to inquire into the *governance of ecosystem services*. The objective is not to provide a state-of-the-art treatise, but rather to propose a limited set of contingent domains of inquiry that merit particular attention. Chapter XXX (reference to the 2nd governance chapter 13) will build on the presently identified domains and provide a series of approaches to the complex issues at hand.

Ecosystem Services. http://dx.doi.org/10.1016/B978-0-12-419964-4.00005-6

Indeed, the project of inquiring into how the governance of ecosystem services is institutionally arranged raises a considerable number of questions that quite directly link to entire potential domains of enquiry. The very fact that the governance of ecosystem services is a project relevant for descriptive as well as prescriptive approaches not only doubles the number of perspectives to be taken into account, but also involves quite fundamental investigations into the linking of prescription to comprehension. First-hand questions would include:

- What is and what should be governed: for example, the interplay between nature and society of course, but by precise problem framing and with which policy options in sight and on top of which policy priorities.
- What is and what should be a relevant unit of governance: for example, the stakes of actors and their distributed power relationships, but along which definitions of benefits and burdens, and against which understanding of equity.
- Who is and who should merit representation in governance arrangements: for example, the sociopolitical groups of actors, but along which definitions to qualify as a stakeholder and against which expectations and voice principles.
- When are and when should actors interact with governance processes: for example, the moments of issue- and problem framing are strategic, as are moments of process design, policy interpretation, definition of policy options, prioritization, and evaluation, but along which understanding of the rules of interaction and on the basis of which informational basis.

Subsequently, we introduce a series of domains of inquiry as an attempt to provide some order—eventually even priority—in the wider struggle to enhance the comprehension of governance arrangements of ecosystem services. We will first (Section 2) open a building site around the conceptualizations of governance. This is followed (Section 3) by a call to develop inquiry on the parallel understandings of particular modes of governance, and how inquiry can be initiated at the level of knowledge and power (Section 4).

2. WHAT IS GOVERNANCE?

While historically and conceptually leaning on governing or government, "governance" or "governmentality" encompasses a wider, more inclusive, and extensive set of perspectives on the study of institutional arrangements [6–11]. Andrew Jordan mobilizes Rosenau [12. p.4] to give an account of the differentiation between both conceptualizations: "Both [governance and governing/government] refer to purposive behaviour, to goal oriented activities, to systems of rule; but 'government' suggests activities that are backed by formal authority ... whereas 'governance' refers to activities backed by shared goals that may or may not derive from legal or formally prescribed responsibilities.... 'Governance', in other words, is a more encompassing phenomenon than 'government'. It embraces governmental institutions, but it also subsumes informal, non-governmental mechanisms ... whereby those persons and organizations within its purview move ahead, satisfy their need and fulfil their wants.

In other terms, whereas "governing" refers quite explicitly to a theory of public intervention that favors public authority-based (aka governmental) steering of societies, "governance" refers to the societal processes of acting and interacting in decision making, in implementation, and in evaluation of public policy. Governance hence accounts for a less linear, less one-directional, less instrumental, less substantive, less exclusive relationship between public authorities and their subjects. Lee [6] refers to governance as social coordination that aims at "solving social problems by coordinating (the) interactions of various actors." As a direct consequence, governance cannot be limited to inquiry into formalized public institutions or into how the work of policy makers and their formalized organizations impacts their subjects. Indeed, governance entails comprehending the dynamics and processes of the sociopolitical configuration of the public good. It assumes paying equal attention, for instance, to issues of corporate management or to indigenous communities and households, because the "subjects" have become a comprehensive network of diverse actors that play a role in, are influenced by, and at the end of the day co-define a complex set of governance activities and outcomes. This co-definition does indeed call for spanning the inquiry of governance across different sociospatial scales and should be regardless of administrative boundaries—in other words, it should be extended into the inquiry of "multilevel governance."

More profoundly even, governance furthermore asks to expand our attention from the sole inquiry of the interactions between social actors (i.e., the classical authority/subject interactions) into the roles played by and imposed on objects and artifacts, technology, science, and infrastructure. In other words, no-human animals, and even nonliving beings, are fundamentally to be included into the inquiry of governance to a level where inquiry can no longer be based on an obsolete ruler–subject–object divide, but could, for instance, be developing into the inquiry of actor–network interactions.

In this sense, the shifts from government to governance and from ecosystems/ species as the object of steering to ecosystem services drastically opens the set of domains of inquiry in a considerable number of ways into recognizing more complexity, more processes, more actors, more natures of actors, more dynamics, more networks, more scales and times. A number of approaches—even schools and epistemologies—propose ways to close down and bring priority into the seamless web of interactions and processes that an inquiry of governance entails; one practice-oriented perspective is to distinguish between concurring modes.

3. THE PRACTICE OF GOVERNANCE

An inquiry of the governance of ecosystem services can be organized along a series of characterizations of the practice of governance, which, to a certain extent, coincide with prevalent schools in the discipline of public management.

In this sense, authors such as Kitthananan [9] and Lee [6] identify a diversity of governance modes that individually define their own perspectives on inquiry into socio-institutional arrangements. Such modes of governance include, for instance: [1] *New Public Management*, focusing on the processes created by the introduction of (adapted private-sector) management methods into government bureaucracies; [2] *Good Governance*, focusing on the processes created by the introduction of ethical, moral standards against which to measure public and democratic institutions and investigating criteria such as transparency, accountability, and participation, and in a wider sense attempting to redefine the operationalization of democracy or even of societal empowerment; [3] (*self-organizing*) *Network Governance*, paying attention not primarily to formalized governmental institutions as main governance actors, but where the emphasis is put on the complex web of interactions created by a network of organizations, social norms, resources, and competences that might be largely autonomous from the classical state-defined actors of governance.

An inquiry into the practice of governance has been raising questions on the nature of the *structure* that organizes the governance mechanisms, as well as of the *processes* prevailing therein. Kitthananan [9], among others,[1] identifies four types of structures: hierarchies (as within public authorities), markets (as resource-allocating machines), networks (as the web of organized societal stakes and voices), and communities (as localized levels providing the entry points for citizen involvement). Kitthananan's process perspective asks to inquire into the outcomes and types of the interactions between sociopolitical actors, which can be viewed from a steering or a coordination perspective, respectively, putting public authorities in a more or less directive position within social processes.

With respect to the place, importance, and right-to-initiative of public authorities, both Kitthananan and Lee distinguish somewhat binary between old (state-centered steering) and new (society-centered coordination) modes of governance, Initially, this distinction seems to blur somewhat into the more traditional, fundamental distinction between government and governmentality. In effect, the everyday practice of governance runs only imperfectly along the lines of demarcation of the above-described modes of governance; the large numbers of existing state-steered networks are perfect examples of the combination of a (new) network-based mode of governance grounded in a traditional (old) hierarchical right-to-initiative of the state. Such realities show the obvious limitations of developing an inquiry into governance from the perspective of practices. In the example given above, the old versus new divide has been more accurately conceptualized as existing forms of civic epistemologies [13], which help to characterize modes of governance against their respective sociopolitical, historical, cultural backgrounds that form the basics of state interventionism, of

---

1. For example, Lee [6] distinguishes three structural arrangements of what he addresses as social coordination: markets, hierarchies, and democracy.

the organization of institutions, their relationship to citizens and stakeholders, and their recourse to policy instruments.

As a conclusion, the inquiry of the governance of ecosystem services along modes of governance may perhaps provide us with a series of labels that we could attach to distinctive public-private arrangements, but it qualifies only imperfectly as a means of helping to define the relevant domains of our inquiry.

4. KNOWLEDGE: DIVERSITY, ETHICS, AND POWER

The invitation to discuss governance requires a fundamental shift to observation of "underlying objects" of study. Developing a coherent and encompassing program of study of these objects is beyond the scope of the present introduction, but we will attempt to shed some light on it here.

The knowledge base for governance is an important object in this sense. Many scientific challenges remain regarding the complex relation between the natural environment and humans [14]. The question of which and whose knowledge is perceived to be relevant is a crucial one when inquiry is made into the emergence and development of complex social interactions (i.e., governance arrangements). Relevance in turn demands asking questions regarding, for example, the respective roles of technocratic, expert-based knowledge as well as of local and lay knowledge. Consequently, perception of relevance demands asking questions, for instance, at the level of how social debate on knowledge/information can enhance the understanding of the complex socioecological issues. Analytical-deliberative approaches toward knowledge construction, in which scientific analysis is procedurally coupled with social deliberation [15–18], are intended to help actors configure their very own knowledge base by handing over the decision of what information is relevant or not relevant to the very group of actors. Processes such as these are not meant to entail a fundamental questioning of science-based expertise, but rather to permit opening the potential information-base to alternatives. Facing complexity and uncertainties, deliberative knowledge construction has been seen as basic to the operationalization of governance arrangements by attributing inherent value-laden choices to proto-democratic institutions.

Kemp et al. [8] show that recognizing and internalizing diversity of information and knowledge could be an important domain of inquiry for understanding the wider configurations of governance arrangements of complex issues, because such mechanisms are a source of collective learning and hence are at the basis for adaptation and reorganization. As Lebel et al. [19] put it: "The capacity to build and maintain social and ecological diversity is important as a source of renewal and reorganization following major crises."

Embracing knowledge diversity has been considered a strategy for taking into account the many faces of complexity, and notably organizing critical mass. In turn, Keune [18] stresses that "critical complexification," and more generally "critical complexity thinking," admit that limits of knowledge are inherent to complexity, necessitating a reduction that should be accompanied by critical reflection on the normative basis for such simplifications. For Cilliers [20. p.259], "knowledge is provisional. We cannot make purely objective and final claims about our complex world. We have to make choices and thus we cannot escape the normative or ethical domain." Kunneman [21. p.134] points out that "it is impossible to prove the validity of normative judgments with the help of empirical arguments. In twentieth century philosophy an unbridgeable gap between 'is' and 'ought' has opened up…. Instead of bridging the gap between 'is' and 'ought,' however, this line of critique has strengthened the idea that ethical perspectives cannot be justified on rational grounds." In this sense, knowledge-providing mechanisms that build—even implicitly—on a singular track of "science speaking truth to power" tend to underestimate the necessary tribute to the social and ethical complexity of the governance of ecosystem services.

The foreclosure of knowledge-providing mechanisms is an obvious issue of power that needs to be subjected to closer enquiry in socioecological governance thinking [22–29]. The development of this particular program of inquiry is itself difficult and potentially complex, and might in turn ask for its own deliberative mechanisms. Smith and Stirling [3] pose an initial set of questions along which to organize that particular development: "How do challenging bottom up governance initiatives confront the deeply structural forms of economic power vested in current global patterns of system reproduction? How are different bodies of knowledge and interests in social-ecological systems negotiated? How is consent achieved, and how is dissent reconciled? To what extent are plural development pathways tolerated, and how is dialogue between their advocates and constituents maintained? How should these problem-focused, adaptive, and reflexive governance activities link to the more general-purpose and formal institutions of political authority and democracy? What alternate forms of direct democracy can be brought into adaptive governance and transition management?"

REFERENCES

1. Vatn, A. 2009. An institutional analysis of methods for environmental appraisal. *Ecological Economics* 68, 2207–2215.
2. Norgaard, R.B. 2010. Ecosystem services: From eye-opening metaphor to complexity blinder. *Ecological Economics* 69(6), 1221–1227.
3. Smith, A., and Stirling, A. 2010. The politics of social-ecological resilience and sustainable socio-technical transitions. *Ecology and Society* 15(1), 11.
4. Wilkinson, C. 2011. Social-ecological resilience: Insights and issues for planning theory. *Planning Theory* 11(2),148–169.
5. Voß, J-P., and Bornemann, B. 2011. The politics of reflexive governance: Challenges for designing adaptive management and transition management. *Ecology and Society* 16(2), 9.
6. Lee, M. 2003. Conceptualizing the New Governance: A New Institution of Social Coordination. Proceedings of the Institutional Analysis and Development Mini-Conference, May 3–5, Workshop in Political Theory and Policy Analysis, Indiana University, Bloomington.
7. Hooghe, L., and Marks, G. 2003. Unravelling the central state, but how? Types of multi-level governance. *The American Political Science Review* 97, 233–234.

8. Kemp, R., Parto, S., and Gibson, R.B. 2005. Governance for sustainable development: Moving from theory to practice. *International Journal of Sustainable Development* 8(1/2), 12–30.
9. Kitthananan, A. 2006. Conceptualizing Governance: A Review. *Journal of Societal and Social Policy* 5/3, 1–19.
10. Jordan, A. 2008. The governance of sustainable development: Taking stock and looking forwards. *Environment and Planning C* 26(1), 17–33.
11. Kluvánková-Oravská, T., Chobotová, V., Banaszak, I., Slavikova, L., and Trifunovova, S. 2009. From Government to Governance for Biodiversity: The Perspective of Central and Eastern European Transition Countries. *Environmental Policy Governance* 19, 186–196.
12. Rosenau, J. 1992. Governance, order and change in world politics. In *Governance without Government*. (Rosenau, J., and Czempiel, E. O., eds). Cambridge: Cambridge University Press.
13. Jasanoff, S. 2005. *Designs on Nature: Science and Democracy in Europe and the* United States. Princeton, NJ: Princeton University Press.
14. Liu, J., Dietz, T., Carpenter, S.R., Alberti, M., Folke, C., Moran, E., Pell, A.N., Deadman, P., Kratz, T., Lubchenco, J., Ostrom, E., Ouyang, Z., Provencher, W., Redman, C. L., Schneider, S.H., and Taylor, W.W. 2007. Complexity of coupled human and natural systems. *Science* 317(5844), 1513–1516.
15. Stern, P., and Fineberg, H. (eds.) 1996. Understanding risk: Information decisions in a democratic society. Washington, DC: National Research Council, National Academy Press.
16. Haines-Young, R. 2011. Exploring ecosystem services issues across diverse knowledge domain using Bayesian Belief Networks. *Progress in Physical Geography* 35(5), 685–704.
17. Fish, R. 2011. Environmental decision making and an Ecosystems approach: Some challenges from the perspective of social science. *Progress in Physical Geography* 35(5), 671–680.
18. Keune, H. 2012. Critical complexity in environmental health practice: Simplify and complexify. *Environmental Health* 11(S19), 1–10.
19. Lebel, L., Anderies, J.M, Campbell, B., Folke, C., Hatfield-Dodds, S., Hughes, T. P., and Wilson, J. 2006. Governance and the capacity to manage resilience in regional social-ecological systems. *Ecology and Society* 11(1): 19. http://www.ecologyandsociety.org/vol/iss/art/19.
20. Cilliers, P. 2005. Complexity, deconstruction and relativism. *Theory, Culture and Society* 22, 255–267.
21. Kunneman, H. 2010. Ethical complexity. In: *Complexity, Difference and Identity*. (Cilliers, P., and Preiser, R., eds.). Dordrecht: Springer, pp.131–164.
22. Kemp, R., and Loorbach, D. 2003. Governance for sustainability through transition management. Proceedings of the Open Meeting of the Human Dimensions of Global Environmental Change Research Community, October 16–19, Montreal, Canada.
23. Lemos, M.C., and Agrawal, A. 2006. Environmental governance. *Annual Review of Environment and Resources* 31, 297–325.
24. Wittmer, H., Rauschmayer, F., and Klauer, B. 2006. How to select instruments for the resolution of environmental conflicts? *Land Use Policy* 23, 1–9.
25. Biermann, F. 2007. Earth system governance as a crosscutting theme of global change research. *Global Environmental Change* 17, 326–337.
26. Vatn, A. 2009. An institutional analysis of methods for environmental appraisal. *Ecological Economics* 68, 2207–2215.
27. Smith, A., and Stirling, A. 2010. The politics of social-ecological resilience and sustainable socio-technical transitions. *Ecology and Society* 15(1), 11.
28. Wilkinson, C. 2011. Social-ecological resilience: Insights and issues for planning theory. *Planning Theory* 11(2), 148–169.
29. Voß, J-P., and Bornemann, B. 2011. The politics of reflexive governance: Challenges for designing adaptive management and transition management. *Ecology and Society* 16(2), 9.

Part II

Ecosystem Services: Conceptual Reflections

Shutterstock.com / © Jan Martin Will

Monetary Valuation of Ecosystem Services: Unresolvable Problems with the Standard Economic Model

John Gowdy and Philippe C. Baveye

Rensselaer Polytechnic Institute, Troy, New York

John Gowdy, Professor of Economics, Rensselaer Polytechnic Institute, Troy, New York USA. John Gowdy's current research includes ecosystem valuation, behavioral economics, and evolutionary economics. He was a member of the TEEB (The Economics of Ecosystems and Biodiversity) initiative and is currently working on the topic of human ultrasociality with an interdisciplinary group of social cientists and evolutionary biologists.

Ecosystem Services. http://dx.doi.org/10.1016/B978-0-12-419964-4.00006-8

Philippe C. Baveye, Professor of Environmental Engineering, Rensselaer Polytechnic Institute, Troy, New York, USA

The monetary valuation or "Monetization" of Ecosystem Services (MES) and its stepchild, Payments for Ecosystem Services (PES), have taken the environmental policy field by storm in the years following the influential paper by Costanza et al. [5], "The Value of the World's Ecosystem Services and Natural Capital." Thousands of academic articles have been devoted to MES and, more and more governments and international organizations are implementing policies based on MES and PES. Although the motivation for ecosystem valuation exercises is laudable, and the political value of these studies is quite positive because they usually show high economic values for nature's services, most of the discussion and implementation of the valuation paradigm has taken place in an intellectual vacuum. Measuring the economic value of ecosystem services is usually presented as a novel idea emerging out of the blue in 1997. In fact, as early as 1864 George Perkins Marsh discussed the economic value of nature, including the waste disposal and pest control services the world provides. More relevant to the current MES debate is the work of a number of authors in the 1970s [7, 8, 13, 15, 16] who pointed out the limitations of the monetary valuation of nature. These critiques have largely been ignored, most likely because they cannot be answered in the standard economic model underlying MES and PES.

The standard economic framework is essentially a financial investment model [9]. It answers the question: "How should a rational individual at a point in time allocate her economic resources so as to maximize the present value of the flow of future income?" For an individual, it is logical to discount the future, to reduce all investments to a common metric, to be self-regarding, and to focus on maximizing income. Problems arise when that model is extended beyond its original narrow scope. The crux of the problem is that the well-being of society is more than the additive sum of the discounted financial assets of a collection of individuals at a point in time. This simple observation was the basis of early criticisms of attempts to monetize environmental features, and it is still valid. Unresolvable difficulties exist with the assumptions of the standard valuation framework. These include the following.

DISCOUNTING THE FUTURE

For a number of reasons, including impatience, a finite life span, and the presence of investment opportunities, an individual will prefer to receive a given monetary sum today rather than at some point in the future. The use of a financial model to evaluate the flow of nature's services requires that a lower value be placed on the future; that is, a positive discount rate is assumed. Indeed, if a zero discount rate is used, any service of nature will have a near-infinite value since it generates positive values forever. However, for the human species putting a lower value on the future is untenable. Why should the well-being of those in the future count less than the well-being of those living in the present? This line of reasoning led a number of economists to call for considering the natural world as a stock to be maintained rather than as a source of flows to be used for GDP maximization [2, 3, 6, 7]. An impressive list of economists has rejected the notion of discounting the future well-being of society based on ethical considerations alone.

THE SOURCES OF UTILITY ARE COMMENSURABLE

In order to represent the myriad services of nature in an economic model, the services must be reduced to a common monetary metric, reducing all features of the natural world to a single monetary numéraire. All objects of economic value must have some common attribute that makes them comparable and tradable with one another. Yet this line of reasoning flies in the face of human experience. E. Odum [13] and Georgescu-Roegen [7] wrote eloquently about the failure of the market to account for natural features for which there are no substitute or that were subject to irreversible change. Westman ([16], p. 960) asks: "What is the value to societies, present and future, of the inspirations that flowed from Wordsworth's poetry, and indirectly from nature? These questions seem safely relegated to the realm of the unanswerable because they deal with qualities upon which our society has not placed a quantitative value." The same observation applies to all single-valued measures such as Marx's labor theory of value and energy theories of value as advocated by H. Odum [14] and Costanza [4]. Ghiselin [8] described cost-benefit analyses of environmental services as the "commensuration of the incommensurable."

HUMANS ARE NARROWLY RATIONAL AND SELF-REGARDING

To assume that properly pricing nature is an adequate way to ensure that society optimally allocates nature's services, it must also be assumed that humans rationally and consistently respond to price signals. Furthermore, market decisions must be self-regarding to ensure that society's welfare is the sum of individual choices. But as Knetsch [10] points out, according to well-established findings from behavioral economics, the standard valuation framework denies the very behavior that makes us human. Observed human behavior is generally empathetic and other-regarding. Nelson ([12], p. 264) puts it succinctly: "Defining economics as the study of *rational choice*, neoclassical economics treats human physical bodies, their needs, and their evolved actual psychology of thought and action as rather irrelevant. The notion that humans are created as rational decision-makers is, from a physical anthropology point of view, just as ludicrous as the notion that humans were created on the sixth day." Understanding other-regarding behavior is one key to formulating effective economic policies having complex and long-lasting consequences. But taking this into account is impossible in a standard cost benefit framework. For example, changes in individual income cannot be meaningfully aggregated if an increase in person A's income decreases the well-being of person B."

Raising the price of ecosystems services as an incentive to conserve them is a reasonable policy tool in many circumstances. But the price system does not exist in a vacuum. The belief that "correct" prices lead to the rational allocation of the services of nature is based on an array of underlying assumptions that have proved to be logically inconsistent and wildly at odds with actual human behavior. These assumptions lie at the heart of welfare economics, a subject unfamiliar to most biologists and one that is disappearing from standard economics curricula [1]. Until economists and biologists begin to appreciate the theoretical difficulties with MES and PES, they will continue to be disappointed with the policy results.

Fortunately, other frameworks exist for assessing ecosystem values without forcing all them into the straightjacket of neoclassical economic theory. The TEEB (The Economics of Ecosystems and Biodiversity) [13] initiative stressed the importance of multiple valuation approaches. Economic values such as ecotourism can be expressed in monetary units, noneconomic benefits to human society can be quantified using a variety of measures (health or well-being indexes, for example), and such things as the value of biodiversity to ecosystems can be described in detail even if they cannot be quantified. The fact that this approach does not end up with a single number to compare all policies should be seen as an advantage, not a drawback.

REFERENCES

1. Atkinson, A. 2001. The strange disappearance of welfare economics. *Kyklos* 54, 193–206.
2. Ayres, R., and A. Kneese. 1969. Production, consumption and externalities. *American Economic Review* 59, 282–297.
3. Boulding, K. 1966. The economics of the coming spaceship earth. In H. Jarrett (ed.), *Environmental Quality in a Growing Economy*. Baltimore, MD: Johns Hopkins University Press, 3–14.
4. Costanza, R. 1980. Embodied energy and economic valuation. *Science* 210, 1219–1224.
5. Costanza, R., et al. 1997. The value of the world's ecosystem services and natural capital. *Nature* 387, 253–260.

6. Daly, H. 1968. On economics as a life science. *Journal of Political Economy* 76, 392–406.
7. Georgescu-Roegen, N. 1975. Energy and economic myths. *Southern Economic Journal* 41, 347–381.
8. Ghiselin, J. 1977. Perils of the orderly mind: Cost-benefit analysis and other logical pitfalls. *Journal of Environmental Management* 1, 295–299.
9. Gowdy, J., Howarth, R., and Tisdell, C. 2010. Discounting, ethics, and options for maintaining biodiversity and ecosystem services. In *The Economics of Ecosystems and Biodiversity: Ecological and Economic Foundations* (Pushpam Kumar, ed.), pp. 257–283. London: Earthscan.
10. Knetsch, J. 2005. Gains, losses and the US-EPA Economic Analysis Guidelines: A hazardous product? *Environmental and Resource Economics* 32, 91–112.
11. Kumar, P. (ed.). 2010. *The Economics of Ecosystems and Biodiversity: Ecological and Economic Foundations.* London: Earthscan.
12. Nelson, J. 2006. *Economics for Humans.* Chicago: University of Chicago Press.
13. Odum, E.P. 1979. Rebuttal of "Economic Value of Natural Coastal Wetlands." *Coastal Zone Management Journal* 5, 231–237.
14. Odum, H.T. 1971. *Environment, Power and Society.* New York: John Wiley & Sons.
15. Shabman, L., and Batie, S. 1978. Economic value of coastal wetlands: A Critique. *Coastal Zone Management Journal* 4, 231–247.
16. Westman, W. 1977. How much are nature's services worth? *Science* 197, 960–964.

Chapter 7

Biodiversity and Ecosystem Services: Opposed Visions, Opposed Paradigms

Martin Sharman

Independent expert, Avenue des Orangers

Martin Sharman: I am a member of the currently dominant mammalian species on my planet, Homo sapiens. Most of my ancestors reached Europe some 35000 years ago. Like most of the descendants of those people, I am affiliated with a group whose skin changes colour in the sun since it contains low levels of melanin. We lack an epicanthal fold and are lactose tolerant. Like most of the males in this group I possess potentially abundant facial hair and suffer male pattern baldness. Within this group, I might affiliate myself with those who tend not to affiliate themselves with anything much, only I don't care enough to do so.

In 1982 the United Nations' World Charter of Nature stated that "mankind is a part of nature, and life depends on the uninterrupted functioning of natural systems."

This statement is as true today as it was then—or even 5000 years earlier when the Vedas remarked: "So long as this land has fields, forests and fountains, the Earth will survive, sustaining you and the generations that follow you." Five thousand years ago, however, it was perhaps not as widely understood as it was in 1982 that humans emanate from and are part of nature.

The premise of some advocates of ecosystem services [1,2] is that since biodiversity loss continues unabated, conservation is not working. Biodiversity loss is seen as a problem—something has gone wrong, perhaps that "current

Ecosystem Services. http://dx.doi.org/10.1016/B978-0-12-419964-4.00007-X

policies and economic systems do not incorporate the values of biodiversity effectively in either the political or the market systems [3]." By working on what went wrong, and why, a solution to the problem offers itself; we must add the motive of self-interest. The beneficiaries of ecosystem services will look after the ecosystems that provide them, and this will in turn reduce the rate of loss of biodiversity and reverse the degradation of ecosystems.

This paper focuses only on the concept of ecosystem services, although payment for ecosystem services [4,5] and other financial incentives to protect nature [6] are contentious in their own right [7].

The vision behind the concept of ecosystem services is that if people understand the utility of nature, they will value and protect it. By definition, therefore, it extends its protection only to those elements of biodiversity that benefit humans. Protection of all the rest is left to those humans who gain benefit from the "intrinsic" or "existence" value of biodiversity [8,9].

The paradigm of ecosystem services is apparent from the noun "service". It expresses a master-servant relationship; nature is there to serve human needs, whether for goods, culture or recreation [10]. Ecosystem services improve the prospects for human well-being. Properly managed, they may contribute to economic growth and create jobs—an idea that is sometimes expressed as "nature-based solutions" [11] —and their study advances scientific understanding.

In opposition to the cold equations and hermetic vocabulary of ecosystem services, and its vapid, morally vacant and profoundly meaningless "intrinsic" or "existence" values of biodiversity, one might place Albert Schweitzer's emotionally engaging and consequential phrase, "Ehrfurcht vor dem Leben," [12] or "reverence for life".

Those who promote the concept of reverence for life, or biodiversity, also have a vision: If people understand the true relationship between humans and nature, we can possibly agree that we face a predicament. We cannot solve biodiversity loss and be done with it; we have to respond to it and live with it [13]. This vision has a major difficulty that prevents its easy acceptance: We do not want a predicament; rather, we want problems we can solve.

The paradigm of biodiversity is that a proper framing of the human predicament places our species on a planet on whose living fabric we depend. In this framework, biodiversity loss is not a problem, but one important symptom of a predicament to which there is no solution—far less, a solution free of pain. No amount of ecosystem services can extract us from the shrinking gap between the Scylla of a spherical, and therefore limited, planet and the Charybdis of ever-growing human demands.

If we trust that ecosystem services are "the way forward for conservation" [14], we are gambling that the value of services provided by the intact ecosystem is greater than the value of a system transformed for other exploitative purposes [15]. Furthermore, we are gambling that the ecosystem service with a high-value in the short term happens not to degrade any high value service for future generations. Moreover, although the intended message of the ecosystem service was originally that nature is essential for human well-being, the message received may only be that nature can be coerced into providing services that allow humans to continue their unsustainable ways. "Biodiversity should be seen not as a problem," says the International Union for the Conservation of Nature, "but as an opportunity to help achieve broader societal goals. [16]"

Underlying this objection to ecosystem services is the vision that the relationship between humans and nature is one of human dependency rather than nature's servitude. This dependency is expressed in many, sometimes mutually conflicting, ways, of which we might briefly examine three—the ethical, the spiritual and the material.

The ethical understanding of dependency is simply put. "There is something overspecialized about an ethic, held by the dominant class of *Homo sapiens*, that regards the welfare of only one of several million species as an object and beneficiary of duty. [17]" Humans, in developing the power to destroy, have created a responsibility to protect and cherish. This ethical argument can be extended in many directions; for example, if ecosystem services are defined as benefiting humans, then different ecosystems provide lesser or greater benefits. This suggests that we might abandon ecosystems that provide no obvious benefit, and we can instead focus on those ecosystems that provide high returns. From an ethical perspective, such anthropocentric valuation is invalid; humans have no right, even if they have the power, to decide what to save and what to destroy on the grounds of (almost always short-term) human well-being. Biodiversity is not collapsing because of a failure of conservation. On the contrary, biodiversity is doing best where it is conserved by effective legislation, but by a recent capsizing of the historic moral courage that gazetted the nature reserves in the first place. The fifth columnists here are "ecosystem service advocates [who] are finding allies and enjoying traction in places where ethical arguments for biodiversity conservation are given short shrift. [18]"

One manifestation of the spiritual understanding came in October 2012 when Bolivia enacted a law[1] designed to give legal standing to Pachamama—Mother Earth. This decree follows an earlier law that accords Mother Earth a personality and a Bolivian-led debate on the issue in the United Nations. A laconic comment from one non-Bolivian diplomat sums up the dismissive attitude of the General Assembly toward this spiritual view: "the concept 'Mother Earth'," he said, "is not universally accepted. [19]" Underlying this concept, however, is a spiritual relationship with nature that is perhaps not universal but that is shared by many people far beyond the borders of Bolivia [20]. To these people our spiritual mother, Earth, is not a servant. She may be capricious, cruel, and callous, but she is our mother on whom we depend. Biodiversity is not collapsing

---

1. Ley Marco de la Madre Tierra y Desarrollo Integral para Vivir Bien. Ley N° 300, 15 de octubre de 2012

because conservation has failed, but because when we treat nature with disdain, she is not endlessly bountiful but patient, and remorseless.

The material understanding of the dependency of humans on nature relates, in essence, to the failure of anaerobic decomposers in the Carboniferous era to fully exploit the results of above-swamp productivity. Cultural and technological adaptations, including agriculture and the use of fossil fuel, have allowed the human population to expand into every corner of the Earth, avoiding the natural mechanisms of pestilence and famine that limit the populations of most creatures and out-competing almost all other extended genomes. The human mechanism of genocide has also failed to limit our numbers. We have maintained our inexorable population growth by greatly altering the biosphere. The material understanding of dependency would claim, therefore, that the underlying premise of ecosystem services is wrong. Biodiversity is collapsing because humans extract food, fibre and vegetable oil from ecosystems simplified to the point of monoculture.

This anthropocentric framing of the relationship between humans and the rest of the living world is anathema to many people [21]. Even the misleading metaphors [22,23] used in association with ecosystem services—system, service, function, stocks of natural capital, indicators, targets, resources, goods, benefits—seem not to describe something unique, miraculous, and endlessly wonderful, but something base, mundane, and replaceable. But more importantly, it is far from evident that the dominant position that our species has achieved gives it the right to view nature as a servant.

Furthermore, just as empires can be built by working slaves to death, so short-term human well-being may be widely provided, not by maintaining ecosystem services but by continuing to extract them at such a rate as to drive ever faster the precipitous decline in biodiversity [24]. One might also suspect that the idea of ecosystem services expresses a hubristic (and hence dangerous) misapprehension of the relationship between humans and nature, to which we may give, in our turn, the last laugh.

The concept of ecosystem services seems to be conceived and construed in the same exploitative mindset that has led to the loss of biodiversity in the first place. Moreover, it is not clear that the poor and powerless will benefit from this concept; it is more likely that the powerful, the multinationals, the billionaires will use it to their advantage, making ecosystem services a veil that only half-conceals a taboo trade-off [25]. From this perspective ecosystem services are not, then, are a sneaky not way to get people to care for nature but to subdue and subjugate nature for new purposes—to extend yet further humankind's crass exploitation of the ineffable.

If some scientists find ecosystem services a source of interesting questions for research [26], others are more doubtful—pointing out, for example, that an ecosystem service is not a fixed, immutable object of study, but a "cultural product deriving from a system of beliefs and values, symbolically expressed within particular knowledge systems that relate to particular patterns of behaviour and practice, all of which are contested" [27].

Is the path from here to a workable future really as easy as "ecosystem services"? Are ecosystem services even a paving stone in the path? Or does the concept lend itself to wishful thinking, Polyanna's win-win world and denial, helping us not to notice that biodiversity loss is a symptom of a predicament?

We like to deal with problems one at a time. Poverty is solved by economic growth. Economic stagnation is solved by innovation. Climate change is solved by renewable energy. Biodiversity loss is solved by ecosystem services. Unfortunately, in real life, all these "problems" interact and the solution to one only aggravates the other. By framing issues piecemeal we make it difficult to see that they are all facets of a predicament—human-induced environmental change is making the future increasingly uncertain. Since nature does not negotiate, and humans do little to diminish their aggregate demands, no argument is likely to prevent further biodiversity loss—even if we have always known that we depend on bounteous nature.

REFERENCES

1. Balmford A. et al. (2002) Economic Reasons for Conserving Wild Nature. Science 297: 950–953
2. Kumar Duraiappah A. et al. (2005) Ecosystems and human well-being: Biodiversity Synthesis. Millennium Ecosystem Assessment. World Resources Institute, Washington, DC.
3. Ash N. et al. (2007) Chapter 5 Biodiversity of 4th report in the Global Environment Outlook. United Nations Environment Programme.
4. Gómez-Baggethun E. et al. (2010) The history of ecosystem services in economic theory and practice: From early notions to markets and payment schemes. Ecological Economics 69(6):1209–1218
5. Redford K.H & W.M. Adams (2009) Payment for Ecosystem Services and the Challenge of Saving Nature. Conservation Biology 23(4):785–787
6. Vatn A. et al. (2011) Can markets protect biodiversity? An evaluation of different financial mechanisms Noragric Report 60. Department of International Environment and Development Studies, Norwegian University of Life Sciences (UMB)
7. Sullivan S. (2013) Banking Nature? The Spectacular Financialisation of Environmental Conservation. Antipode 45(1):198–217
8. Skroch M. & López-Hoffman L. (2009) Saving Nature under the Big Tent of Ecosystem Services: a Response to Adams and Redford. Conservation Biology 24(1):325–327
9. Reyers B. et al. (2012) Finding Common Ground for Biodiversity and Ecosystem Services. BioScience 62(5) 503–507
10. Haines-Young R. & Potschin M. (2010) The links between biodiversity, ecosystem services and human well-being. In: Raffaelli, D. & C. Frid (eds.): Ecosystem Ecology: a new synthesis. BES Ecological Reviews Series, CUP, Cambridge
11. IUCN (2012) The IUCN Programme 2013-16. Third Draft, Revised January 2012. Available at http://cmsdata.iucn.org/downloads/iucn_programme_2013_16_third_draft_january_2012.pdf
12. Brown P.G. (2004) Are there any natural resources? Politics and the Life Sciences 23(1):11–20
13. Mora C. & Sale P. (2011) Ongoing global biodiversity loss and the need to move beyond protected areas. Marine Ecology Progress Series 434: 251–266
14. Armsworth P.R. et al. (2007.) Ecosystem-service science and the way forward for conservation. Conservation Biology 21(6):1383–1384

15. Adams W.M. & K.H. Redford (2009) Ecosystem Services and Conservation: a Reply to Skroch and López-Hoffman. Conservation Biology 24(1):328–329

16. IUCN (2012) Nature+: The Jeju Declaration for a New Era of Conservation, Sustainability and Nature-based Solutions. IUCN World Conservation Congress, September 2012

17. Rolston H. (1991) Environmental Ethics: Values in and Duties to the Natural World. In Bormann F.H. & S.R. Kellert (Eds.) The Broken Circle: Ecology, Economics, Ethics. Yale University Press.

18. Armsworth P.R. et al. (2007) Ecosystem-service science and the way forward for conservation. Conservation Biology 21(6):1383–1384

19. Wachtel J. (2011) U.N. Prepares to Debate Whether 'Mother Earth' Deserves Human Rights Status. FoxNews.com. Accessed at http://www.foxnews.com/world/2011/04/18/prepares-debate-rights-mother-earth/

20. Wilson E.O. (1984) Biophilia: the human bond with other species, Cambridge, Harvard University Press.

21. Botkin D.B. (2000) No Man's Garden: Thoreau And A New Vision For Civilization And Nature. Island Press.

22. Vira B. & W.M. Adams (2009) Ecosystem services and conservation strategy: beware the silver bullet. Conservation Letters 2:158–162

23. Norgaard R.B. (2010) Ecosystem services: From eye-opening metaphor to complexity blinder. Ecological Economics 69:1219–1227

24. Raudsepp-Hearne C. (2010) Untangling the environmentalist's paradox: why is human well-being increasing as ecosystem services degrade? BioScience 60(8):576–589

25. Fiske A.P. & P.E. Tetlock (1997) Taboo trade-offs: reactions to transactions that transgress the spheres of justice. Political Psychology 18(2):255–297

26. Anton C. et al. (2010) Research needs for incorporating the ecosystem service approach into EU biodiversity conservation policy. Biodivers Conserv 19:2979–2994

27. MacDonald I.K. (2004) Conservation as Cultural and Political Practice. Policy Matters 13:6–71

Chapter 8

Earth System Services—A Global Science Perspective on Ecosystem Services

Sarah Cornell

Stockholm Resilience Centre

Sarah E. Cornell - Stockholm Resilience Centre. I lead a transdisciplinary global sustainability research group at SRC, researching the "safe and just operating space for humanity". We use models and observational evidence of global change to explore options, informing responses by policy, business and communities.

It is worthwhile to reflect on the pace of advances in knowledge about environmental change. The field of Earth system science is just a little over 50 years old. This area of research analyzes the interlinked dynamics of the physical and living "components" of Earth, bridging local to global-scale processes. The first satellite orbit of Earth occurred in 1957, marking the beginning of Earth observation, which in turn enabled the first global climate models to be developed and validated. In 1969, James Lovelock [1] first described the whole planet as "the maximum unit of life" and proposed the paradigm-shifting Gaia hypothesis: Earth's biota is shaped by and adapted to its physical environment, and at the same time changes it. These interactions create the "geosphere–biosphere" feedbacks that play a vital role in determining climatic and ecological conditions on Earth. The science that initially aimed to understand the cosmos "out there" has enabled

Ecosystem Services. http://dx.doi.org/10.1016/B978-0-12-419964-4.00008-1

us to fundamentally reconceptualize the planetary ecosystem, and it increasingly challenges us, as human beings, to rethink our place within it.

These last 50 years or so have also seen the most rapid pace of change in the human world [2]. Humanity's transformative power over our environment started to grow at the start of the Industrial Revolution, fueled by new technologies and new forms of energy. Around the middle of the 20th century, human population, innovation, trade, and resource use all underwent a "Great Acceleration." This escalation in the use of natural resources has triggered significant changes in the biophysical Earth system. The human species has now become the prime driver of environmental change, pushing planet Earth into what is becoming widely viewed as a new epoch [3, 4] – the *Anthropocene.*

Viewing the whole Earth, including its human inhabitants, as a complex and dynamic system brings new perspectives to decision making about natural resource use. Economic theory and practices have long accommodated many aspects of natural resource use and management, including issues of scarcity and the management of "the commons." [5] A large body of policy and academic literature addresses current patterns of consumption and production, documenting unsustainable trends from both local and global perspectives [6]. Much of this work has started from a tacit—and in some places an explicit—assumption of fixed limits to natural resources. John Dryzek has summarized the history of the "limits" discourse [7]. The most widely known work in this area is the Club of Rome's 1972 Limits to Growth report, which analyzed economic pathways subject to supply-side constraints. Related approaches include the widely used footprint calculations, and many sustainability indicators and metrics used by nations, businesses, and intergovernmental bodies.

Relatively recently, focus has shifted toward seeing the natural environment not just in terms of fixed capital stocks of resources, but in terms of the many processes and functions that underpin services to humanity [8, 9]. The TEEB Report (2010) describes a "utility cascade"—the structures and processes of an ecosystem determine its functioning, which determine the services it provides to humanity, which determine the value of that ecosystem. "Ecosystem services" originated as an ecological idea (specifically, as a strong argument for greater attention to ecological conservation). It evolved rapidly into an economic concept, and greater efforts to monetize ecosystem values followed [10]. The TEEB Report makes clear that not all valuation effort necessarily leads to market exchanges of ecosystem services, arguing that there is an essential place in decision making for the simple recognition of ecosystem values. Nevertheless, valuation studies enable the pricing of nonmarket services, which now underpins many cap-and-trade and payments schemes. They also underpin Pigouvian taxes, an attempt to better incorporate the cost of environmental externalities.

As the value of well-functioning ecosystems becomes recognized in society's decision making (or at least, becomes a more prominent feature in the rhetoric about environmental decisions), attention is turning to the dynamic processes of ecosystems, as well as the interacting processes of social and economic change. The capacity of ecosystems to maintain these dynamics in the context of change presents an entirely different kind of "limit." However, dynamic behavior presents major challenges to the scientific understanding of the world, as well as to policy responses to many contemporary sustainability issues. When natural resources were only extracted and consumed locally, problems of resource depletion and undesirable environmental impacts tended to be visible to the users. Economic development involved the worldwide expansion of extraction and trade, unfortunately leaving a trail of "exported" social-ecological problems in its wake. Yet these problems are not merely the sum of local impacts. The science of the last 50 years informs us about global dynamics, involving links and feedback between the physical climate system, land and marine ecosystems, and the fundamental chemistry of land, oceans, atmosphere, and life. Society's choices influence these processes and feedbacks, not just Earth's structures.

In 2009, a Swedish professor, Johan Rockström, with a team of 28 co-authors, published an article [11] arguing that the fundamental processes and functioning of these linked components of the Earth system set nonnegotiable constraints for human activities. They described a set of "planetary boundaries," globally quantifiable processes or conditions of the Earth system, associated with the "desirable" state of the Holocene (the geological period in which human societies have developed and thrived). The boundary processes that Rockström and colleagues identified are profoundly interdependent. This implies that transgressing thresholds of change in any one process may shift the "safe operating margins" for other boundaries, or even trigger cascading effects through the planetary ecosystem, which could be catastrophic for human societies. For example, changes in land cover affect the hydrological cycle, alter local and regional climate, perturb the cycling of essential nutrient elements (notably nitrogen and phosphorus), and can have serious effects on biodiversity and ecosystem resilience. Rockström and colleagues—and many other commentators—highlight the need for better and more adaptive approaches to environmental stewardship, at all scales from local to global.

In today's world, knowledge about ecosystem functioning translates into realizable material benefits, and knowledge of global ecological dynamics brings new perspectives on risks and opportunities for the human enterprise. The predictive power of Earth system science means that this field of scholarship plays an increasingly important role in global policy. However, it also brings new risks that are too poorly understood and responsibilities that have not yet been admitted. One problem is that people's needs are largely invisible from this global viewpoint, so that in the arenas of global environmental change science and policy, many feel that issues of unfairness and inequity do not get the attention they need [12]. "Meeting people's needs" lies at the heart of sustainability, so better links are needed between global-scale insights and the needs of the individual. Another set of problems arises because not all aspects of global change are equally predictable, but the power of quantification can hide some of these differences.

Richard Norgaard (2010) drew attention to some of the problems that arise as complex systems become simplified and abstracted for the purposes of analysis and decision making [13]. We, as a global society, tend to have a very short-term view: Ecosystems are important because they supply resources and services that are useful to humans right now. Longer-term perspectives are very hard to comprehend, and the problems become even worse when our scientific understanding must also rely on the simplification of Earth's complexity. It is a wide-open question for society, poised as we are at the start of the Anthropocene. Can we obtain the ecosystem services we need from ecosystems that function differently on a global scale from the ones with which our societies coevolved? We know that degraded ecosystems have reduced functionality, but what about human-controlled or even artificial ecosystems? What trade-offs can we get away with in the natural world?

We, as a global change science community, should urge global change researchers to focus more on what we know, how we know it, and how we transmit that knowledge. It is especially important to pause and to reflect, in a world of rapid social and environmental change. Knowledge is certainly needed for management of our global resources and for stewardship of Earth's processes. But wisdom and humility are also needed as we travel along the sustainability pathway, for the sharing of resources in ways that are fair and inclusive.

REFERENCES

1. Lovelock, J.E., and C.E. Giffin. 1969. Planetary atmospheres: Compositional and other changes associated with the presence of life. *Advances in the Astronautical Sciences* 25, 179–193.
2. Steffen, W., A. Sanderson, J. Jäger, P.D. Tyson, B. Moore III, P.A. Matson, K. Richardson, F. Oldfield, H.J. Schellnhuber, B.L. Turner II, and R.J. Watson. 2004. *Global Change and the Earth System: A Planet under Pressure*. Heidelberg: Springer Verlag.
3. Steffen, W., P.J. Crutzen, and J.R. McNeill. 2007. The Anthropocene: Are humans now overwhelming the great forces of nature? *Ambio* 36, 614–621.
4. Zalasiewicz, J., M. Williams, A. Haywood, and M. Ellis. 2011. The Anthropocene: A new epoch of geological time? *Philosophical Transactions of the Royal Society* 369: 1938, 835–841.
5. Ostrom, E. 1999. Revisiting the commons: Local lessons, global challenges. *Science* 284, 5412, 278–282.
6. For an overview, UNEP. 2010. *The ABC of SCP*. Paris: United Nations Environment Program. www.uneptie.org/scp/marrakech/pdf/ABC%20of%20SCP%20-%20Clarifying%20Concepts%20on%20SCP.pdf
7. Dryzek, J. 2005. *The Politics of the Earth: Environmental Discourses*. Oxford: Oxford University Press.
8. Millennium Ecosystem Assessment. 2005. *Ecosystems and Human Well-being: Synthesis*. Washington, DC: Island Press.
9. TEEB. 2010. *The Economics of Ecosystems and Biodiversity: Ecological and Economic Foundations*. Washington, DC: Island Press.

10. de Groot, R., L. Brander, S. van der Ploeg, et al. 2012. Global estimates of the value of ecosystems and their services in monetary units. *Ecosystem Services* 1(1), 50–61.
11. Rockström, J., W. Steffen, K. Noone, et al. 2009. Planetary boundaries: Exploring the safe operating space for humanity. *Ecology and Society* 14(2), 32.
12. For example, Raworth, K. 2012. *A safe and just space.* Oxfam Discussion Paper. www.oxfam. org/en/policy/safe-and-just-space-humanity.
13. Norgaard, R. 2010. Ecosystem services: From eye-opening metaphor to complexity blinder. *Ecological Economics* 69, 1219–1227.

Chapter 9

Ecosystem Services in a Societal Context

Joachim H. Spangenberg

Helmholtz Centre for Environmental Research—UFZ, Department Community Ecology

Joachim H. Spangenberg - Helmholtz Centre for Environment Research. Several research projects on ecosystem services and related issues: in LEGATO we analyse the interaction of provisioning, regulating and socio-cultural services in Asian rice agriculture under the conditions of development and climate change. In APPEAL we analyse the biocontrol services in European agriculture; in both cases my work is focussed on monetary and non-monetary valuation. EJOLT deals with conflicts around service provision and access worldwide, ENRI with environmental rights based on human rights, and EO-Miners with equitable information access in Europe, South Africa and Central Asia.

A NEW AND USEFUL CONCEPT

EcoSystem Services (ESS, in other publications also abbreviated ES) is a stunning new concept, a metaphor useful for raising public awareness of the crucial role ecosystems and their biodiversity play in maintaining the quality of our everyday life. Basically, it is not about monetizing the value of ecosystems (an exercise that can be useful or even necessary in specific circumstances, like fixing compensation payments in court cases). It is rather about the basic role ecosystems play in human societies.

Ecosystem Services. http://dx.doi.org/10.1016/B978-0-12-419964-4.00009-3

There are many schools of thought and strands of discussion in the field, testifying to how new, that is, not yet consolidated, the field is and how creative the researchers and activists involved are. However, to avoid omissions and confusion, a closer look at the process of service generation is recommended.

A CLOSER LOOK AT THE CASCADE

Ecosystems consist of a multitude of interacting elements and processes, which on the system level constitute ecosystem functions, a natural phenomenon. The first step of human involvement is use value attribution, that is, humans recognizing the potential usability of a system and its functions. This intellectual step, without any physical intervention, changes the character of the system from a natural phenomenon to a societal resource, constituting the Ecosystem Service Potential ESP. The step is a crucial one as different groups of society may recognize different potential uses that can be complementary or mutually exclusive. For instance, a forest can be used for leisure, learning, fruit collection or logging, fertile soil for planting grain, flowers, or as a nature reserve. Any such activity is a mobilization of (part of) the potential and creates the ecosystem service. It usually requires the investment of human labor, time, and resources, and often money as a means to provide them.

FROM ECOSYSTEMS TO VALUE: A LONG CASCADE

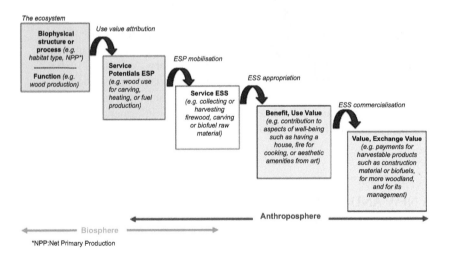

Then we have fruit, wood, grain, and so on growing in the field or forest, but not yet a benefit to humans. This requires appropriation of the ESS (collection, harvesting, and the like), including processing to transform the raw material into a useful product such as flour, firewood, and timber. Of course, appropriation and processing require labor, time, and resources again, and it is a matter of power and social interaction as to who can appropriate a service and who is excluded, with or without compensation. If directly used as in a subsistence economy, the ESS provides a use value or human benefit.

In our societies, however, most of the grain harvested is not consumed by the farmer, but sold. This is the process of ESS commercialization providing monetary exchange value, the main driver of ecosystem management. Additional conflicts may arise over the distribution of the monetary value created.

WHAT IS A SERVICE? NO FINAL CONSUMPTION, NO FREE GIFT OF NATURE

Some authors define the final consumption (bread, furniture, firewood) as the ESS. However, because, for instance, a maize harvest can be used for food, feed, or biofuels, depending on market prices, in this conceptualization the market (which so far plays no role in the ESS discourses) will decide which service the ecosystem provides, and the result would still be unknown at the time of harvest. The farmer selling his crop would lose the benefits, as the goods bought with the money earned are not allocated to the ESS that made it possible to purchase them. I consider that unsatisfactory and prefer to define the crop on the field as the actual service, generating benefits and income, which in turn provides additional benefits (but that is economics, no longer ESS analysis).

As this stepwise description shows, ESS and their benefits are not predefined by EcoSystem Functions ESF nor free gifts of nature, but which ESS is realized depends on the outcome of social discourses, conflicts, and compromises, which can change over time (de- and reforestation, draining and re-wetting swamps). Each ESS to be provided requires investments of labor, time, and resources, and often money as well. Thus it is society, not nature, that decides which ESS are available: Nature provides a potential for diverse services, and society makes the choice, based on the prevailing policy priorities. Thus it is a matter of policy orientation how the concept is implemented and which ESS are provided in practice, whether it is taken as a means to define exploitable resources or to promote ecosystem health (different use-value attributions), whether it serves the commercial interest or local stakeholders' needs value attribution and ESS mobilization), and whether the local stakeholders are compensated for the loss of otherwise available services and whether the environment is used sustainably or in contrast exploited and degraded.

ES IN A POLICY CONTEXT

Although the ESS concept provides valuable means to make people—including decision makers—aware of the importance of ecosystems and thus contribute to their conservation, parts of its current applications are of dubious value in that respect, and other applications are outright dangerous.

On the one hand, the concept might be used as the basis for an extended nature and biodiversity protection approach, integrating the multiple functions and services of natural and human-made landscapes into a broader conservation concept. Then it could serve to raise awareness regarding the risks of exploiting one service at the expense of others, like maximizing yields by pesticide use in agriculture at the expense of services such as freshwater provision or recreation. On the other hand, if applied under a standard economic framework, it could limit the perception of value to economically viable services. When the value of services is monetized, substitutability and commodification tend to threaten ecosystem integrity.

Payments for ecosystem services are a case in point: If designed as behavior-changing incentives, they tend to support conservation. However, if they are designed to internalize the value of so far free services, not only do social tensions about access to them arise, but it is also not certain whether they are effective (i.e., if they do actually change the behavior of agents), or whether they legitimize continuing unsustainable practices. In particular, if ESS are perceived as "free gifts of nature" and the societal context is ignored, power-based decisions are masked as natural facts—a biologism similar to Social Darwinism.

ES IN A GROWTH CONTEXT

In the past, economic growth has tended to undermine ecosystem health and service provision. The current focus on green, integrative, or sustainable growth claims to overcome this link and allow continued growth without overstressing the environment's capacity to provide services. However, we are already beyond the limits; the task is not to stabilize the impacts but to significantly reduce them to remain within the planetary boundaries. Furthermore, while delinkage is possible in the short run, and permanently so for selected economic activities, it is an illusion for the longer term development of the economy as a whole. The reason is simple: Imagine that to preserve the provision of ESS, resource consumption is cut by half today (a rather heroic assumption). Then, assuming the 3% GDP growth rate that the EU strives for materializes, the same unsustainable level of resource use will be reached again in about 30 years. In a growth context, ESS management offers at best temporary relief.

ESS IN A SUSTAINABILITY CONTEXT

According to the *full definition* of the Brundtland Commission (WCED, 1987, p. 43)

Sustainable development is development that meets the needs of the present without compromising the ability of future generations to meet their own needs. It contains within it two key concepts:

1. The concept of "needs," in particular the essential needs of the world's poor, to which overriding priority should be given, and
2. The idea of limitations imposed by the state of technology and social organization on the environment's ability to meet present and future needs.

In ESS parlance, this statement can be read to mean that a sufficient availability of ESS should be guaranteed for all members of the present and future generations, that in the case of distributional conflicts (use value attribution, ESS appropriation) the needs of the poor should have priority (e.g., indigenous land management over commercial exploitation) and that ESS mobilization must not reduce the ESP.

In such a context, the ESS concept can help understand the diverse ESP attributed to ecosystems by different stakeholders and prepare the ground for solving conflicts by deliberations and well-informed democratic decision making. It can promote cultural pluralism (different use value attributions are of equal value), and indicate the need for societal intervention (the investment in mobilization and appropriation public of ESS). It indicates distributional conflict potentials that require political solutions. Use value/benefits of ESS accrue to agents long before commercialization. This demonstrates the limits of market-based valuations and the necessity to take nonmarket values appropriately into account.

CONCLUSION

The ecosystem service concept is an innovative means to clarify views on the interaction of nature and of humankind. Its applications and impacts depend on the political context as much as on the natural one. Its full potential can best be exploited in a sustainability context, while in a neoliberal growth context it may be used to unsustainable ends.

Chapter 10

The Value of the Ecosystem Services Concept in Economic and Biodiversity Policy

Leon C. Braat

Alterra, Wageningen University and Research, Wageningen

Leon C. Braat - Alterra, Wageningen University and Resreach. Project leader of European scale DG ENV project "Mapping and Assessment of Ecosystems and their Services"; senior researcher in "The Economics of Ecosystems and Biodiversity - NL" and the "Digital Atlas of Natural Capital", both for the Dutch National Government, and in EU Framework 7 projects BESAFE and ROBIN; Chairman of the Management Board of the European Biodiversity network "ALTER-Net" and Editor-in-Chief of the international Elsevier Scientific Journal Ecosystem Services.

INTRODUCTION

Every human being is making value judgments all the time. Most of the assigned values to goods, services, people, or religious rituals have to do with the survival of humans as individuals, many have to do with the welfare and well-being of individuals and social groups—these are often called economic valuations—and some with the ethical considerations humans make about other people's and other species' rights to live.

Economic valuations are widely misunderstood to be equal to monetary valuation, while money is only one metric to assess values to our welfare and

Ecosystem Services. http://dx.doi.org/10.1016/B978-0-12-419964-4.00010-X

well-being. Energy would be doing much better (see, e.g., [15]). Also, because mainstream economics, which dominates the political processes and newspapers, has been based on market transaction mostly since the late 18th century, many, mostly the economically ignorant, consider economic valuation to be equal to include only such market goods and services. Furthermore, the argument goes that biodiversity can and should not be traded in markets, and thus should not be valued through money metrics, and therefore not be economically valued at all—as if biodiversity has no meaning to our economic welfare and individual and social well-being.

Refraining from putting explicitly assigned, quantitative values on nature and biodiversity has been the strategy of conservationists for more than a century. In spite of considerable success in establishing protected area and protective laws and raising awareness among the general public, the net result is a continued loss of biodiversity worldwide, and unsustainable use of ecosystems and their services. It is, in short, a strategy that has to be strengthened with concepts that enlighten the powers that be, whose decision paradigm is based on mainstream economics about the real and economic value to the human species of nature, ecosystems, natural capital, and the goods and services they produce. Of course, we should not stop at copying the neoclassical economic concepts and methods, but we must extend the model to what has been known as ecological-economics since the early 1990s (see, e.g., [3]).

An interesting view on valuation processes is proposed in TEEB Foundations [16], where it is described as a form of regulatory adaptation via positive and negative feedback in an economic system. In this view, the valuation of changes in biodiversity, natural capital, and ecosystem services then becomes a logical and necessary element of the sustainable development policy cycle. Economic valuation that includes a broadly market- and nonmarket-based overview of the benefits and costs may well contribute to adjusting economic policies and environmental regulations to the natural science-based knowledge of productive and carrying capacity of ecosystems and the welfare and well-being requirements of human beings.

ECOSYSTEM SERVICES

A bridging concept (see [1]) is available; it is called ecosystem services, and it has matured sufficiently over the last few decades to play a decisive role in the political arena where survival, welfare and well-being decisions are made and become the rational basis for decision making. In the European Union this ostentatiously happened with the signing of the EU Biodiversity Strategy in 2011 [10], where ecosystem services has become the cornerstone for sustainable economic development. The concept need not replace the ethical drive of people, which aims to conserve species and habitats via an appeal to decision makers' conscience. Informal education of the current powerful politicians is, however, necessary, as the generations that are still receiving formal education are not yet influential enough to make a difference and stop the loss of biodiversity.

The thesis of the present chapter is in fact that the design of sustainable development policies can be much better informed (than is currently practiced) by properly using economic valuation and accounting exercises, which include the nonmarketed services, and wisely using expressions of value in monetary terms to make public decision making transparent. To try to stop biodiversity from being entered in the economic valuation equations is effectively giving it a value of zero. And that cannot be the goal.

A recent summary of the history of the concept of ecosystem services is given in [1]. It involves the utilitarian framing of those ecosystem functions that are considered beneficial to individual humans, social groups, economic sectors, and society at large, in terms of economic goods and services. This type of thinking has origins in different corners of society. Some have propelled the concept in order to increase public interest in conservation of biodiversity. Others have seen an opportunity to support the sustainable development debate, but it has also been picked up in mainstream economics as a topic for environmental economic research.

The definitions of the concept have evolved, with varying attention to the ecological basis (ecosystem services are the conditions and processes through which natural ecosystems, and the species that make them up, sustain and fulfill human life [4] or the economic use (ecosystem services are the benefits human populations derive, directly or indirectly, from ecosystem functions; see [4]) to a bridging definition by the TEEB project (ecosystem services are the direct and indirect contributions of ecosystems to human well-being—[16]; see Figure 10-1).

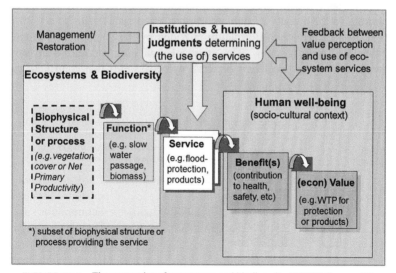

FIGURE 10-1   The economics of ecosystems and biodiversity (TEEB) diagram [7].

According to most sources, the term *ecosystem services* was coined in 1981 by Paul and Anne Ehrlich [12], but the process of bridging the gaps between ecology and economics, and between the domains of nature conservation and economic development, and the landing in the political arenas took a few decades longer. In the 1970s and 1980s, a growing number of environmentally aware authors started to frame ecological concerns in economic terms in order to stress societal dependence on natural ecosystems and raise public interest in biodiversity conservation. The rationale behind the use of the ecosystem service concept was for some creative conservationists to demonstrate how the disappearance of biodiversity directly affects ecosystem functions that underpin critical services for human well-being. The TEEB study, building on, among other things, the Cost of Policy Inaction study initiated by the European Commission [2], has brought ecosystem services into the policy arena, with a clear economic connotation to what used to be (and still is) called biodiversity policy (www.teebweb.org). With increasing research on the role of ecosystem services in public and private financial flows, the interest of policy makers has turned to both the design of market-based instruments to create economic incentives for conservation, for example, payments for ecosystem services, and to tax and subsidy instruments, for example, to get rid of perverse subsidies in fisheries.

Following H.T. Odum's [15] energy flow approach to complex systems, we argue that in the real world provisioning and cultural ecosystem services are only delivered (and subsequently beneficial and of value) to humans with some investment of energy, for example, labor, by humans (see Figure 10-2). The energy content of the ecosystem services flows is in all such cases a combination of natural (ecosystem processes based) energies with human energies.

Even a basic provisioning service, such as food delivery, requires labor in the form of gathering, hunting or harvesting work. All cultural services by definition involve the activity of human sensory organs and brains to absorb and process, respectively, the information provided by the components, structure, and dynamics of ecosystems. The group of regulating services is diverse in this respect. They all are forms of work by ecosystems that contribute to an

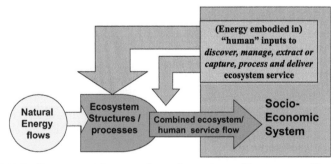

FIGURE 10-2   Ecosystem services as products of ecosystem and human energy (Braat, adapted from Odum [15]).

environment in which humans (and many other species) can live (e.g., climate regulation by carbon sequestration; air pollution capture), buffering extreme events (floods, droughts, erosion) or facilitating other services (pollination), without human labor required in that ecosystem work directly. In the real world, at least in developed economies, human interventions have often reduced the capacity of ecosystems to perform such regulating services, by pollution, over-exploitation, and destruction, so that targeted restoration and management are necessary to reestablish the capacity for regulating services. In economic terms opportunity costs are always involved—for example, by having a forested land not available for urban activities.

THE ECONOMICS OF ECOSYSTEMS AND BIODIVERSITY

With the introduction of *Environmental Impact Assessment* (EIA) [11] and its relative, *Strategic Environmental Assessment* (SEA) [9], the foundations were laid to include nature and biodiversity in economic development planning. Many countries are now extending these decision support instruments and are combining them with traditional *Cost-Benefit Analysis* (CBA) into so-called *Social Cost-Benefit Analysis* (SCBA) or *Sustainability Assessment*. The TEEB procedure, as outlined in the TEEB Synthesis Report [16], could then be seen as the next step in the process of maturation of ecologically based, social, and economic decision making . Braat and De Groot [1] discuss the procedure and the necessary research to make it a full-fledged decision-support tool. A short summary is given here to launch the argument of the necessity of economic valuation.

Identify and Assess: Indicators, Mapping, and Quantification

Maintaining ecosystems capable of delivering multiple services requires a consistent approach to sustaining a considerable level of biodiversity. It is then essential to map the ecological and human systems in the landscapes where ecosystem services are to be assessed. Without precise delineations of system boundaries, the quantification processes will be unreliable, and in human systems ultimately legal consequences of policies require exact property boundaries. Maes et al. [14] give an introduction to and an overview of the challenges of mapping ecosystem services. In assessing trade-offs between alternative uses of ecosystems, the total bundle of ecosystem services provided by different conversion and management states should be included. Economic assessment should be spatially and temporally explicit at scales meaningful for policy formation or interventions, inherently acknowledging that both ecological functioning and economic values are contextual, anthropocentric, individual-based, and time specific. Braat and Ten Brink [2] provide a provocative visualization of the trade-offs between provisioning and other ecosystem services with an increase in intensity of land use.

Estimate and Demonstrate: Valuation and Monetization

Valuation, especially monetary valuation, is sometimes understood to imply that ecosystem services must be privatized and commodified (traded in the market). First, this is not a necessary corollary, but second something that can be countered by demonstrating that public goods and services (and the natural capital they come from) may be better managed in the public domain. EIA does not provide direct insight into welfare gains and losses. Historically, this was thought to be obtained from cost-benefit analyses. But a broader approach was seen to be needed, including the nonmarket aspects of welfare and well-being to the decision-making process. If small changes are the issue, scenario comparison is considered particularly important for monetary valuation based on marginal values. When, however, the proposed land-use change involves nearly complete loss of ecosystems, biodiversity features, and disappearance of ecosystem services, marginal value changes are in fact irrelevant [13].

Capture and Manage the Values

Step 3 in the TEEB procedure is interpreted as being "to capture the values for a sustainable society." To a large extent the third step is represented in the TEEB diagram (Figure 10-1) by the feedback loop from the economics box to the ecological box, and to the services flows, as institutional, policy, and societal response. Protective legislation or voluntary agreements can be appropriate responses where biodiversity values are generally recognized and accepted.

CONCLUSION

It is important to follow the procedure completely from Step 1 through Step 3 to identify, quantify, and valuate all changes in ecosystem services. To choose a priori and arbitrarily to exclude some classes of services makes no sense. A systematic checklist of ecosystem services should lead the selection process. And the capital stocks of those services that are then shown to be prominent should also be part of the analysis. In addition, decision makers also need information about who is affected and where and when the changes will take place. Such a demonstration of economic value may also lead to more efficient use of natural resources. The capturing of course refers to making the "value" in the service actually visible, in some cases cashable and accountable, and generally includes payments for ecosystem services, reform of harmful subsidies, tax breaks for conservation, or creation of a green market economy. The development (or adjustment) of the legal system with respect to rights over natural resources and liability for damage to ecosystem service potential is essential.

REFERENCES

1. Braat, L.C., and R.S. de Groot. 2012. The ecosystem services agenda: Bridging the worlds of natural science and economics, conservation and development, and public and private policy. *Ecosystem Services* 1, 4–15.
2. Braat, L.C., and P. ten Brink (eds.). 2008. The cost of policy inaction: The case of not meeting the 2010 Biodiversity target. Report to the European Commission under contract: ENV.G.1./ ETU/2007/0044; Wageningen, Brussels; Alterra report 1718/ http://ec.europa.eu/environment/ nature/biodiversity/economics/index_en.htm
3. Costanza, R. (ed.). 1991. *Ecological Economics: The Science and Management of Sustainability.* New York: Columbia University Press.
4. Costanza, R., d'Arge, R., de Groot, R., Farber, S., Grasso, M., Hannon, B., Naeem, S., Limburg, K., Paruelo, J., O'Neill, R.V., Raskin, R., Sutton, P., and van den Belt, M. 1997. The value of the world's ecosystem services and natural capital. *Nature* 387, 253–260.
5. Daily, G. 1997. *Nature's Services: Societal Dependence on Natural Ecosystems.* Washington, DC: Island Press.
6. Daly, H., and J. Farley. 2010. *Ecological Economics: Principles and Applications*, 2nd ed. Washington, DC: Island Press.
7. De Groot, R.S., B. Fisher, M. Christie, J. Aronson, L.C. Braat, R. Haines-Young, J. Gowdy, E. Maltby, A. Neuville, S. Polasky, R. Portela, and I. Ring. 2010. Integrating the ecological and economic dimensions in biodiversity and ecosystem service valuation. In TEEB Foundations (Kumar, P., ed.), Chapter 1, pp. 9–40. *The Economics of Ecosystems and Biodiversity (TEEB): Ecological and Economic Foundations.* London: Earthscan. [http://www.teebweb.org]
8. De Groot, R.S., L. Brander, S. van der Ploeg, F, Bernard; L.C. Braat, M. Christie, R. Costanza, N. Crossman, A. Ghermandi, L. Hein, S. Hussain, P. Kumar, A. McVittie, R. Portela, L. C. Rodriguez, P. ten Brink, and P. van Beukering. 2012. Global estimates of the value of ecosystems and their services in monetary terms. *Ecosystem Services* 1:50–61.
9. EC. 2001. SEA Strategic Environmental Assessment Directive 2001/42/EC. Brussels.
10. EC. 2011. Our life insurance, our natural capital: an EU biodiversity strategy to 2020 COM(2011) 244 final. Brussels.
11. EEC. 1985. EIA Directive 85/337/EEC, Environmental Impact Assessment—EIA Directive. Brussels.
12. Ehrlich, P., and A. Ehrlich. 1981. *Extinction: The Causes and Consequences of the Disappearance of Species.* New York: Random House.
13 Farley, J. 2012. Ecosystem services: The economics debate. *Ecosystem Services* 1, 40–49.
14. Maes Joachim, Benis Egoh, Louise Willemen, Camino Liquete, Petteri Vihervaara, Jan Philipp Schagner, Bruna Grizzetti, Evangelia G. Drakou, Alessandra LaNotte, Grazia Zulian, Faycal Bouraoui, Maria Luisa Paracchini, Leon Braat, and Giovanni Bidoglio. 2012, July. Mapping ecosystem services for policy support and decision making in the European Union. *Ecosystem Services* 1(1), 31–39.
15. Odum, H.T. 1983. Systems ecology: and introduction. New York: John Wiley & Sons.
16. TEEB Foundations. 2010. TEEB—*The Economics of Ecosystems and Biodiversity (TEEB): Ecological and Economic Foundations* (P. Kumar, ed.). London: Earthscan.

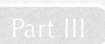

Ecosystem Service Debates

Shutterstock.com / © zhu difeng

Chapter 11

Valuation of ES: Challenges and Policy Use

Inge Liekens and Leo De Nocker

Flemish Research and Technology Organisation (VITO)

Chapter Outline

1. INTRODUCTION

The monetary valuation of ecosystem services (ES) presents a promising approach to highlight the relevance of ES to society and the economy, to serve as an element in the development of cost-effective policy instruments for nature restoration and management, and for use in impact assessments in cost-benefit analysis. Economic valuation may also be useful in developing payments for ecosystem services [1]. However, it is also recognized that considerable challenges still remain if the valuation of ecosystem services is to be fully incorporated into policy appraisal.

Economic valuation requires a three-step approach: first, the goods and services that may be delivered by an ecosystem need to be identified; second, they need to be quantified; and third, they need to be valued [2]. These steps require a thorough understanding of all the mechanisms that affect supply and

Ecosystem Services. http://dx.doi.org/10.1016/B978-0-12-419964-4.00011-1

demand for these goods and services. For applications in specific policy studies, it is very unlikely that one can study all these aspects in detail and for the location and contexts of the study. Consequently, one has to rely on information and data gathered and developed for similar ecosystems in similar contexts. In areas with a lot of well-documented data and information, this will suffice to make reliable estimates. However, for some ecosystems and/or services, understanding and data will very likely be rather limited, and so benefit transfer will become a difficult exercise. In this chapter, we list and discuss a number of important issues that need to be accounted for. First, we discuss issues related to selection and interpretation of information and data. We distinguish between the quantification and valuation steps, although in practice these steps and issues are likely to be interwoven. Finally, we discuss issues related to use of results for policy studies.

2. UNCERTAINTY AND COMPLEXITY IN QUANTIFICATION AND VALUATION

2.1. Uncertainty and Complexity Surrounding Quantification

The quantification of goods and services should reflect the uncertainties in biophysical and ecological sciences, as well as the account for context specificity of numbers or functions used. This implies a strict selection of data and functions suited for benefit transfer, and it requires the reporting of uncertainty boundaries for the results. Improved knowledge and data related to variability, interrelation of ecosystem services (ES), and the importance of thresholds are essential for a more precise selection and use of data and functions [3].

Uncertainty in Ecosystem Functioning

Uncertainty reflects the variability in the delivery of goods and services by ecosystems, but we are not able to specify the factors that explain this variability. Consequently, a better understanding and reporting about this variability is essential to improve quantification and valuation of ES.

Uncertainties relate to which services different ecosystems provide, how these services may change over time, and how changes to ecosystems may affect the quantity and quality of the services they provide (see Chapters 14 and 15, this volume).

Science is only starting to get some insight into these questions and the role of biodiversity in the delivery of services. Robust information is still lacking on how biodiversity contributes to the ecological functions that lead to tangible benefits for society (see Chapter 3, this volume). The uncertainties in ecological sciences can be accounted for in ES valuations, either by means of using the range of values from different studies or by testing the impact of different insights in ecological science on the result of valuations (sensitivity analysis). This implies that the value of ES should be presented as a range of possible values rather than as single-point estimates.

Interrelation of Ecosystem Services

Interrelationships within and between ecosystems means that the delivery and/ or economic value of any one service may depend on its relationship with other services; that is, a change in one service could have an important impact on other services. For example, improving the production service of an ecosystem may decrease regulating services. Ignoring these interrelations by focusing assessments on one single ecosystem service could therefore lead to suboptimal decisions for society. Therefore, monetary valuation must aim to take the total bundle of ecosystem services into account.

Accounting for Thresholds

As the state of an ecosystem deteriorates, the services it provides are likely to diminish. This process may be a gradual one, but in some circumstances thresholds are reached. Ecological thresholds can be defined as breaking points of ecosystems at which the pressures lead to abrupt changes in the ecosystem. Beyond these thresholds, an irreversible change to the ecosystem may occur (e.g., total collapse), resulting in a permanent loss of ecosystem services. Well-known examples are the change of a clear lake into a turbid one, species extinction, and the change from "grassy" systems to bare soil. Economic valuation focuses on marginal changes and values; for this to hold true, the next unit should not tip the ecosystem over a functional threshold [4, 5] into an alternative state. Current knowledge about biodiversity and ecosystem dynamics to forecast these thresholds is insufficient due to the considerable complexity and uncertainty surrounding ecosystem functioning and interrelationships. One of the difficulties is that thresholds are not constant. An example is described by [6] for rangelands. If the grass layer consists of all perennials, the threshold ratio of shrubs (S) to grass (G) (the slow-changing, controlling variable) is higher than if the grass layer consists of only annuals.

2.2. Uncertainty and Complexity in Valuation

Monetary valuation estimates the welfare impact of marginal changes in the supply of ecosystem goods and services. The added value of monetary valuation relates to unveil the "hidden" benefits of the environment into a metric that policy makers and the general public can easily relate to and that allows for aggregation and comparison. The range of goods and services covered requires the use of different methods, including market prices, avoided costs (replacement or repair cost; costs of illness), and revealed and stated preferences [7, Chapter 1.3.].

The use of nonmarket valuation techniques is unavoidable but requires careful application and thoughtful interpretation of results. Consequently, data availability and uncertainty vary significantly between the different ecosystem goods and services considered [7–9]. To use data from literature, benefit transfer methods are developed, but their practical application is hampered by inappropriate basic data [10, 11].

Nonmarket valuation techniques have been well developed for a number of areas and provide good and useful information for purposes such as valuation of health impacts or visits to a forest. In these cases, different methods can be used, including revealed and stated preferences. Stated preferences methods can successfully be applied in areas where people can rely on their own experiences (e.g., related to illness or visits to similar green areas). In addition, the indicators describing their quantity (e.g., number and duration of visits or episodes of a specific health condition) are relatively easy to interpret and communicate. Consequently, there are more and better data related to so-called use values of ecosystems, compared to the nonuse values. For use values such as visits, the available meta-analysis shows that we start to understand the factors that impact these values [12–15] and we can use the results for benefit transfer. For nonuse values, the situation is different. Nonuse values are typically a mix of different types of values, referring to option values, bequest values, and altruistic values. We have much fewer expectations from theory or other disciplines about what factors contribute and we can hardly compare results with other methods. Consequently, for nonuse values, less data is available and it is hard to use in benefit transfer.

Although the number of valuation studies has grown substantially over the last decades, there is still a need to enlarge the number of data and functions suited for benefit transfer and to present the existing information in a better format.

Very often, past studies were not designed for the purpose of benefits transfer or application beyond the specific scope of the study. Moreover, many valuation studies focus on single ecosystems (forests and wetlands, for example, received a lot of attention) or ecosystem services (provisioning) and especially on the most evident ones. For example, the value of wetlands is mostly studied in regions where these habitats are very scarce, not in regions where a lot of substitutes can be found [1]. Some ecosystem services (e.g., cultural services and regulating services such as air quality and noise mitigation) or even ecosystems (e.g., urban ecosystems, grasslands, marine ecosystems) are barely looked at. Although the available data on ecosystem values mostly refer to a limited set of services obtained within a specific context, they may be misinterpreted as indicators for total value and for an indicator that is relevant in a wider range of contexts. These misunderstandings can be limited by a more precise presentation of results that show the limits of services included and the range of context for which it is relevant.

If new studies would result in value functions suited for benefit transfer and if they could present their results more precisely, the uncertainty in the benefit transfer process could significantly be reduced and the risk of misinterpretation of data in policy appraisals would be limited.

Option Value and Insurance Value

Valuation research has produced data and information related to the current use of the ecosystem (see also Chapter 2, this volume), for example, food production, climate regulation, and recreational value. For some of the important arguments to protect ecosystems, data and information are scarce. This relates to some future option values and insurance values.

The option value can be defined as the value people put on future use of known and unknown services. Even if an ecosystem or some component of it currently generates no value, its option value may still be significant [8] For example, the future may bring human diseases or agricultural pests that are unknown today. This means that today's biodiversity has an option value insofar as the variety of existing plants may already contain a cure against the as yet unknown disease, or a biological control of the as yet unknown pest [9].

The insurance value is the value of ensuring that no regime shift in the ecosystem takes place that will have irreversible negative consequences for human well-being. The insurance value is the value of the ecosystem's capacity to maintain a sustained supply of benefits (ecosystem resilience). In order for ecosystems to deliver services, some minimum safe scale of organization and operation is needed before services can be provided [3]. Minimum "safety standards" are hard to define due to the complex nature of the processes. To measure this insurance value in economic terms, one needs to account for the risk preferences of society. The more risk averse a society is, the more weight it will place on strategies to maintain or improve ecosystem resilience. Although a lot of information on the willingness to pay for risk aversion is available, it is not obvious to apply that information to value ecological resilience. For some of the important arguments to protect ecosystems (e.g., related to insurance and option values or to biodiversity), the good or service valued is much more difficult to describe and unequivocal indicators to measure their quantity are less available or hard to communicate. Consequently, it is much more difficult to elicit the preferences of people in these areas, even if people are for example, familiar with the concept of risk and the benefits of risk-avoiding behaviour.

Moreover, the valuation of resilience goes beyond the marginal approach of most valuation methods. The data or functions for economic valuation originate from studies made in specific contexts related to socioeconomic conditions and the supply and scarcity of the specific good or service. These data inform us about the value of marginal changes of supply in similar conditions, but they hardly tell us anything about their value in cases where supply drops significantly and scarcity increases.

Nonmonetary Valuation Techniques

Further investigation is needed on to determine the extent to which nonmonetary valuation techniques can help to overcome these obstacles. Nonmonetary valuation techniques also aim to estimate preferences of the citizens to value goods and services, but the indicator is not expressed in a monetary value and it does not reflect preferences defined under budget constraints. Civic valuation can elicit information about preferences through either revealed behavior (e.g., outcome of referenda) or stated valuations (deliberative processes such as citizen juries, roundtables, and Participatory Rural Appraisals) [16, 17]. An example of the use of public referenda is given in [18] where preferences related to climate regulation (CO_2 abatement) are estimated based on the outcome of Swiss referenda.

Citizen jury is a stated preference method. It derives information about preferences through a deliberative process that helps individuals understand and assess trade-offs among multiple attributes [19]. In such deliberative processes, values emerge from a communicative social process [20, 21]. This method is often used to rank options for specific decisions and to promote a common understanding and ranking, but the experiences related to ecological valuations are limited [16] (see also Chapter 1, this volume).

Individual and Collective Deliberations

Economic valuation methods can be based on individual or collective deliberations. Economic valuation with stated preferences typically uses collective deliberation methods (e.g., focus groups) to develop surveys, but the data-gathering phase is usually based on individual (or family) deliberations. It works well only if people have some predefined values that can easily be generated in the process initiated with the survey. Although collective deliberations or discussions are possible, it is not common to use it. In addition, guidelines promote individual deliberations to avoid socially desirable answers and to account for differences in budget constraints among the population. Studies aim to understand not only average value in society but also heterogeneity in values.

Deliberative processes use more often collective deliberations, especially if the objective is to derive common understandings and rankings. This has proven to be very useful for decision making for local problems. Nevertheless, proper implementation of these methods remains challenging. These methods are time consuming and labor intensive. Ensuring the representativeness of the affected population could also be a limiting factor. Moreover, it is not straightforward to apply these techniques to value resilience or future options in a way that the results can be used as general indicators in the more formal process of economic appraisal [2].

A comprehensive report from the United States Environmental Protection Agency (EPA) proposes an approach of multiple methods (e.g., ecological methods, cost-based methods, willingness to pay methods, deliberative valuation) to help capture different dimensions of biodiversity and ecosystems services, as well as the perspectives of different stakeholders. The underlying idea is that an integrated and multidimensional approach will be more likely to capture the full range of contributions, thus the broader value of biodiversity and ecosystems, including values that may be context specific (global, national, regional, local) [16].

2.3. A Multidisciplinary Approach to Reduce Uncertainties

Addressing the current gaps in knowledge requires a coordinated, multidisciplinary approach. Too often, ecological, economic, and social studies are carried out separately. As a result, the most reliable ecological and economic information cannot be brought together. Instead, a new kind of interdisciplinary and transdisciplinary approach is needed to build understanding of social-ecological systems. To understand changes in ecosystem services, the interactions of social and ecological constituents of the earth system must be considered. Discipline-bound approaches that hold one component constant while varying the other lead to incomplete and incorrect answers. Although many important questions of basic interdisciplinary science must be addressed, there might be a need for networked, place-based, long-term social-ecological research [22].

Under these circumstances, uncertain future losses need to be taken into account. One such method is to conduct sensitivity analysis in testing how sensitive the outcomes are to changes in values and to assumptions used in valuing ecosystem services. In addition, there may be a need for decision making to consider precautionary approaches in the face of considerable uncertainty and close to thresholds.

3. CHALLENGES IN USING MONETARY VALUES FOR POLICY APPRAISAL

A major critique of early ecosystem services valuation work was the rudimentary treatment of ecological systems at the scale of the biomes and the extrapolation of site-specific values across the entire globe [23, 24]. At the other end of the spectrum, the utility of plot-scale experiments for policy formation is questionable. Advanced research in ecosystem services is using a more balanced trade-off and has focused on spatially explicit economic and ecological models. The trend has been to move away from standard lookup tables assuming constant €/ha values for a similar ecosystem type across Europe [25–27] but still allowing feasible benefit transfer beyond the plot scale. This research is needed to aggregate different data.

3.1. Spatial Explicitness and Distance Decay

Aggregation refers primarily to multiplying a monetary unit value with the appropriate quantity (e.g., kg N-removal/ha multiplied with a €/kg , number of households multiplied with the willingness to pay per household). The quantification requires both spatial explicitness of the factors influencing the biophysical processes and the number of beneficiaries. Important site characteristics with an impact on valuation include the type of ecosystem, area, and integrity of the ecosystem. For example, the extent to which a wetland will have an impact on the water quality of a nearby watershed depends on the soil texture, soil moisture, and load of nutrients entering the wetland [28]. Benefit transfer methods therefore need to account for differences in site characteristics. In addition, it needs to take into account socioeconomic data and spatial information.

In aggregating the number of beneficiaries, the analyst needs to assess what the size of the market is for the ecosystem service, that is, identify the population that holds values for the ecosystem service. In conducting benefits transfer (BT), it is also important to control for differences in the characteristics of beneficiaries between the study and policy sites (e.g., spatial location and socioeconomic parameters). A recent improvement in valuation is accounting for distance decay effects between beneficiaries and ecosystems in determining the market size for an ecosystem service and in aggregating per-person values across the relevant population (e.g. [29]). It should be noted that the market size and rate of distance decay are likely to vary across different ecosystem services from the same ecosystem. For example, beneficiaries may be willing to travel a large distance to view unique fauna (distance decay of value is low, and people in a wide geographic area hold values for the ecosystem and species of interest), whereas beneficiaries may not travel far to access clean water for swimming (distance decay of value is high due to the availability of substitute sites for swimming, and only people within a short distance of the ecosystem hold values for maintaining water quality to allow swimming). Aggregation of transferred values across beneficiaries without accounting for distance decay and substitutes will affect the results significantly. An illustrative example can be found in Bateman et al. [29], who compare different aggregation methods and assess the effect of neglecting distance effects. It is also important to account for differences in site context such as availability of substitute and complementary ecosystems and related services [30]

3.2. Double-Counting

In assessing trade-offs between alternative uses of ecosystems, the total bundle of ecosystem services provided by different conversion and management states should be considered. "In this case aggregation means adding up the different services of the same ecosystem in order to calculate the total economic value of the ecosystem" [7, p. 232]. Aggregation of data for individual services requires checking for potential double-counting. In valuing the different ecosystem services, it is best to focus on the end products (benefits) in order to avoid adding supporting or intermediate services values to the value of the final good or benefit. For example, the value of a pollination service, which is already embodied in the market price of a crop, should not be counted separately unless the value of its input to the crop is deducted [31]. A classification scheme recommended by Fisher, Turner and Morling [5] helps to avoid the problem by drawing a clear distinction between intermediate services, final services, and benefits, with only the benefits being subject to economic valuation.

Second, the interrelation between the final goods or services needs to be examined. As long as the ecosystem services are entirely independent, adding up the values is possible. However, ecosystem services can be mutually exclusive, interacting, or integral [4]. The interaction of ecosystem services and values can also depend on their relative geographical position, for instance, with substitutes that are spatially dependent. For example, estimations of health benefits of living in an area with more open and green space is likely to overlap partly with estimates of benefits for recreation, but knowledge and data do not provide clear guidance on how to correct for this potential overlap. Knowledge on interlinkages between ecosystem services is indeed not always there, so adding up should integrate an analysis of risks and uncertainties, acknowledging the limitations of knowledge on ecosystems and their services and on their importance to human well-being.

3.3. Spatial Scale

When adding up a bundle of ecosystem services, one needs to make sure that the beneficiaries of a certain good or service are well identified. Spatial scale is recognized as an important issue to the transfer of ecosystem service values [32]. The services that ecosystems provide can be both on- and off-site. For example, a forest might provide recreational opportunities (on-site), downstream flood prevention (local off-site), and climate regulation (global off-site). On the demand side, beneficiaries of ecosystem services also vary in terms of their location relative to the ecosystem service(s) in question. Although many ecosystem services may be appropriated locally, beneficiaries also receive manifold services at a wider geographical scale. Local agents tend to attach higher values to provisioning services than national or global agents, who attach more value to regulating or cultural services. Different stakeholders often attach different values to ecosystem services, depending on cultural background and the impact the service has on their living conditions [32, 33]. Furthermore, goods with wider spillovers are more "public" in nature and require contributions from a more diverse set of donors. For this reason, different types of ecosystem services are valued differently as the spatial scale of the analysis varies [32, 34].

The majority of valuation studies do not account for different scales of ecosystem services, although considering spatial and temporal scales and stakeholders enhances the ability of ecosystem service valuation studies to support decision making. The formulation of management plans that are acceptable to all stakeholders requires balancing their interests at different scales [32]. Spatial scale is also highly relevant to the issue of distance decay, fordepending on a local or global scale, the number of beneficiaries will change and the value people put on a certain ecosystem services will drop faster with distance (e.g., local agriculture versus high biodiversity of the rainforest).

3.4. Upscaling

The challenges we have discussed in this chapter relate to the valuation of individual ecosystem sites. Scaling up refers to "the use of benefit transfer to estimate the value of an entire stock of an ecosystem or provision of all ecosystem services within a large geographic area" [8, p. 237]. In addition to the other challenges mentioned above, scaling-up values requires accounting for the possible nonconstancy of marginal values across the stock of an ecosystem. Simply multiplying a constant per-unit value by the total quantity of ecosystem service provision may over- or underestimate total value. For example, using four studies on the monetary value of nutrient cycling by coastal systems, TEEB [8] found a range of 170 to 30 451 $/ha.year.

Many ecosystem services values have nonconstant returns to scale. Some ecosystem service values exhibit diminishing returns to scale; that is, adding another ha to a large ecosystem (1000 ha) increases the total value of ecosystem services less than an additional ha to a smaller ecosystem (e.g., 10 ha) [35] Diminishing returns may occur either because of underlying ecological relationships (e.g., whether there are more marshes, how lower the value of an additional marsh could be for flood control) or because of declining marginal utility to users of services (e.g., as vacationers have already a large number of forests in their neighbourhood, their value for an additional forest will be lower than when no forests at all are nearby). In contrast, other ecosystem services such as habitat provision may exhibit increasing returns to scale over some range. For example, if the dominant goal is to maintain a viable population of some large predator, habitats too small to do so may have limited value until they reach a size large enough to be capable of supporting a viable population.

"Appropriate adjustments to marginal values to account for large-scale changes in ecosystem service provision need to be made, for example by using estimated elasticities of value with respect to ecosystem scarcity" [8, p. 237]. But for some ecosystem services (e.g., services for which no adequate substitutes are available), estimations on the demand curves are hard to predict, again because knowledge about the dynamics of the functioning is missing.

4. CONCLUSION

Many challenges in economic valuation remain, ranging from methodological challenges in economics and valuation studies to challenges for a better coupling of economic information with ecological information and challenges related to produce more and better data [8]. Challenges remain concerning the difficulty in accounting for interlinkages between different ecosystem services, spatial scales and distance decay, and upscaling. Properly used, however, economic valuation based on sound biophysical data can deliver robust estimates of ecosystem service values that are extremely useful for policy.

There is a need for studies and methods to account more systematically for all the impacts on ecosystems and their services. Investment in multi/transdisciplinary studies is necessary to better understand the links between ecosystem functioning, the delivery of ecosystem services, and the values these generate for society. New valuation studies should aim to produce and report results in a way that can be used for benefit transfer.

The long-term challenge for those involved in policy appraisal is to fully take into account all the impacts on ecosystems and ecosystem services. However, current information and methods can already be useful, as the "perfect" ecosystem service valuation may not be necessary for many appraisal purposes [3]. Practical appraisals need to compare the relative magnitude of changes in the provision of ecosystem services across different alternatives; this can be possible even with limited availability and precision of scientific and economic information. In these cases, it is important to report the constraints of the valuation exercise. Attention should be paid to uncertainty concerning estimates of environmental effects (e.g., timing, magnitude, and significance); assumptions entailed in the transfer of economic values or functions; and the potential significance of any incomplete information or nonmonetized impacts.

REFERENCES

1. Markandya, A. 2011. Challenges in the Economic Valuation of Ecosystem Services. BEES Workshop IV: Ecosystem Services and Economic Valuation, May 18, 2011.
2. Gantioler, S., Rayment, M., Bassi, S., Kettunen, M., McConville, A., Landgrebe, R., Gerdes, H., and ten Brink, P. 2010. Costs and socio-economic benefits associated with the Natura 2000 Network. Final report to the European Commission, DG Environment on Contract ENV.B.2/SER/2008/0038. Institute for European Environmental Policy/GHK /Ecologic, Brussels.
3. DEFRA. 2007. An introductory guide to valuing ecosystem service. London: Department for Environment, Food and Rural Affairs.
4. Turner, R.K., Paavola, J., Cooper, P., Farber, Stephen, Jessamy, V., and Georgiou, S. 2003. Valuing nature: Lessons learned and future research directions. *Ecological Economics* 46, 492–510.
5. Fisher, B., Turner, R.K., and Morling, P. 2009. Defining and classifying ecosystem services for decision making. *Ecological Economics* 68, 643–653.
6. Walker, B.H. 1993. Rangeland ecology—understanding and managing change. *Ambio* 22(2–3), 80–87.
7. Turner, R. K., Georgiou, S., and Fisher, B. 2008. *Valuing Ecosystem Services: The Case of Multi-Functional Wetlands*. London: Earthscan, 229 p.
8. TEEB. 2012. *The Economics of Ecosystems and Biodiversity: Ecological and Economic Foundations* (P. Kumar, ed.). Abingdon and NewYork: Routledge.
9. Barbier, E.B. 2009. Ecosystems as natural assets. *Foundations and Trends in Microeconomics* 4(8), 611–681.
10. Brouwer, R. 2000. Environmental value transfer: State of the art and future prospects. *Ecological Economics* 32, 137–152.
11. Brouwer, R., and Bateman, I.J. 2005.The temporal stability and transferability of models of willingness to pay for flood control and wetland conservation. *Water Resources Research* 41(3):W03017.
12. Bateman, I.J., Carson, R., Day, B., Hanemann, W.M., Hanley, N., Hett, T., et al. 2002. Economic valuation with stated preference techniques: A manual. Cheltenham: Edward Elgar.
13. Brander, Luke M., and Koetse, M.J. 2011, October. The value of urban open space: Meta-analyses of contingent valuation and hedonic pricing results. *Journal of Environmental Management* 92(10), 2763–2773.

14. Zandersen, M., and Tol, R.S.J. 2009, January. A Meta-analysis of forest recreation values in Europe. *Journal of Forest Economics* 15(1–2), 109–130.

15. Sen, A. , Darnell, A., Bateman, I., Munday, P., Crowe, A., Brander, L., Raychaudhuri, J., Lovett, A., Provins, A., and Foden, J. 2012. Economic assessment of the recreational value of ecosystems in Great Britain. Cserge working paper.

16. EPA. 2009. Valuing the protection of ecological systems and services, a report of the EPA Science advisory Committee, EPA-SAB-09-012.

17. Chambers, R. 1991. Shortcut and participatory methods for gaining social information for projects. In *Putting People First: Sociological Variables in Rural Development* (Cernea, M.M., ed.), 2nd ed., pp. 515–537. Baltimore, MD: Johns Hopkins University Press.

18. Heck, T. 2004. Public preferences for CO2 control revealed in referenda in Switzerland. In *European Commission (2004)*, New Elements for the Assessment of External Costs from Energy Technologies (NewExt). *Final Report to the European Commission, DG Research, Technological Development and Demonstration (RTD). WP3. Contract No: ENG1-CT2000-00129.*

19. Shmelev, S.E. 2008. Multicriteria analysis of biodiversity compensation schemes: review of theory and practice with a focus on integrating socioeconomic and ecological information. Environment Europe, UK.

20. O'Connor, M. 2000. The VALSE project: An introduction. *Ecological Economics* 34, 165–174.

21. Zografos, C., and Paavola, J. 2008. Critical perspectives on human action and deliberative ecological economics. In *Deliberative Ecological Economics* (Zografos, C., and Howarth, R.B., eds.). Delhi: Oxford University Press, pp. 146–166.

22. Carpenter, Stephen R., Harold A. Mooney, John Agard, Doris Capistrano, Ruth S. DeFries, Sandra Díaz, Thomas Dietz, Anantha K. Duraiappah, Alfred Oteng-Yeboah, Henrique Miguel Pereira, Charles Perrings, Walter V. Reid, José Sarukhan, Robert J. Scholes, and Anne Whyte. 2009. Science for managing ecosystem services: Beyond the Millennium Ecosystem Assessment. *PNAS* 106(5), 1305–1312.

23. Bockstael, N.E., Freeman, A.M., Kopp, R.J., Portney, P.R., and Smith, V.K. 2000. On measuring economic values for nature. *Environmental Science and Technology* 34(8), 1384–1389.

24. Naidoo, R., A. Balmford, R. Costanza, B. Fisher, R. E. Green, B. Lehner, T. R. Malcolm, and T. H. Ricketts. 2008. Global mapping of ecosystem services and conservation priorities. *PNAS* 105, 9495–9500.

25. Barbier, E.B., Koch, E.W., Silliman, B.R.,Hacker,S.D., Wolanski, E., Primavera, J., Granek, E.F., Polasky, S., Aswani, S., Cramer, L.A.,Stoms, D.M., Kennedy, C.J., Bael, D., Kappel, C.V., Perillo, G.M.E., and Reed, D.J. 2008. Coastal ecosystem based management with nonlinear ecological functions and values, *Science* 319, 321–323.

26. Naidoo, R., and Ricketts, T. 2006. Mapping economic costs and benefits of conservation. *Plos Biol.* 4, 2153–2164.

27. Bateman, I.J., Lovett, A.A., and Brainard, J.S. 2003. *Applied Environmental Economics: A GIS-Approach to Cost-Benefit Analysis.* Cambridge: Cambridge University Press.

28. Liekens, I., Schaafsma, M., Staes, J., De Nocker, L., Brouwer, R., and Meire, P. 2009. Economische waarderingsstudie van ecosysteemdiensten voor MKBA. Studie in opdracht van LNE, afdeling milieu-, natuur- en energiebeleid, VITO, 2009/RMA/R308.

29. Bateman, I. J., Georgiou, S. and Lake, I. 2006. The aggregation of environmental benefit values: Welfare measures, distance decay and BTB. *BioScience* 56(4), 311–325.

30. Schaafsma, M. 2010. Spatial effects in stated preference studies for environmental valuation. PhD thesis, VU University, Amsterdam.

31. Morse-Jones, S., Luisetti, T., Turner, R.K., and Fisher, B. 2010. Ecosystem valuation: Some principles and a partial application. *Environmetrics* 22(5), 675–685. http://onlinelibrary.wiley.com/doi/10.1002/env.1073/abstract

32. Hein L., van Koppen, K., De Groot, R.S., and van Ierland, E.C. 2006. Spatial scales, stakeholders and the valuation of ecosystem services. *Ecological Economics* 57, 209–228.

33. Kremen, C., Niles, J.O., Dalton, M.G., Daily, G.C., Ehrlich, P.R, Fay, J.P., Grewal, D., and Guillery, R.P. 2000. Economic incentives for rain forest conservation across scales. *Science,* 288, 1828–1832.

34. Martín-López, B. 2007. Bases socio-ecológicas para la valoración económica de los servicios generados por la biodiversidad: implicaciones en las políticas de conservación. PhD-dissertation, Universidad Autónoma de Madrid.

35. Brander, L.M., R.J.G.M. Florax. and J.E. Vermaat. 2006. The empirics of wetland valuation: A comprehensive summary and a meta-analysis of the literature. *Environmental and Resource Economics* 33, 223–250.

Chapter 12

Ecosystem Services in Belgian Environmental Policy Making: Expectations and Challenges Linked to the Conceptualization and Valuation of Ecosystem Services

Tom Bauler and Nathalie Pipart

Université Libre de Bruxelles, Institut de Gestion de l'Environnement et d'Aménagement du Territoire, Centre d'Etudes du Développement Durable

1. INTRODUCTION

Ecosystem services (ES) are "the benefits people obtain from ecosystems" [1]; in other words, ES are "the aspects of ecosystems utilized actively or passively to produce human well-being" [2]. In 2005, the Millennium Ecosystem

Ecosystem Services. http://dx.doi.org/10.1016/B978-0-12-419964-4.00012-3

Assessment (MEA) identified the loss of biodiversity as a major threat to eco-systems as well as to human well-being: "the conceptual framework for the MEA places human well-being as the central focus, while recognizing that bio-diversity and ecosystems also have intrinsic value and that people take decisions concerning ecosystems based on consideration of well-being as well as intrinsic value" [3]. In this respect, the MEA sought to improve the knowledge base for decisionmaking by developing indicators of ecosystem conditions, ecosystem services, and human well-being. Similarly, the TEEB processes and publica-tions [4] (TEEB—The Economics of Ecosystems and Biodiversity) want to guide policy actors to acknowledge the economic impacts of biodiversity and ecosystem services loss, and to draw attention to the global economic benefits of biodiversity and the contribution of ecosystem services to human well-being (see Chapter 1, this volume). TEEB reports typically collect evidence on case studies showing that pro-conservation choices are also a matter of economic common sense.

In addition to the conceptual framework of ES, developed under the umbrel-las of the Convention on Biological Diversity (CBD) and the MEA, initiatives [5] such as TEEB pursue the idea that management is founded on measurement and monitoring, and actively engage with the development of a set of policy instruments and tools that allow assignment of values to ES. TEEB processes make the assumption [6] that the economic language is a potentially very pow-erful translation of nature's values in a way that will improve the recognition of these values in policy-making processes. Hence, TEEB and similar efforts are providing for the basic knowledge that should allow policy makers to conceive of the (economic) consequences of their (environmental) policies on ES and consequently on human well-being. These evaluations and assessments of the economic aspects of ES do not always use economic language (i.e., become valuations)[1]. TEEB reports, for instance, affirm that (monetary) valuation of ES is not useful and applicable in all situations and that "economic valuation has its limits and can ever only be one input to the decision process" [7].

An ES approach seems to entail two different, but intertwined, mecha-nisms: (1) the adoption of the conceptual framework of ES, as a particular perspective to analyze the linkages between people and the environment; and, (2) the development of and experimentation with a set of ES (e)valua-tion tools and instruments in order to reduce complexities through the adop-tion of economic/monetary values wherever suitable. The present chapter explores both mechanisms at the level of Belgian environmental policy

---

1. In the following, we establish a semiotic difference between evaluations (i.e., the generic termi-nology for attempts into assessing societies and their development) and valuations (i.e., assessments that rely on monetization). In order to account for the fact that ES can be assessed by a multitude of methodologies using economic language or not, relying on monetization techniques or not, we will make use of "(e)valuation," which allows stressing typographically the value component in any form of assessment.

making in order to give an overview of the level of institutionalization of ES into Belgium'senvironmental policy making. We focus on the following two different layers of institutional adoption: (1) the occurrence in official policy documents and environmental policy makers' knowledge of the ES concept; and (2) the expectations and challenges that environmental policy makers perceive to be linked to the translation of ES into economic language.

Section 2 of this chapter presents an account of our empirical exploration of the institutional adoption of the concept of ES (2.1), and subsequently of the perceptions of policy actors of (e)valuation tools (2.2). We will then discuss, in Section 3, both of these empirically revealed adoptions by analyzing the policy actors' explicit and implicit expectations in terms of challenges to be addressed. Section 4 will conclude that an explicit governance of ES (e)valuations should become an integral part of the future ESpolicy framework in Belgium, an issue that is still to be acknowledged by the ongoing, often hasty, governmental processes.

2. THE ADOPTION OF ECOSYSTEM SERVICES IN BELGIAN ENVIRONMENTAL POLICY MAKING

A first layer of adoption of ecosystem services (e.g., the conceptual adoption of ES) can be approached empirically by raising questions such as: how much has the concept been referred to in Belgian environmental policy-making documents, and how do policy actors perceive the strengths of the concept of ES, for instance, when compared to a more traditional conservation terminology?

In this perspective, federal and regional environmental policy plans and legislation were screened[2] during a document analysis to detect the usage made of ES language, that is, taking the factual occurrence of the concept as a sign for its institutional adoption in the environmental policy domain. Where ES occurred in documents and legislation, a set of interviews were conducted with policy actors from regional Belgian environmental authorities in order to improve our understanding of their knowledge of the conceptual underpinnings of ES. The second layer of adoption brings us to explore the perceptions by Belgian environmental policy makers of the translation of the concept of ES into economic language. Accepting the argument of TEEB processes in terms of the attractiveness of the economic language to translate ES for policy processes, in our secondary investigation we will attempt to qualify the demand of Belgian policy actors for (e)valuations of ES: how do selected policy actors perceive the value added of (e)valuations of ES? What is their knowledge of the existence and application of ES (e)valuation tools? Is there a "demand" for monetary

---

2. The document analysis covers all legislation and planning documents as referred to in the official databases of the three Belgian regions (Brussels, Flanders, and Wallonia) as well as those of the federal Belgian level until December 2011.

valuation? What are their expectations in terms of value added against other (non-economic) assessment approaches of ES?

2.1. The ES Concept in Policy Documents and ES Knowledge by Policy Actors

An inventorization[3] of the occurrence of the words *ecosystem service(s)* in Belgian policy and legal documents reveals only a few factual occurrences of the term *service(s)* and even fewer occurrences of *ecosystem service(s)*. Where the terms do occur, they are used mainly for descriptive purposes and less to motivate a shift in the underlying conceptual basis for the policy as such.

On national and federal levels, the ES concept appears in a series of planning/strategizing documents. The National Biodiversity Strategy (NBS) adopts the ecosystem approach, and ES developed in the MEA as a guiding principle and rationale. The NBS uses several linked terms (i.e., ecosystem services, ecological services, environmental services), but without suggesting whether they have a similar or a different meaning. The Federal Plan for the Integration of Biodiversity in four key federal sectors is part of the federal implementation of Belgium's NBS. Among the actions planned is an exploration of the feasibility of the economic (e)valuation of the Belgian ecosystem services, including a call to integrate socioeconomic sciences in biodiversity research. The ES concept is also identified in three royal decrees (all from 2007), which transpose the European directive (2004/35/CE) on environmental liability regarding the prevention and compensation of environmental damages.

At the regional level, the occurrence of the ES concept reveals a somewhat differentiated adoption between the three regions. In the Brussels Region, no explicit reference to the term *ecosystem services* could be identified in policy documents. In the Walloon Region, the word *services* appears in a series of documents and legislation, most notably in the Code de l'Environnement (CdE), where services are defined as "the functions provided by a natural resource to the benefit of another natural resource or the public"[8] and where it is used as a conceptual reference to conceive of damage to natural resources (i.e., degradation of services). Furthermore, the CdE stipulates that in case of damage, priority should be given to actions that provide services of the same type, quality, and quantity as those that were lost. When not possible, the competent authorities can prescribe evaluation methods, including monetary valuation, to determine abatement costs equivalent to the estimated value of the services lost. In Flanders more references

---

3. We proceeded as follows: the term *ecosystem services* and its French and Dutch translations were searched for (1) in the official Belgian journal (*Belgisch Staatsblad/Le Moniteur Belge*); (2) in federal and regional environmental legislation; and (3) in federal and regional policy plans and policy-related publications (such as the regional states-of-the-environment). In a second turn, the inventorization was expanded to the term *services* when referring to ecosystems and to environmental or ecological services.

to the ES concept are occurring, but above all they reveal a different nature of adoption. Indeed, the term *ecosysteemdiensten* can be found 18 times in the Flemish Environmental Policy Plan 2011–2015 (FEPP), and the reference to ES exceeds the more descriptive nature prevailing in the Walloon and federal documents. The FEPP recognizes ecosystem services as a being a concept in itself, distinct from the biodiversity concept. It announces that ES will be "operationalized" [9] for Flanders, notably through an inventory/assessment of ES in Flanders and through the development of a series of policy tools [10]. Interestingly, focus is given to the monetary valuation of ES as an input for cost-benefit analysis; consequently, valuation studies of a series of ecosystem services in Flanders are announced as well as their integration into an Internet-based decision-aiding tool.

One possible key to the differentiation in the nature of the percolation into environmental administrations might simply be a different level of knowledge and/or interpretation of the ES concept by the policy actors themselves. Thus, we pursued our document-based inventorization with a face-to-face exploration of the knowledge/interpretation of ES by interviewing a selection of environmental policy actors linked to authorship of the respective policy documents.

Factually, every interviewed policy actor was easily able to give a short definition of ecosystem services, which most defined roughly as "the services provided by nature to society or humans." A few of them emphasized the broad spectrum of the concept and its inherent complexity: "We can make it very broad, we can put everything under this concept." Interestingly, some of the interviewees directly referred to economic/monetary valuation when asked about the ES concept. Finally, interviewees estimated that the concept is well known throughout their respective environmental administrations.

In addition to equaling ES as a concept, the policy actors referred to several other framings of what ES is: argument, reflection, current of thought, communication tool. They acknowledge that the ES concept is foremost a "marketing and communication concept," a way to communicate about and raise people's awareness of the societal interest of ecosystems. Some interviewees saw in ES an opportunity to develop an argument in favor of "less visible services," like those provided by soils or wetlands. ES were seen to be targeted mainly to "non-biologists" (natural sciences being the discipline of most of our interviewees), a term that was broadly used to qualify politicians, citizens, and economic actors. The persuasive force of ES was seen as promising mainly because ES would allow rendering a wider, more encompassing perspective on environmental issues and broadening the reflection on the different aspects to consider in land management and land use. It was mentioned that ES would induce a more multidisciplinary approach that could highlight complexities and (environmental) values that current decision rationales do not take into account. However, as one of the interviewees emphasized, ES is a recent concept that might be only a buzzword: He points out that, historically, in environmental policy making there have been a whole series of semantic changes, which in the end have not led to a fundamental change in policy formulation.

The innovativeness of the ES concept appears to be perceived in diverging ways. Some interviewees see it as a promising alternative avenue for communicating about nature. ES is acknowledged to speak more directly than the concept of biodiversity to different publics who lack disciplinary training or expertise on issues related to ecosystem, because "it is a way to tell in their language [the public] that everything in nature can be useful in some way for the people." Quite differently, other interviewees felt that the ES concept has in reality underpinned environmental policy making for a long time. Some interviewees expressed the idea that they have "always" either unconsciously or indirectly used equivalent anthropocentric approaches, even if the exact terminology of ES is recent.

The interviews with policy makers also reveal that the ES concept seems to have found its way into Belgian environmental policy making through two simultaneously operating dynamics. From the federal policy level, interviewees observe a top-down percolation onto regional and local authorities, which is sparked by the involvement of federal policy makers in institutional processes such as the MEA and CBD. For instance, Article 6 of the Convention states the obligation of its parties to elaborate a national biodiversity strategy. After the findings of the MEA had been endorsed in 2006, parties to the CBD were encouraged [10] to make use of the MEA's conceptual ES framework and methodologies in their environmental impact and strategy assessments. Simultaneously, since the TEEB initiative has been endorsed by the European Union, it is gaining a significant influence among the member states as well, mainly through the considerable dissemination efforts done through the TEEB reports.

In parallel, a bottom-up (from nongovernmental actors to policy actors) percolation can be observed: Scientific bodies and public administrations, universities, research centers, and consultancy firms increasingly publish on topics or frame projects and reports relating directly to ES; they trigger discussions on the economics of ecosystem; and they lobby for the use of ES to structure research efforts and collaborate in related research projects with the scientific or environmental administrations. The entire BEES-processes (BElgian Ecosystem Services) and their annexes are perfect examples of research actors rallying behind ES as a joint framework of thought, and consequently allowing the concept to gain credibility for the policy actors.

2.2. The Place of ES Tools: Representations of Monetary (E)valuations of ES

One of the acknowledged advantages of the ES conceptualization of nature–society interlinkages, as set out in the literature, is to facilitate the conception of assessments of nature/biodiversity. Measuring the services provided by nature to human well-being appears to be at the very forefront even of the concept of ES. Although many authors and international policy-making institutions fully acknowledge the necessity of methodological pluralism when striving toward such assessments of ES, the force of persuasion and institutional integration of the ES concept is often intrinsically linked to assessing ES in an economic, monetary language. The presumed facilitation of the uptake of nature into policy decisions through use of the ES concept appears to render economic valuations of ES have a major attractive force, if not a necessity. This ambiguity of the operationalization of ES, far from being a new principle, has been synthesized as follows: "Contending views in this controversy range from the support of valuation and market solutions as core strategies to solve present environmental problems (which from this perspective are framed as market failures) (Engel et al., 2008; Heal et al., 2005) to an outright rejection of utilitarian rationales for conservation (Child, 2009; McCauley, 2006). In between, there is a strategic endorsement of valuation as a pragmatic and transitory short-term tool to communicate the value of biodiversity using a language that reflects dominant political and economic views (Daily et al., 2009; de Groot et al., 2002). This strategic endorsement of valuation has become part of an increasingly dominant position as the environmental movement attempts to look for novel conservation strategies where traditional ones have failed to halt biodiversity and habitat loss (Armsworth et al., 2007; Daily et al., 2009)." [11]. Valuing nature and ES in monetary terms is controversial [12] on at least two different levels. First, economic valuation is anthropocentric per se, as it focuses on the interest of ecosystems for humans, and consequently, as it refers to the consequences for human well-being and not to intrinsic values [13]. The utilitarian language that underlies economics, and hence monetary valuation, hardly allows for more differentiated languages of evaluation to emerge. Second, valuations of ES rely on an important, irreducible amount of methodological assumptions linked to the very fact that one values in monetary terms "goods and services" thatare (mostly) out of the markets; that is, they don't wear a price tag.

This strategic endorsement of economic valuation when adopting ES as an underlying policy concept appears to be a matter of concern with the interviewed policy actors. Discussions with interviewees on (e)valuation of ES, even if special care was taken to keep the space widely open to any form of assessment or (e)valuation, focused almost exclusively on monetary valuation. Among the opportunities of monetary ES (e)valuations that were explicitly identified are classical issues such as facilitated communication (and persuasion) to politicians by adopting their language of predilection. Citizens and the public at large were also identified as potential privileged target groups for such a simplified, synthetic language of valuation. Several of the environmental policy makers positively acknowledged the prospects of conducting cost-benefit analyses on issues of environmental management and of providing in the future for nature conservation and management with a least-cost approach. Monetarization was seen as a way to facilitate and objectify[4]—priority setting at the level of intradepartmental processes (e.g., trade-offs within the environmental policy portfolio) and at the level of inter-department processes (e.g. giving a potentially higher priority to environmental issues as compared to other governmental portfolios). However, several of the interviewed actors expressed resistance, or at least anxiety, especially to this line of argumentation. Finally, a number of interviewees cited the potentially facilitated linkages that would emerge from ES (e)valuations with economic policy instruments, which should become more prominent in environmental governance than they are currently. Occasionally, this turn toward more economic policy instruments was mentioned to be conditioned by the uptake of ES values into national accounting, and to a wider extent into the formulation and configuration of policy indicators.

3. CHALLENGES OF ES-BASED POLICY MAKING: A DISCUSSION OF MONETARY VALUATION

From the empirical exploration of these two layers of institutional adoption (i.e. at the level of the concept and the (e)valuation), an overarching questioning arises on how policy actors see the future of ES-based policy making. In particular, valuing nature and its services in monetary terms, and subsequently linking these to the definition and configuration of policy instruments (such as subsidies, for instance), indeed touches a sensitive cord with many of the interviewed policy actors. An implicit strategic endorsement of monetary ES valuations appears to be the widely shared position among Belgian environmental policy makers: Criticism of monetary ES valuations is recurrent, and some of their limits are even acknowledged as being irreducible per se. The present section synthesizes some of the issues raised by policy makers as a series of challenges, whose resolution or societal discussion might be conditional for the future of ES-based policy making in Belgium. The feasibility and legitimacy of monetary valuation remain uncertain for all the interviewed policy actors, and should thus be qualified as a necessary topic for debate. In parallel, questions related to the manageability and necessity of monetary valuations were less shared among policy actors, but nevertheless appear to be fundamental issues of concern for some of them.

Such reserved opinions—and the calls we heard for organized debate and discussion—may simply correspond to a prudent civil servant position, considering that most of the policy actors have not yet experienced concretely with ES valuations. In the following section, we expose—and to a certain extent might exacerbate—the more fundamental remarks, resistances, and objections raised

---

4. With many of our interviewees, it appeared through their line of argumentations that they strongly assume that ES valuation efforts will attribute particularly important values to nature. Hence, that objectivation would unavoidably point toward a prioritization of nature against other sociopolitical goals.

by Belgian environmental policy actors *against* monetary valuations. Virtually none of the issues raised are unique to Belgian environmental policy actors; all of them can be identified in the existing literature on monetary (e)valuation, often even beyond the mere environmental field of study. However, the discussion of the captured statements might help configure the terms of reference of future multi-actor discussion on the limits of the application of ES valuations to Belgian environmental policy making. In other words, such discussion might facilitate the necessary step from implicit to explicit strategic endorsement of ES (e)valuations.

3.1. Monetary Valuation as a Necessity?

Adopting ES (e)valuations as the yardstick to conceive of and prioritize nature conservation and management is fundamentally different in its outcomes than what environmental policy actors experience today. Although the aforementioned strategic endorsement of ES (e)valuations could lead to beneficial governmental prioritization of nature, it would necessarily change the outcomes of policy decisions in a way that the interviewed policy actors cannot easily apprehend today. Current nature conservation in Belgium was said to target the completion of ecological networks and the protection of (red-listed) species and ecosystems. The selection and management of Natura 2000 areas, one policy that was repetitively mentioned, are based on biological criteria, and no proxy (i.e., economic or monetary) assessments were acknowledged by interviewees to be able to outperform the direct ecological ones. ES (e)valuations could, on the level of specific decisions and policies, have a series of perverse and surprising effects, notably when comparing the richness (economic and biological) of specific ecosystems to some of the currently protected areas of presumably lesser economic and biological richness (e.g., chalky grasslands, which are degraded ecosystems that are maintained and protected because they are habitats for patrimonial species). One of the interviewees pointed to a relatively obvious but currently unanswered question: How are the logic of ES (and their (e)valuation) and the one based on the protection of (endangered) species and conservation of ecological networks fitting together? Both are fundamentally different logics for decision making, with repercussions that go far beyond the mere semantic change that is advocated for at the level of using ES as a communicational concept. The outcomes in terms of policy choices are unclear, but as the risks of pernicious effects possible with current nature conservation choices are evaluated as real, some of the policy actors advocate limiting ES conceptualizations for communicating, and hence refraining from the development of tools for ES (e)valuation.

More generally, policy makers might not give as much importance to economic information as had been assumed. While long-standing studies [14] of the impacts of (scientific) information on decision situations have raised a number of profound questions on the implementation of evidenced-based policy making, relatively few attempts have been made to investigate the impact of economic valuations on environmental decision making. The discussion of the influence of cost-benefit analyses in the environmental policy domain lead, however, to conclusions such as "(a)lthough policy processes take a variety of forms, decisions are typically taken without referring to Net Present Values (NPV[5]). NPV calculations resulttypically in loss of information instead of informing complex decisions" (Arild Vatn, personal communication, 2012).

3.2. The Feasibility of Monetary Valuation

When switching from the discourse on necessity to issues related to the feasibility of ES (e)valuations, a relatively unanimous questioning of the robustness of the methods used to value nature emerged from the interviews. Among the aspects mentioned in the interviews were the inherent complexity and lack of knowledge of the linkages between biodiversity and ES. More fundamentally, the capability of economic methodologies to capture values in a robust and precise (i.e., scientific) way was questioned. In this regard, the subjectivities (at the level of methodological choices and their implementation) of the valuation approaches were seen as an entry point to manipulations, or at least for errors. Case study approaches, that is, small-scale, spatially, and temporally focalized monetary valuations, were generally acknowledged to be preferable to large-scale (e.g., regional), complex valuation efforts. With regard to the aforementioned advantages to develop ES (e)valuation to disentangle trade-offs, an interesting opinion was introduced showing the necessity to develop unambiguous rules of application for ES (e)valuations: If ES (e)valuations were "unapt to capture the value of this ecosystem or of nature," but could capture the relative importance of a specific service within a particular ecosystem, then monetary valuation should be limited to use as an intra-environmental policy tool and should not be used to compare environmental policies to competing (e.g., socioeconomic) policy portfolios.

3.3. Monetary Valuation and Legitimacy

The specificity and uniqueness of ecosystem- or nature-related management decisions, as compared to social or economic policy problems, was repetitively recognized, following an argument by Ludwig [15] that "few of us require much convincing that economic values belong to a different sphere than personal and social values." The responsibility of human beings as ecosystem managers might thus imply not approaching questions of the environment in the same way we deal with human systems (e.g., the social and economic spheres). Intrinsic values of nature and ES leave us with a decision space that includes not only

---

5. *Net Present Value* is a standard economic calculus that discounts future inflows and outflows of a project (e.g., investment) to their present value with the help of applying a discount rate.

differentiated value judgments on policy options, but also encompass fundamentally different value perspectives that are not reducible to a common unit (such as monetary valuation implies).

Commodification has been defined as follows: "the expansion of market trade to previously non-market areas; it involves the conceptual and operational treatment of goods and services as objects meant for trading." [16]. It has not been explicitly mentioned by policy actors, but a strand of the revealed comments clearly refer to it. The difference between value and price–that is, between economic approaches to valuation and market approaches to governance—can be thin, or at least rendered more fragile through the very fact of engaging with monetary (e)valuation tools. ES valuation might facilitate the emergence of market-based governance schemes of nature conservation (specifically mentioned were payments for ecosystem services—PES). Economic and market-based policy instruments, potentially even the appearance of environmental finance, was identified as a relatively logical step, once the monetary, value-seeking (e)valuations would have been more widely spread. One of the mentioned consequences would be changes in the attribution of decision capacities. Strategic endorsement might bring economic actors to the forefront of environmental governance decision arenas. Although many interviewees saw an opportunity in opening policy making on environmental issues to private economic actors, some fears were expressed because of the consequent change in decision logics. Decisions on the common, environmental good might no longer be solely taken by society through its democratic representatives, but more directly influenced by economic, private interests. A certain depoliticization of environmental issues was felt, that is, that the "fundamental fallacy in application of economic criteria to determine our policies is the substitution of techniques for judgment" [17]. This evolution might trigger questions with regard to the legitimacy of the deciding actors. It might also modify the rationale at stake: implicitly setting economics and markets as an end rather than as a means.

4. PERSPECTIVES: GOVERNANCE OF ES AND GOVERNANCE WITH ES

In this chapter, we attempted to present a clearer picture of how Belgian environmental policy actors view the ES concept and ES (e)valuation. The percolation of the concept is obvious, even if it is not yet widely inserted into policy documents. The related drive toward ES (e)valuation does, however, raise concerns that should lead to taking a careful look at the broader implications of the semantic and paradigmatic changes that ES has introduced into environmental policy arenas. The concepts are not neutral: They carry their own representation. Conceptualizing ecosystem functions as services points for many environmental policy actors toward a framing of the environment in economic terms, and economic approaches induce a series of underlying assumptions, providing an entry for new actors to join the policy arena. Policy concepts such as ES are not solitary either, but rather display linkages with related policy instruments and tools that also are not neutral. For instance, policy instruments and tools have been framed [18] as institutions, which carry their own set of values, principles of actions, and belief systems. The fact that policy actors appear to redirect the debate toward the specific issues related to monetary valuation of ES—although ES assessments can also be social or biophysical, qualitative or quantitative—might be taken as a sign that they recognize these mechanisms of framing. The linkages between the ES concept and ES tools are thus indeed not solely analytical, but are a matter of preoccupations by policy actors. In this sense, our exploration could be understood as a call to develop at the Belgian level a "Governance with Ecosystem Services," that is, to explicitly address via a policy–science–society process of deliberation the implications of switching nature governance toward a governance of ES. A series of scientifically led initiatives are trying to structure and trigger such a process, also via the BEES project which is at the basis of the present book. It appears to be necessary—indeed urgent—that environmental policy actors develop their own processes of deliberation on the issue of adopting the ES concept and the linked tools.

ACKNOWLEDGMENT

We are particularly thankful to Jennifer Hauck (UFZ-Leipzig) for her comments on an earlier version of the chapter.

REFERENCES

1. Millennium Ecosystem Assessment (MEA). 2005. *Ecosystems and Human Well-being: Synthesis Reports.* Washington, DC: Island Press.
2. Fisher, B., Turner, R.K., and Morling, P. 2009. Defining and classifying ecosystem services for decision making. *Ecological Economics* 68(3), 643–653.
3. Millennium Ecosystem Assessment (MEA). 2005. *Ecosystems and Human Well-being: Synthesis Reports.* Washington, DC: Island Press.
4. The Economics of Ecosystems and Biodiversity (TEEB). 2009. *Responding to the value of nature. Summary for National and International Policy Makers.* UNEP, Nairobi. The Economics of Ecosystems and Biodiversity (TEEB). 2010. *Mainstreaming the Economics of Nature: A synthesis of the approach, conclusions and recommendations of TEEB.* Nairobi: UNEP.
5. UK National Ecosystem Assessment. 2011. *The UK National Ecosystem Assessment: Synthesis of the Key Findings.* Cambridge: UNEP-WCMC.
6. The Economics of Ecosystems and Biodiversity (TEEB). 2009. *Responding to the Value of Nature. Summary for National and International Policy Makers.* Nairobi: UNEP.
7. The Economics of Ecosystems and Biodiversity (TEEB). 2010. Mainstreaming the Economics of Nature: A Synthesis of the Approach, Conclusions and Recommendations of TEEB. Nairobi: UNEP.
8. Code de l'Environnement sur les dispositions communes et générales (Livre Ier, Titre II). http://environnement.wallonie.be/legis/menucode.htm. Accessed April 25, 2012.
9. Vlaamse Overheid. 2011. *Milieubeleidsplan 2011–2015.* Brussels.

10. Schlesser, M. 2007. Belgium's National Biodiversity Strategy: Jow is it linked to the Millenium Ecosystem Assessment? In Bourdeau, P., and Zaccaï, E., *Millenium Ecosystem Assessment: Implications for Belgium.* Conference Proceedings, Brussels, October 27, 2006.

11. Gomez-Baggethun, E., and Ruiz-Pérez, M. 2011. Economic valuation and the commodification of ecosystem services. *Progress in Physical Geography* 35(5), 613–628. Citation comprises the following references: Engel, S., Pagiola, S., Wunder, S. 2008. Designing payments for environmental services in theory and practice: An overview of the issues. Ecological Economics 65: 663–674. Heal, G. M., Barbier, E.B., Boyle K.J., Covich A.P., Gloss S.P., Hershner C.H., et al. (2005) Valuing Ecosystem Services: Toward Better Environmental Decision Making, Washington, DC: The National Academies Press. Child, M.F., 2009. The Thoreau ideal as unifying thread in the conservation movement. Conservation Biology 23: 241–243. McCauley, D.J. 2006. Selling out on nature. Nature 443(7107): 27–28. Daily, G.C., Polasky, S., Goldstein, J., Kareiva, P.M., Mooney, H.A., Pejchar, L. 2009. Ecosystem services in decision making: Time to deliver. Frontiers in Ecology and the Environment 7: 21–28. de Groot, R.S., Wilson, M., Boumans, R. 2002. A typology for the description, classification and valuation of ecosystem functions, goods and services. Ecological Economics 41: 393–408. Armsworth, P.R., Chan, K.M.A., Daily, G.C., Kremen, C., Ricketts, T.H., Sanjayan, M.A. 2007. Ecosystem-service science and the way forward for conservation. Conservation Biology 21: 1383–1384.

12. Spash, C. L. 2008. Deliberative monetary valuation and the evidence for a new value theory. *Land Economics* 84(3), 469–488.

13. Turner, R.K., Paavola, J., Cooper, P., Farber, S., Jessamy, V., and Georgiou, S. 2003. Valuing nature: Lessons learned and future research directions. *Ecological Economics* 46(3), 493–510. Salles J.-M. 2011. Valuing biodiversity and ecosystem services: Why put economic value on Nature? *Comptes Rendus Biologies* 334(5–6), 469–482.

14. Weiss, C.H. 1977. Research for policy's sake: The enlightenment function of social research. *Policy Analysis* 3 (4), 531–545. Henry, G.T., and Mark M.M. (2003). Beyond Use: Understanding Evaluation's Influence on Attitudes and Actions. *American Journal of Evaluation* 24(3), 293–314. Kingdon, J.W. 1984. *Agendas, Alternatives and Public Policies.* Boston: Little, Brown.

15. Ludwig, D. 2000. Limitations of economic valuation of ecosystems. *Ecosystems* 3(1), 31–35.

16. Gomez-Baggethun, E., and Ruiz-Pérez, M. 2011. Economic valuation and the commodification of ecosystem services. *Progress in Physical Geography* 35 (5), 613–628.

17. Ludwig, D. 2000. Limitations of economic valuation of ecosystems. *Ecosystems* 3(1), 31–35.

18. Lascoumes, P., and Le Galès, P. 2007. Introduction: Understanding public policy through its instruments—from the nature of instruments to the sociology of public policy instrumentation. *Governance* 20(1), 1–21.

Chapter 13

Ecosystem Services Governance: Managing Complexity?

Hans Keune[1,2,3,4],Tom Bauler[5] and Heidi Wittmer[6]

[1]*Belgian Biodiversity Platform,* [2]*Research Institute for Nature and Forest (INBO),* [3]*Faculty of Applied Economics – University of Antwerp,* [4]*naXys, Namur Center for Complex Systems – University of Namur,* [5]*Université Libre de Bruxelles, Institut de Gestion de l'Environnement et d'Aménagement du Territoire, Centre d'Etudes du Développement Durable,* [6]*Helmholtz Centre for Environmental Research—UFZ Division of Social Sciences, Department of Environmental Politics*

1. FRAMING ECOSYSTEM GOVERNANCE

1.1. What Is (Ecosystem) Governance?

To unambiguously define *governance* is a challenge. According to Jordan [1], it is one "of the most essentially contested terms in the entire social sciences." But perhaps to define it narrowly would also be a pity, As with other rather vaguely defined buzzwords like sustainable development, by defining it too narrowly one runs the risk of excluding, rather than including, a diversity of those actors or groups one

Ecosystem Services. http://dx.doi.org/10.1016/B978-0-12-419964-4.00013-5

would want to involve and to join forces with, and so one might lose an audience [2, 1]. In this chapter we present governance mainly as an important challenge, open for debate, with options to invent and choose from, and with practical relevance for ecosystem services, but without an unambiguous clear definition or roadmap. More on governance specifics are presented in Chapter 1.4 [3]; here we elaborate on ecosystem services and how governance issues can be framed. Attention for and development of governance in real-world practice is urgently needed according to several scholars in the dynamic field of ecosystem services [4–8]. These scholars realize that the "speaking truth to power" strategy (as coined by Jasanoff [9]) is an oversimplification of the social complexity of ecosystem governance issues [10]. But what is this social complexity? How can we frame and picture it?

1.2. Bring in the Social Cloud

With the introduction of the concept of *ecosystem services*, proponents of nature and biodiversity conservation (Cowling et al. [4] call it *a mission-oriented discipline*) aim to clarify to others and to society at large the importance of nature and biodiversity for humankind. They do this for good reason and with good intentions: Indeed, one may rightfully argue that nature and biodiversity are in many respects beneficial to human beings and societies, and the value of nature and biodiversity in this regard is often not taken into account when we organize our lives and govern our societies. This has led us far away from appreciation of nature and biodiversity, and even is leading to unprecedented destruction: We run the risk of destroying not only nature, but also important natural foundations for our well-being with it. Through clarifying the importance of nature for society, the conservation community hopes to convince others of the need for and urgency of taking conservation actions. A dominant strategy for this purpose seems to be "one-way conviction" by means of establishing an evidence base. The question that arises is whether this is realistic. In trying to connect this body of evidence and conviction to "others" and their way of doing things (individual or household or company behavior, governance in different sectors and at different levels), this largely one-way communicative approach seems to underestimate its limitations: The issue of ecosystem services is only partly a knowledge (or scientific or truth) issue; it is also a social issue, an issue of social debate.

Strikingly, some efforts to connect ecosystem expertise to governance issues are examples of what we may call *bring in the social cloud*: The bigger part of the picture of reality is built around an ecosystem functional perspective, to which, usually at a later stage, some sort of sociopolitical or social institutional "stuff" is attached—as a cloud in the bigger landscape. Though stressing the importance of this cloud, it is almost treated as some form of externality, being outside of or vaguely linked to what really matters in the reality of this model. This is illustrated in Figure 13-1. Although some governance elements are incorporated in the main part of the model, the sociopolitical questions ("*core*

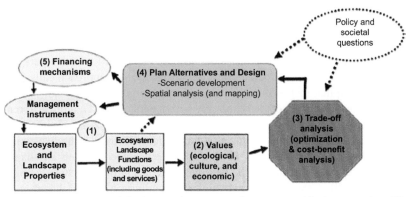

FIGURE 13-1 Framework for integrated assessment of ecosystem and landscape services. [6]

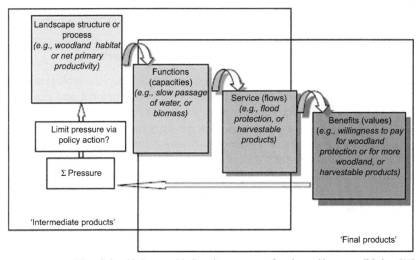

FIGURE 13-2 The relationship between biodiversity, ecosystem function and human well-being. [11]

business" of *governance*), and with it the process of *how to do governance*, is somewhat loosely related and in fact almost put in brackets by using dotted lines.

Often this social cloud is also treated as a kind of endpoint in a rather linear one-way visualization of complex reality: from biodiversity to human well-being. (See, e.g., Figure 13-2.)

The point here is not to state these models are without relevance. On the contrary, they are, have been, and will remain very important in visualizing the issue of ecosystem services and have contributed significantly to putting ecosystem services on the agenda of many policy makers and other societal actors. Recently some countries have invested in or are planning to invest in what has become almost a must for policy makers: "*you don't have a TEEB yet?*" TEEB (*The Economics of Ecosystems and Biodiversity;* [12]) in its first volume [13], offers a framework for such assessments where decision making is constituted by a box informed mainly by the values of ecosystem services. The need to include decision making as a social process is picked up in Hussain et al. [14], in the TEEB volume for local policy, where a stepwise approach for ecosystem appraisal is recommended that starts out with an identification of the perspectives of all stakeholders involved. Such a transdisciplinary approach calls for a much more explicit inclusion of the social cloud. As long, however, as the social dimension is not conceptualized explicitly within a framework, a more differentiated consideration of social processes will largely depend on who is in charge of conducting an assessment. Typically, this is dominated by natural scientists or economists. At the same time, we claim that the way these models are shaped is almost a natural outcome of the fact that the concept of ecosystem services is rooted mainly within the more ecological disciplines. Although this more or less opens up Pandora's box by linking ecosystems to human well-being, still the social complexity side of ecosystem services governance is in clear need of more integrated approaches, of more reality-proof frameworks.

1.3. Integrated Ecosystem Governance Frameworks

Carpenter et al. (15; Figure 13-3) present a combined picture of ecological and social processes, including actors and actions and institutional and ecological dynamics across multiple scales. The governance of ecosystem services is addressed in an integrated picture (Figure 13-4). Strikingly these pictures are multidirectional and open up the picture of complexity to both *substance* (what is the subject of governance?) and *process* (how do we do governance?).

Carpenter et al. [15] contribute to picturing the complexity of ecosystem services governance in a more complete and integrated manner, pleading for a new type of science that will allow evaluation of the effectiveness of governance practices. Bodin and Crona ([16]; Figure 13-5) show how different actors in different societal sectors may relate to each other and to nature in many different ways, forming social networks relevant for dealing with issues concerning natural resource governance. Cowling et al. ([4]; Figure 13-6) combine both process phases, stakeholder engagement, scales, and ecological system states in an operational model of ecosystem governance, and place social assessment on an equal footing with biophysical assessment at the start of operations. As such, they take the more integrated approach a step further than Carpenter et al. [15] and Bodin and Crona [16] by making it operational in a stepwise approach: from assessment to management. Cowling et al. [4] propagate an adaptive management approach of learning by using the most effective strategy for dealing with complexity (see also Section 2.3.1 of this chapter). Key to this approach is the institutionalization of such an approach in learning organizations: organizations that have the capacity, authority and flexibility to respond adequately to actions in practice that are often characterized by constantly changing circumstances. In the next section we will focus on this action situation from an institutional perspective.

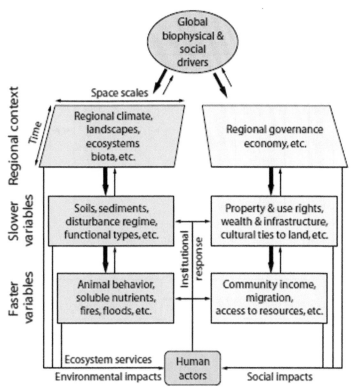

FIGURE 13-3   Conceptual framework for integrated analysis of coupled social–ecological systems. [11]

1.4. Institutional Aspects of Ecosystem Governance

Several of the above-mentioned frameworks allude to the need to integrate the ecological picture of reality with an institutional picture. According to Daily et al. [5], science on ecosystem services should not only further develop, but should be integrated into decision making (individuals, households, companies, governments), without which "the value of nature will remain little more than an interesting idea, represented in scattered, local, and idiosyncratic efforts," Daily et al. continue: "Without institutional change, communities may well continue to carry on with behaviours that are widely known to be harmful to society over the long term (e.g. overfishing, high use of fossil fuels). p.22" According to Norgaard [17], "we need new global institutions and far more resources devoted to environmental governance. The flurry of enthusiasm for optimizing the economy by including ecosystem services has blinded us to the more important question of how we are going to make the substantial institutional changes to significantly reduce human pressure on ecosystems. p.1220" He continues, stating "we should be focusing on the combination of markets and institutions that best reach social goals, p.1225" thus hinting at a combination of governance approaches.

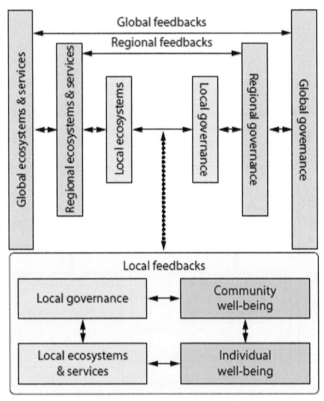

FIGURE 13-4   Governance and ecosystem services are interlinked at multiple scales. [15]

Vatn [18], from the perspective of institutional economics, introduces the concept of *value articulating institution* for environmental appraisal, stressing the social process of valuation in which we need to clarify "rules concerning a) who should participate and in which capacity, b) what is considered data and which form data should take, and c) rules about how a conclusion is reached. p.2207" The data/knowledge issue is a crucial one, as Wilkinson [19] also acknowledges: "How human–nature relations are conceptualized significantly informs the basis for governance of social-ecological systems. It informs what and whose knowledge matters and to what end this knowledge is put. p.10"

According to Lee [20], the Institutional Analysis and Development (IAD) framework by Ostrom [21] can be useful in conceptualizing and analyzing governance. By institution Ostrom [21, 22] means "the prescriptions that humans use to organize all forms of repetitive and structured interactions. [21, p.3]" Central in the IAD framework is the *action situation* (Figure 13-7) that leads to interactions and outcomes.

In Figure 13-8 Ostrom [22] presents "a common set of variables used to describe the structure of an action situation [which] includes (i) the set of actors, (ii) the specific positions to be filled by participants, (iii) the set of allowable

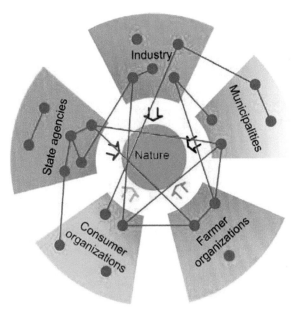

FIGURE 13-5   Different sectors of society involved in the use and management of the natural environment. [16]

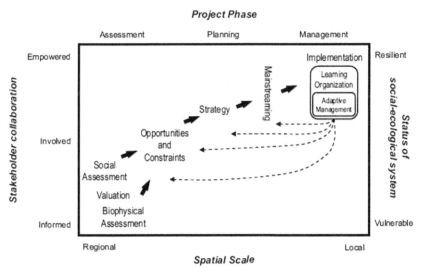

FIGURE 13-6   An operational model for implementing the safeguarding of ecosystem services. [4]

actions and their linkage to outcomes, (iv) the potential outcomes that are linked to individual sequences of actions, (v) the level of control each participant has over choice, (vi) the information available to participants about the structure of the action situation, and (vii) the costs and benefits—which serve as incentives

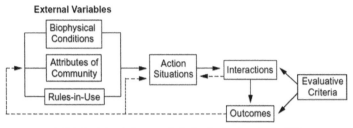

FIGURE 13-7   A framework for institutional analysis. [21]

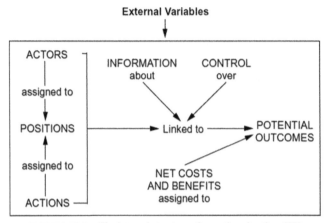

FIGURE 13-8   The internal structure of an action situation. [22]

and deterrents—assigned to actions and outcomes. p.11" In Figure 13-9 action situations are embedded within social-ecological systems.

Pahl-Wostl ([23]; Figure 13-10) frames several strategic options that can play a role in (ecosystem) governance. (1) The *actors* involved and their role in the governance process, distinguishing state and nonstate actors. (2). *Institutional arrangements,* distinguishing formal and informal approaches. (3) Typologies of combinations of (1) and (2) are given: traditional top-down hierarchical arrangements, voluntary "market" arrangements, and more egalitarian bottom-up network arrangements. According to Pahl-Wostl, governance regimes that are characterized by more complexity and diversity of these strategic options have a higher adaptive capacity. A key element of this adaptive capacity is social and societal learning (see also Section 2.3.1). Moreover, Pahl-Wostl stresses the importance of context specificity: Whatever framework or model is developed or applied, it should be specific only to a certain extent, as to allow case-specific room for adaptive maneuver.

The question Duit and Galaz [24] try to answer is similar to the title of this chapter: Can complex adaptive systems (for a definition, see Section 2.3.1) be governed? They link governance typologies to adaptive (problem-solving)

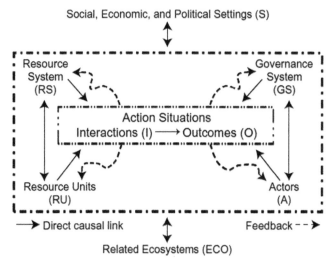

FIGURE 13-9   Action situations embedded in broader social-ecological systems. [22]

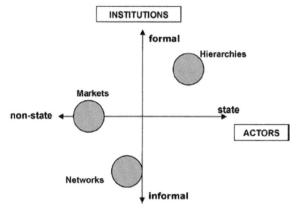

FIGURE 13-10   Difference of governance modes of bureaucratic hierarchies, markets and net-works regarding the degree of formality of institutions and the importance of state and non-state actors. [23]

capacity in the face of ecological change (Figure 13-11). The focus on problem-solving capacity of governance types in the face of complexity is urgently needed according to Duit and Galaz [24]: Until now there has been too much focus on merely characterizing governance types. Based on their analysis, it is concluded that a robust type of governance for complex adaptive systems is dependent on "resolving the fundamental tension between institutional stability and flexibility. p.329" On the one hand, institutional flexibility is needed in order to be able to respond adequately to change and to prevent path dependency and other threats of rigidity. Stability, on the other hand, is needed in order to provide institutional structures and capacity capable of (collectively) addressing issues.

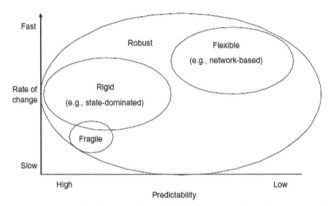

FIGURE 13-11   Adaptive capacity and different types of complex change. [24]

Next we present several governance approaches that we consider relevant to ecosystem services governance, sketching the diversity of options to choose from or indeed combine. By no means is this an exhaustive overview, but we hope to illustrate diversity and room for maneuver and inspire debate.

2. ECOSYSTEM GOVERNANCE APPROACHES: SOME EXAMPLES

2.1. Economics and Ecological Governance

There are many theories and views on economics and environmental/ecological governance. We cannot present an extensive overview here of all the economic methods and concepts relevant for ecosystem services governance. For this we refer the reader to TEEB, to other chapters in this book, and to the vast amount of relevant literature. We will only briefly introduce two modes of framing economics and ecological governance to show that here too choices can be made: environmental and ecological economics. Being two major streams of economic thought relevant to integrating the environmental and nature lens into the visor of economics, they fundamentally disagree on how this should be done [25, 26]. The left side of Figure 13-12 illustrates how the economy can be seen as a sphere different from but equal in size to the environment and society, overlapping to some extent but mainly independent from each other [26]. The right side of Figure 13-12 presents the economy as a subsystem of society, which is a subsystem of the environment, all being interdependent. This places economic activity within the social sphere, showing that social relations are crucial, which is also important from a governance perspective.

From a sustainability perspective, Pearce and Atkinson [25] describe the two different approaches as weak sustainability and strong sustainability—the left and right side of Figure 13-12, representing, respectively, the views from *environmental economics* and *ecological economics*. A crucial difference between the two approaches concerns the view on nature as capital. Capital, a

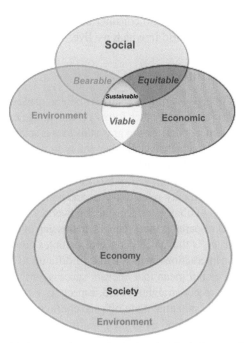

FIGURE 13-12   Environmental economics (top) and ecological economics (bottom) [26]

core economic concept capturing/measuring wealth, is considered to be substitutable in the weak sustainability approach; that is, all forms of capital can be substituted by other forms of capital: human-made capital (including human capital and social capital) and natural capital, the total amount of capital to remain equal. This approach is called weak because it may in principle define a society with vast decrease of natural capital to be sustainable, as long as the decrease of natural capital is compensated by an increase in human-made capital, such as technology or savings. Proponents of the strong sustainability approach criticize this view, stating that natural capital is not to be substituted by human-made capital. The difference between the two approaches has important implications for governance, as differences both in sustainability framing and monitoring of society's well-being will lead to radically different governance assumptions and actions. The internationally dominant "economic growth" governance strategy from a sustainability perspective can be considered a strategy of weak sustainability, where in principle declining stocks of natural capital are supposed to be substituted by technological advancement, for example.

To conclude, we want to stress that both views have one important aspect in common: Both *environmental economics* and *ecological economics* distinguish the environment, the social (society), and the economy as different spheres of reality. Alternatively, we might also consider them as different lenses focusing on different aspects of the world in which we live.

2.2. Science Policy Interface

An important aspect of ecosystem governance is policy uptake of relevant knowledge, ecosystem knowledge, and other forms of knowledge considered relevant for policy making. A concept that tries to grasp this idea is science-policy interface (SPI): "relations between scientists and other actors in the policy process which allow for exchanges, co-evolution, and joint construction of knowledge with the aim of enriching decision-making" [27, p.807 28]. According to Neßhöver [29], Interactions between biodiversity-related science and policy-making are still poorly understood. The Science–Policy Interfaces for Biodiversity: Research, Action and Learning [30] FP7 EU research project focuses on this gap of knowledge [28–30]. An important conclusion is that an SPI concerns complex constellations of many factors and nonlinear interactions, which makes it difficult to pinpoint best practices as such. This does not mean that relevant success attributes of SPIs cannot be detected, but they are not straightforward, especially not for combinations of attributes. We can identify potential trade-offs and conflicts among the SPI success attributes, some of which remind us of the ones discussed earlier in this chapter, such as the formal–informal or stable–flexible dichotomies. Their relative importance is heavily context-dependent, varying according to the type of problem, the stage in the policy cycle, and various other factors. Considering these factors helps to determine which attributes to emphasize. There is no single recipe for success, but rather a suite of features that need to be taken into account in a context-dependent way. Moreover, the successful impacts of SPIs on whatever goal they aim to achieve depends on many features within and outside the SPI.

An international example of an SPI relevant to ecosystem services governance is the Intergovernmental Platform on Biodiversity and Ecosystem Services (IPBES; [31]). IPBES was established in 2012 in Panama at a meeting of over 90 countries, after several years of intense international negotiation. IPBES is described as "a two-way interface between the scientific community and policy makers that aims to build capacity for and strengthen the use of science in policy making." The international community agreed on four main functions of IPBES: knowledge generation; regular and timely assessments; support policy formulation and implementation; and capacity building. Still a lot of important issues concerning the functioning of IPBES have to be negotiated. The BElgium Ecosystem Services community (BEES; [32]) is a Belgian example of an SPI relevant to ecosystem services governance. It was established in 2012 by a variety of mainly science and policy actors, building on the emerging community within the BEES project of which this book is one outcome. The BEES community is an open and flexible network that will serve as an interface between different societal sectors, open to all potentially interested organizations. Its functioning is inspired by the concept of *community of practice* [33]: a network made up of individuals and organizations that share an interest and practice, who come together to address a specific challenge and further each other's goals and objectives in a specific topic area.

2.3. Reflexive Governance

Traditional modes of governance have increasingly been criticized for their incapability to take into account complexities involved in governing complex systems such as the socioecological and sociotechnical, resulting in a lack of awareness and control of solving the problems caused by our modern industrial society: ecological degradation, environmental pollution, and so on. *Reflexive governance* makes us reflect on these unwanted and unforeseen negative effects of modern societies and tries to find new modes of governance in which this *reflexiveness* is incorporated in two ways (after Grin [34], p. 1). (1) What kind of society do we want? (2) How do we realize it? For *reflexive governance* Voß and Kemp [35, 36] stress the importance of several elements. (1) *Integrated transdisciplinary knowledge production*: the need to combine (integrate different perspectives on complexity, both different forms of expertise (disciplines) and other social groups. (2) *Experiments and adaptivity of strategies and institutions*: unpredictable and unforeseen complexity and dynamics require responsiveness and adaptability based on experience and learning. (3) Anticipation of long-term systems effects of measures: constant awareness of possible long-term effects of current actions. (4) Iterative participatory goal formulation: Societal goals demand incorporating social actors, values and their dynamics. (5) Interactive strategy development: Interaction and involvement of a diversity of actors and their knowledge are key to strategy implementation in practice.

Smith and Stirling [37] and Voß and Bornemann [38] present two reflexive modes of governance relevant to our discussion here: adaptive management and transition management. The first, *adaptive management*, is rooted in the field of socioecological system research, and the second in the field of science and technology studies. The first seems logical to discuss here, whereas the second may surprise our readers as perhaps no bigger contrast may exist than between nature and human-made technology and as both approaches stem from completely different fields of expertise. Still we think lessons may be learned for governance of ecosystem services in looking at both. Smith and Stirling [37] in fact state that they belong together: "technologies help constitute social-ecological systems. . . . social-ecological system' extends beyond ecosystem services to include technological use of natural resources such as minerals."

Adaptive Management

An important stream of ecosystem governance is *adaptive management* [4, 10, 19, 23, 24, 37–42], *adaptive co-management* being a more participatory exponent [39, 40]. The core idea of adaptive management is that due

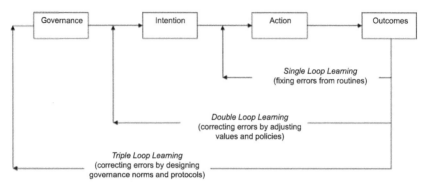

FIGURE 13-13 A multiple-loop learning framework for environmental and resource management. [40]

to complexity, both ecological and social, figuring out beforehand "the" best management approach seems impossible. Crucial in the approach of adaptive management is learning by doing, experimenting, and adjusting management practice based on experience and knowledge gained along the way. This is pictured in Figure 13-13, distinguishing different levels of learning and adaptation: at the concrete action level, at the policy level, and more fundamentally at the governance level. According to Pahl-Wostl [23], one of the main challenges for adaptive management is to take learning processes of ecosystem governance further than merely single-loop learning in order to accomplish the structural change needed to overcome systemic problems. Armitage et al. [40] speak of a paradox of learning: Though considered a crucial element and an important normative goal of adaptive management, learning itself seems somewhat underrepresented in practice and research when it comes to critical reflection, understanding, and capacity building, especially with respect to more participatory approaches such as adaptive co-management. When applying adaptive and participatory approaches to ecosystem services management, this should therefore not be taken for granted, but is in need of careful consideration and research.

Carpenter et al. [41] highlight the relevance of *scenarios* for adaptive management in which actions may be considered informative experiments in the face of (largely unpredictable) complexity. Moreover, scenarios may be used in more vision-driven approaches such as *backcasting* [37]. In *backcasting* [42, 43], the core idea is not to predict the future, but to define a route toward a desirable future: How can the future we envision as desirable be realized by actions starting today? What are the opportunities, and what are the barriers?

Duit and Galaz [24] point out that not only the governance of complex social-ecological systems may be considered in an adaptive manner, but indeed from a complexity perspective, the complex system it aims to govern may be considered to be adaptive as well. Duit and Galaz characterize these *complex adaptive systems* as follows: "we often fail to recognize that these and other

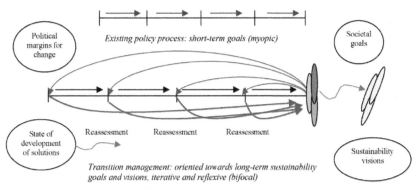

Political
margins for
change

Existing policy process: short-term goals (myopic)

Societal
goals

State of
development
of solutions

Reassessment   Reassessment   Reassessment

Sustainability
visions

*Transition management: oriented towards long-term sustainability
goals and visions, iterative and reflexive (bifocal)*

FIGURE 13-14   Current policy versus transition management. [44]

cross-level drivers of change do not add up in a linear, predictable manner. On the contrary, insights from the last decades of empirical and theoretical research on complex adaptive systems clearly show that biophysical as well as man-made systems are characterized by both positive and negative feedback loops operating over a range of spatial and temporal scales. This results in developments over time, characterized by periods of incremental change followed by fast and often irreversible change and "surprises" with immense consequences for economies, vital ecosystems, and human welfare. p. 311–312"

2.4. Transition Management

There are many similarities between *adaptive management* and *transition management* [44, 37]. Main differences probably are the different topical fields in which both approaches emerged and the more sociotechnical and systemic focus of transition management. A picture of *transition management* is presented in Figure 13-14. Transition management focuses on systemic change oriented at sustainable development. This systemic and to some extent long-term perspective does not exclude incremental steps, nor does it necessarily try to radically break with past policy. According to Kemp and Loorbach [44]: "It puts these policies in a different, longer-term perspective and tries to better align specific policies. Mathematically one could say that transition management = current policies + long-term vision + vertical and horizontal coordination of policies + portfolio-management + process management" p.12.

3. HYBRIDIZATION

Apart from the emergence of new modes of governance, over the years awareness has grown that using singular ideal types of (new) governance approaches will not suffice and that hybrid approaches are needed. According to Daily et al. [5]: "There is no magic recipe for initiating change, and it makes sense

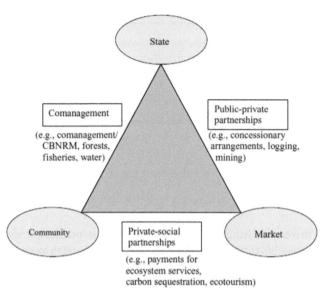

FIGURE 13-15   Mechanisms and strategies of environmental governance. [45]

to experiment with a wide variety of possible mechanisms" and "The complexity of social change, and the diversity of values and decisions facing stakeholders . . . highlight the need for a multi-pronged approach. p.26" Norgaard [17] calls for "methodological pluralism. p.1222" Duit and Galaz [24] discuss how combinations of governance systems on different levels buffer or weaken the capacity to govern complex adaptive systems.

In their review of environmental governance, Lemos and Agrawi [46] specifically address hybridization as a new mode of governance that is emerging more and more. They identify three major forms (Figure 13-15): "co-management (between state agencies and communities), public-private partnerships (between state agencies and market actors), and private-social partnerships (between market actors and communities). p.311" Regarding ecosystem governance, Lemos and Agrawi [46] stress the importance of involving market, state and community actors. They also flag the limitations of such hybrid approaches, especially from the point of view of democracy, equity, and long-term sustainability. A focus mainly on (natural) resource efficiency is one of the most disputed issues.

On a more pragmatic level, Wittmer et al. [47] approach these issues from a slightly different but related angle. They propose to use the criteria presented in Table 13-1 to characterize both the decision situation and the methods concerning the need/ability to deal with complexity, the need/ability to support social dynamics, the legitimacy of the process, and the costs incurred. Such an approach of matching decision situations with adequate methods could also be applied to questions of governance of ecosystem

TABLE 13-1 Criteria for Evaluating Decision Aid Methods [47]

Information	Coping with complexity
	Integrating different types of information
	Coping with uncertainity
Legitimacy	Legal compatibility and integrating procedural knowledge
	Accountability
	Inclusion/representation
	Transparency of rules and assumptions to insiders and outsiders
Social dynamics	Respect/relationship
	Changing behaviour, changing perspectives/learning
	Agency/empowermwnt
	Facilitate convergence or illustrate diversity
Costs	Cost-effectiveness
	Costs of the method
	Decision failure costs

services, particularly where new governance structures are set up. Without providing a fixed recipe, these criteria can help to break down some of the rather abstract concepts to concrete questions and choices in a particular decision setting.

The spiral briefs also offer hands-on practical tips on how to structure and approach some of the challenges encountered at the science–policy interface concerning biodiversity (spiral-project.eu/content/documents#jump2briefs).

4. CONCLUSION

The title of this chapter poses both a topic and a question regarding ecosystem services governance: respectively, *management of complexity* and the question of whether we are in fact *able to manage complexity*. In other words, can we steer society, can we construct institutions that help govern social-ecological systems to whatever ends we choose, such as enhancing ecosystem services? And, if we conclude that we believe we can, can we prove this, can we not only make it happen, but also show it in order to convince others that yes we can indeed? As described above, the new modes of governance have criticized the traditional modes of not being able to fulfil their promises. Alternative modes have emerged, but they themselves can be criticized, perhaps on different grounds (e.g., [10, 40, 48–49]), but in essence by pointing out similarly: you believe you can, but can you really make it happen?

We think it is important to realize that the governance challenge is enormous due to complexity and that whatever governance strategy we choose or develop, inherently has limitations and imperfections. We also think it is important to realize that indeed one *chooses* an approach, in a process of what De Schutter [50] calls *institutional constructivism*. This very much connects to what those engaged in science and technology studies (beautifully described by Bijker [51]) call *the social construction of technologies*: Technical artifacts are considered as social constructions in the sense that the meanings relevant social groups attribute to a technical artifact will shape that artifact and its use. We may add here the *social construction of nature* [52], which is part of ecosystem governance: We choose how we perceive and deal with nature, we attribute meaning to it, and we shape it in a sense to our beliefs about nature. The question then is, of course, if nature listens to us, and perhaps more importantly: does society?

In this chapter, we presented food for thought for those interested in ecosystem or socioecological system governance, without presenting clear-cut recipes or roadmaps. We discussed different ways of framing socio-ecosystem governance and some ecosystem governance approaches. Clearly, a key common denominator and challenge is complexity. The system or systems we aim to govern are complex. But also the governance processes are inherently complex and therefore difficult to capture or steer in terms of effectiveness. Clearly, some key attributes and challenges of governance arrangements are identified, and to some extent they are shared by the authors we discussed. First, taking into account complexity is an important step as such: not only ecological complexity, but indeed also social complexity, and best in an integrated manner. This challenge still seems prominent in the field of ecosystem services. Second is a turn from characterizing governance arrangements to assessing their problem-solving potential, or in other words their effectiveness. Third, adaptive management or learning by doing approaches seem to be favored by most authors. Fourth, in order to deal with complexity and also adapt to changes, balances between several dichotomous traits such as formal–informal, stable–flexible, and top-down–bottom-up have to be found. Fifth, governance regimes that are characterized by more complexity and diversity of the above mentioned traits may have a higher adaptive capacity. Sixth, it is important to take context dependency into account, both when designing, applying, and adapting, andwhen assessing governance processes. Finally, we explicitly want to refer to practice: Governance is about doing, and it is about real life, not just about academic discussions. And, importantly, *ecosystem governance is open for discussion*.

REFERENCES

1. Jordan, A. 2008. The governance of sustainable development: Taking stock and looking forwards. *Environment and Planning C: Government and Policy* 26, 17–33.
2. Hajer, M. 1995. *The Politics of Environmental Discourse*. Oxford University Press.
3. Section 1.4 of Chapter 5 in this book.
4. Cowling, R.M., Egoh, B., Knight, A.T., O'Farrell, P.J., Reyers, B., Rouget, M., Roux, D.J., Welz, A., and Wilhelm-Rechman, A. 2008. An operational model for mainstreaming ecosystem services for implementation. *PNAS* 105(28), 9483–9488.

5. Daily, G.C., Polasky, S., Goldstein, J., Kareiva, P.M., Mooney, H.A., Pejchar, L., Ricketts, T. H., Salzman, J., and Shallenberger, R. 2009. Ecosystem services in decision making: Time to deliver. *Frontiers in Ecology and the Environment* 7(1), 21–28.

6. de Groot, R.S., Alkemade, R., Braat, L., Hein, L., and Willemen, L. 2010a. Challenges in integrating the concept of ecosystem services and values in landscape planning, management and decision making. *Ecological Complexity* 7, 260–272.

7. Haines-Young, R. 2011. Exploring ecosystem services issues across diverse knowledge domain using Bayesian Belief Networks. *Progress in Physical Geography* 35(5), 685–704.

8. Fish, R. 2011. Environmental decision making and an ecosystems approach: Some challenges from the perspective of social science. *Progress in Physical Geography* 35(5), 71–680.

9. Jasanoff, S. 1990. *The Fifth Branch. Science Advisers as Policymakers.* Cambridge, MA: Harvard University Press.

10. Lebel, L., Anderies, J.M., Campbell, B., Folke, C., Hatfield-Dodds, S., Hughes, T.P., and Wilson, J. 2006. *Governance and the Capacity to Manage Resilience in Regional Social-Ecological Systems,* Marine Sciences Faculty Scholarship. Paper 52, http://digitalcommons. library.umaine.edu/sms_facpub/52.

11. Haines-Young, R., and Potschin, M. 2010. The links between biodiversity, ecosystem services and human well-being. Chapter 6 in: Raffaelli, D., and C. Frid (eds.), *Ecosystem Ecology: A New Synthesis.* BES Ecological Reviews Series. Cambridge: Cambridge University Press.

12. *TEEB: The Economics of Ecosystems and Biodiversity,* http://www.teebweb.org

13. de Groot, R.S., Fisher, B., Christie, M., Aronson, J., Braat, L., Gowdy, J., Haines-Young, R., Maltby, E., Neuville, A., Polasky, S., Portela, R., and Ring, I. 2010b. Integrating the ecological and economic dimensions in biodiversity and ecosystem service valuation. In Kumar, P. (ed.), *The Economics of Ecosystems and Biodiversity Ecological and Economic Foundations,* pp. 9–40. London: Earthscan.

14. Hussain, S., Wittmer, H., Berghöfer, A., and Gundimeda, H. 2012. In Wittmer, H., and H. Gundimeda (eds.), *The Economics of Ecosystems and Biodiversity in Local and Regional Policy and Management,* pp. 35–53. London: Earthscan.

15. Carpenter, S.R., et al. 2009. Science for managing ecosystem services: Beyond the Millennium Ecosystem Assessment. *PNAS* 106(5), 1305–1312.

16. Bodin, Ö., and Crona, B.I. 2009. The role of social networks in natural resource governance: What relational patterns make a difference? *Global Environmental Change* 19, 366–374.

17. Norgaard, R.B. 2010. Ecosystem services: From eye-opening metaphor to complexity blinder'. In: *Ecological Economics* 69(6),1221–1227.

18. Vatn, A. 2009. An institutional analysis of methods for environmental appraisal. In: *Ecological Economics* 68, 2207–2215.

19. Wilkinson, C. 2011, May. Social-ecological resilience: Insights and issues for planning theory, *Planning Theory* 11 (2), 148–169.

20. Lee, M. 2003. *Conceptualizing the New Governance: A New Institution of Social Coordination,* Paper presented at the Institutional Analysis and Development Mini-Conference, May 3 and 5, 2003, Workshop in Political Theory and Policy Analysis, Indiana University, Bloomington, Indiana.

21. Ostrom, E. 2005. *Understanding Institutional Diversity.* Princeton, NJ: Princeton University Press.

22. Ostrom, E. 2011. February. Background on the Institutional Analysis and Development Framework. *Policy Studies Journal* 39(1), 7–27.

23. Pahl-Wostl, C. 2009. A conceptual framework for analysing adaptive capacity and multi-level learning processes in resource governance regimes. *Global Environmental Change* 19 (2009), 354–365.

24. Duit, A., and Galaz, V. 2008, July. Governance and Complexity—Emerging Issues for Governance Theory. *Governance: An International Journal of Policy, Administration, and Institutions* 21 (3), 311–335.

25. Pearce, D., and Atkinson, G. 1998. The concept of sustainable development: An evaluation of its usefulness ten years after Brundtland. CSERGE Working Paper PA 98-02. Centre for Social and Economic Research on the Global Environment. University College London and University of East Anglia.

26. Scott-Cato, M. 2008. *Green Economics: An Introduction to Theory, Policy and Practice.* London: Routledge.

27. Van den, Hove, S. 2007. A rationale for science–policy interfaces. *Futures* 39(7), 807–826.

28. Sarkki, S., Niemelä, J., Tinch, R., and the SPIRAL team. 2012. *Criteria for Science-Policy Interfaces and Their Linkages to Instruments and Mechanisms for Encouraging Behaviour that Reduces Negative Human Impacts on Biodiversity*, Science-Policy Interfaces for Biodiversity: Research, Action and Learning (SPIRAL; FP7 EU research project) http://www.spiral-project. eu/sites/default/files/SPIRAL_3-1.pdf

29. Neßhöver, C. 2012. *Study on Landscape of Science-Policy Interfaces*, Science-Policy Interfaces for Biodiversity: Research, Action and Learning (SPIRAL; FP7 EU research project) http:// www.spiral-project.eu/sites/default/files/SPIRAL_1-2.pdf

30. SPIRAL: Science-Policy Interfaces for Biodiversity: Research, Action and Learning, http:// www.spiral-project.eu/

31. IPBES: Intergovernmental Platform on Biodiversity and Ecosystem Services, http://www. ipbes.net/

32. BEES: BElgium Ecosystem Services community, http://www.beescommunity.be

33. Wenger, E., and Snyder, W.M. 2000, January–February. Communities of Practice: The Organizational Frontier. *Harvard Business Review*, 139–145.

34. Grin, J. 2006. Reflexive modernization as a governance issue—or: designing and shaping re-structuration. In Voß J.-P., D. Bauknecht, and R. Kemp (eds.), *Reflexive Governance for Sustainable Development*, pp. 54–81. Cheltenham, UK: Edward Elgar.

35. Voßm, J.-P., and Kemp, R. 2005. Sustainability and reflexive governance: Incorporating feedback into social problem solving. Paper presented at International Human Dimensions Programme on Global Environmental Change (IHDP) Open Meeting. Bonn, October 9–13.

36. Voß, J., and Kemp, R. 2006. Sustainability and reflexive governance: Introduction. In Voß, J.-P., D. Bauknecht, and R. Kemp (eds.), *Reflexive Governance for Sustainable Development*, pp. 3–28. Cheltenham, England: Edward Elgar.

37. Smith, A., and Stirling, A. 2010. The politics of social-ecological resilience and sustainable socio-technical transitions. *Ecology and Society* 15(1), 11.

38. Voß, J., and Bornemann, B. 2011. The politics of reflexive governance: Challenges for designing adaptive management and transition management. *Ecology and Society* 16(2), 9.

39. Armitage, D., Berkes, F., and Doubleday, N. (eds.). 2007. *Adaptive Co-Management: Collaboration, Learning, and Multi-Level Governance*, Vancouver, Canada: UBC Press, University of British Columbia.

40. Armitage, D., Marschke, M., and Plummer, R. 2008. Adaptive co-management and the paradox of learning. *Global Environmental Change* 18 (2008), 86–98.

41. Carpenter, S.R., Bennett, E.M., and Peterson, G.D. 2006. Scenarios for ecosystem services: An overview. *Ecology and Society* 11(1), 29.

42. Kenward, R.E., et al. 2011, March 29. Identifying governance strategies that effectively support ecosystem services, resource sustainability, and biodiversity. *PNAS* 108(13), 5308–5312.

43. Robinson, J. 1982. Energy backcasting: A proposed method of policy analysis. *Energy Policy* 10, 337–344.

44. Quist, J. 2007. *Backcasting for a Sustainable Future: The Impact after 10 Years.* Delft, the Netherlands: Eburon.

45. Kemp, R., and Loorbach, D. 2003. *Governance for sustainability through transition management,* paper for Open Meeting of the Human Dimensions of Global Environmental Change Research Community, http://meritbbs.unimaas.nl/rkemp/Kemp_and_Loorbach.pdf October 16–19, Montreal, Canada.

46. Lemos, M.C., and Agrawal, A. 2006. Environmental governance. *Annual Review of Environmental Resources* 31, 297–325.

47. Wittmer, H., Rauschmayer, F., and Klauer, B. 2006. How to select instruments for the resolution of environmental conflicts? *Land Use Policy* 23, 1–9.

48. Hendriks C.M., and Grin, J. 2007. Contextualising reflexive governance: The politics of Dutch transitions to sustainability. *Journal of Environmental Policy and Planning* 9(3/4), 333–350.

49. Shove, E., and Walker, G. 2007. CAUTION! Transitions ahead: Politics, practice, and sustainable transition management. *Environment and Planning A* 39, 763–770.

50. De Schutter, O., and Deakin, S. 2005. Reflexive governance and the dilemmas of social regulation. General Introduction to: De Schutter O. and Deakin S. (eds.), *Social Rights and Market Forces: Is the Open Coordination of Employment and Social Policies the Future of Social Europe?* Brussels: Bruylant.

51. Bijker, W.E. 1997. *Of Bicycles, Bakelites and Bulbs: Toward a Theory of Sociotechnical Change.* Cambridge, MA: MIT Press.

52. Proctor, J.D. 1998, September. The social construction of nature: Relativist accusations, pragmatist and critical realist responses. *Annals of the Association of American Geographers,* 352–376.

Chapter 14

Ecosystem Service Assessments: Science or Pragmatism?

Sander Jacobs[1,2], Hans Keune[3,4,5,6],Dirk Vrebos[7], Olivier Beauchard[7], Ferdinando Villa[8,9] and Patrick Meire[7]

[1]*Research Institute for Nature and Forest (INBO),* [2]*University of Antwerp. Department of Biology, Ecosystem Management Research Group (ECOBE),* [3]*Belgian Biodiversity Platform,* [4]*Research Institute for Nature and Forest (INBO),* [5]*Faculty of Applied Economics – University of Antwerp,* [6]*naXys, Namur Center for Complex Systems – University of Namur,* [7]*University of Antwerp, Ecosystem Management Research Group (ECOBE),* [8]*Basque Centre for Climate Change (BC3),* [9]*IKERBASQUE, Basque Foundation for Science*

Chapter Outline

1. INTRODUCTION

Ecosystem services (ES) are the benefits humans derive from ecosystems. As the concept of ES gains popularity as an effective framework for management and to facilitate communication of the projected consequences of both global and local change, growing attention is being given to ES in fields as diverse as ecology, agriculture, economics, sociology. and policy science. The ES concept

Ecosystem Services. http://dx.doi.org/10.1016/B978-0-12-419964-4.00014-7

resonates equally strongly within academic, governmental, and corporate discourse. The inherent interdisciplinary nature of ES purportedly gathers actors from many domains and puts recent advances in their fields to practice, providing a sound knowledge base for decision making for resource use in both policy and corporate contexts. But is this really the case?

The Millennium Ecosystem Assessment [1] and the Economics of Ecosystems and Biodiversity [2] reports have served as milestones in creating policy and corporate awareness concerning the dependency of our societies and markets on ecosystems and biodiversity. They demonstrate ecosystem services as a powerful concept to guide anthropocentric, sustainable and equitable natural resource management strategies (e.g., [3]). Nongovernmental organizations, companies, and administrations have now started to embrace the concept and have promoted research concerning their specific regions and services of interest. This has resulted in a high and growing demand for clear, spatially explicit ES accounting from regional and national governments in order to adequately manage natural assets. This need has become the main driver for development of ES research today, firmly establishing ES assessment and valuation as a truly applied activity.

The few scientific contributions that put ES quantification and mapping results to the test (e.g., [4]) have consistently pointed out large inaccuracies and bias in mainstream ES accounting methodologies. Nevertheless, the same straightforward and pragmatic modeling and mapping approaches are used over and over again, and a secondary market of "easy" mapping and valuation tools is emerging quickly. Results from these tools are generally presented both clearly and explicitly, while the uncertainty caused by data scarcity, methodological limitations, and overlooking of the inherently complex ES dynamics is poorly acknowledged, if at all. The need for structured, transparent and comparable approaches—in a word, a *science* of ES—is urgent and clear [5].

A corollary of the fast multiplication of case studies, discussions, and simple assessment tools is a growing skepticism about ES among scientists, policy makers, and practitioners. This challenges the credibility of the ES concept to motivate collaboration between scientific disciplines and to develop sustainable land-use strategies in the long run. Many authors have listed technical/scientific hurdles that need to be overcome (e.g., [6]; see [7] for a review), and argued for a comparative blueprint for Ecosystem Service Assessments [7]. This chapter endorses this debate by:

- Identifying ten main sources of uncertainty that may affect policy and management decisions.
- Proposing three strategies for individual researchers, commissioners, and stakeholders to be mindful of methodological risks and best practices in dissemination or application of results, even under restricted budgets or project goals.
- Arguing for a clear and transparent ES research policy on methodology development, uncertainty acknowledgment, and communication.

2. METHODS

In contrast to earlier contributions discussing challenges for ES research, this chapter focuses on the *diagnosis* of uncertainty. Uncertainty-related problems are mentioned, often under the headings of "challenges" or "questions," in several reviews, editorials, and position papers as well as in the TEEB foundation reports [6–16]. Special issues dedicated to ecosystem services and many case-study reports keep pointing out, more or less explicitly, similar problems in introductions and discussions. Our aim, far from providing yet another literature review, is to detect the main *drivers of uncertainty* in order to enable acknowledgment of complexity, as a first step in decreasing the risks related to application of the ES concept and safeguarding the long-term scientific, societal, and political credibility of ES research.

3. TEN DRIVERS OF UNCERTAINTY

The main sources of uncertainty are represented in all three of the main subsystems involved in the provision, use, and management of ES: the biophysical, the socioeconomic, and the governance. Within these three systems, ten main drivers of uncertainty were distinguished.

3.1. Uncertainty in Our Understanding of the Biophysical System

Data scarcity (a) is an obvious, but often downplayed or underestimated, generator of uncertainty. Globally, only a few species and habitats are considered to be well studied. Data availability and comparability are generally poor, and most of the available studies are not adequate to cope with variability over time and space. Many datasets and databases have limited or no metadata regarding quality and uncertainty. All too often it appears that scientific communities do not acknowledge the importance of including metadata about uncertainty (e.g., [17]).

The same holds for (b) *functional knowledge gaps*. Some ecological processes are considered to occur generally and are very well studied. However the variability in processes on larger scales and interactions among scales are poorly understood. These problems are partially overcome by the development of sophisticated (ecological) modeling methodologies. These can be applied to extrapolate known data/processes over larger spatiotemporal scales, while providing the means to validate the model outputs.

Very often, however, (c) *model validation* is not adequately performed, or application limits of modeling techniques are not respected. Not incorporating spatial autocorrelation, for example, can lead to erroneous conclusions. Moreover, spatial data demand very specific statistics as standard statistics are often inadequate [18]. However, even basic statistical analysis of results is often lacking when applying an Ecosystem Service Assessment.

This is strongly related to the (d) *scaling problem*, which affects even the validity of basic concepts as "the ecosystem" (see, for instance, [19]). For example, in the TEEB documents the concept of the ecosystem is contradictorily defined several times (e.g., [20], p. 44, key message 1 vs. 2). The biosphere is strongly interlinked, and no delineated study site or system is isolated from the influences of other organizational levels [19]. The very choice of research questions influences the scale and resolution of the study and thus the results. Calculating the supply of ES is heavily affected by these choices. In models with lower resolutions, fluxes are often not incorporated as they are too small compared to the model units. As a result, every cell is treated as an isolated system [21, 22]. However, the demand for "truly" distributed models grows drastically as data quality and availability improve [17, 23]. The scaling problem is thus not a problem "to solve", but a driver of uncertainty that should be discussed and accounted for.

Finally, although ecologists have been working on basic questions for many decades, there still exists an important (e) *lack of consensus* concerning principles essential to ecosystem services. A well-known example is the discussion concerning biodiversity and resilient ecological functioning: Although the basic principle remains accepted (biodiversity ensures functioning), technical dissent concerning the mechanisms prevents straightforward application in the ES context (see Chapter 3, this volume).

3.2. Uncertainty Drivers in Understanding and Integrating the Socioeconomic System

Study of the socioeconomics of ecosystems also struggles with many challenges when asked to provide straightforward policy advice concerning ES. Similarly to the biophysical and geospatial study fields, lack of consensus is a very important constraint in formulating ES accounting methodologies. The often controversial discussion between environmental economists and ecological economists exemplifies such unavoidably politically sensitive debates. The ability of monetary parameters to track societal well-being is often contested ([4], Chapter 1, this volume). In the TEEB report, this is well summarized by John Gowdy (box 1 in [24]; see also Chapter 6, this volume).

The (f) *spatiotemporal variability of ES demand* is a crucial driver for uncertainty. The spatial distribution of demand (driven by human population and accessibility) and of the supply of ecosystem services do not necessarily coincide (e.g., flash flood water stored in an upstream wetland offers flood protection in a downstream town). Temporally, demand (and, even more, economical values) are sensitive to changes in socioeconomic dynamics (e.g., in times of crisis people will spend less on recreation), and this restricts the tenability of studies. The high sensitivity of market-derived valuations for biodiversity (e.g., real estate values) and estimates of willingness to pay (e.g., effects of local replacement alternatives) to macroeconomic and social behavioral change also underlines this fundamental problem. Value or function transfer, though often used, is therefore a risky assessment technique.

This brings us to the (g) *methodological shortcomings* that are mostly inherent to quantification of nonuse values, and regulating services strongly related to ecosystem functioning and biodiversity. Restriction of valuation to marginal changes, incompatibilities in secondary units (e.g., €/kg, €/m², €/year), differences of orders of magnitude when valuing a service using different methods, and a general lack of quantitative data hamper the pricing of bundled services and integration with contextual priority setting (see also Chapters 1, 2, and 11, this volume).

The restrictions of Walrasian theorems [25], unpredictable spatiotemporal variations in ES values, and the consequent need for development of nonuse and locally based valuation methodologies require (h) *innovative social science involvement*. Until now, ES science has not fully incorporated social sciences, although most of the aspects summarized here have strong social implications for the broad community and require effective and scientifically correct involvement of it. Moreover, as social survey methods are becoming more popular (e.g., [8]), there is a call for the active involvement of these fields of expertise.

3.3. Uncertainty Drivers in the Governance System

All the drivers of uncertainty discussed this far affect the governance and policy interface of the ES field, where management decisions are taken. TEEB distinguishes several trade-off types ([20]; see also Chapter 20, this volume) which can be considered (i) *socially generated trade-offs*: trade-offs in time (benefit now–cost later); trade-offs in space (win here–lose there), and fairness trade-offs (some win–others lose). These trade-offs have to be clearly acknowledged. The traditional position of both scientists (as objective technicians) and decision makers (as taking an informed decision for the benefit of all) is heavily challenged when implementing ES. In practice, close cooperation among representative stakeholders and transparent scientists from the very onset of ecosystem assessments seems to be the way forward.

A fourth trade-off is the (j) *service trade-off* (manage one service–lose the other) ([20]; see also Chapter 20, this volume). Some of these service trade-offs are evident and linear, others are less clear and predictable. Disentangling the underlying ecological nature and the socioeconomic drivers of these trade-offs is a task that tests the current ES science. Yet, it is exactly what is needed most to optimize, limit, and distribute natural resource use sustainably.

4. THREE PARALLEL STRATEGIES

Our overview gives a cautionary list of some commonly dismissed application risks that impact scientific credibility. Although the literature does acknowledge several problems (mostly formulated as challenges) and suggests research directions to address them, no straightforward tools to *acknowledge* (inherent) uncertainty are proposed. Often, even the classic scientific procedures to assess validity and confidentiality of results do not apply to ES research.

Three parallel strategies for ES researchers are suggested here: (1) continuously enhance fundamental and methodological capacity, (2) acquire innovative estimates to transparently acknowledge and communicate uncertainties, and (3) engage in dialogue with other disciplines (in particular social and political sciences) and stakeholders at different levels to determine relevant questions and methods and consolidate (future) findings.

4.1. Improve Methodologies

Every ecosystem service assessment can be partly dedicated to fundamental and methodological topics. Even under restricted budgets or project goals, at least transparency and comparability with other studies should be aimed at, as stressed by Seppelt et al. [7]. Promising research directions to overcome some of the repeatedly mentioned problems in different fields are continuously developing, for instance:

- The use of derived GIS outputs that link a pixel to its wider environmental context can deal with part of the scaling issues.
- Mapping of ES flows integrate biophysical ES supply potential with socioeconomic value (demand), while providing an explicit spatial dimension [26].
- Probabilistic modeling combine qualitative, quantitative, and expert data and allow for (some) uncertainty, confidentiality, and credibility measurement ([27]; Chapter 20, this volume);
- Census techniques using size and distance-dependent validation for nonuse values offer promising ways of coping with spatiotemporal demand (and value) variability.

Together, the recent scientific evolutions offer reasons for optimism: The ecosystem service concept pushes scientific advance forward and vice versa. This strategy might slow down single applied research projects compared to quick approaches, which are often seen as sufficient to fulfill ad-hoc policy needs. However, scientific quality and long-term credibility is essential in the long run, when decisions that impact on many people's lives have to be made. ES scientists and policy makers have the shared responsibility to improve this quality and the obligation to invest in fundamental improvement. This is the only way to obtain a case-by-case scientific advance.

4.2. Transparently Acknowledge Uncertainty

All new research directions inevitably encounter intellectual, ethical, scientific, and methodological limitations. In every project where processes, functions, potential or actual ES, or benefits are quantified in a spatially and temporally explicit way, uncertainties and unknowns will be inherently part of the final results. Acknowledging and quantifying these uncertainties is essential to minimize the risks for management measures to turn out bad and avoid "paralysis by analysis."

Appropriately applying modeling methods and checking model power and validity are basic practice, but these can be complemented with more innovative confidence evaluations, such as uncertainty acknowledgment based on the developing Intergovernmental Panel on Climate Change (IPCC) communication (see, e.g., [28]). This is not an easy task (e.g., [29]), and many improvements are needed (e.g., [30]). Also, critics might use communication of uncertainties to dismiss conclusions altogether [30], as was painfully demonstrated in the climate change debate. In this regard, transparency on choices made concerning research questions and methodologies is key to advance scientifically in ES research and to honestly inform decisions impacting peoples' lives.

4.3. Organize Transdisciplinarity

Aiming at a sustainable and equitable management of natural resources will be politically challenging. Current resource use practices will be critically evaluated, and many beneficiaries will be confronted. Therefore, an attitude of a priori dialogue between the researchers, policy makers, and stakeholders can serve to consolidate the ecosystem service assessment into its specific social context. This avoids technocratic top-down approaches and the use of analyses as power tools by the governing institutes (e.g., [18]). In this way ecosystem service assessments can stimulate cooperation, broad ownership of future results, and a realization of shared interests in ecosystem protection. This clearly links to the two former strategies mentioned.

5. SCIENCE AND PRAGMATISM

ES scientists are walking the sharp edge between science and pragmatism. Keeping in mind the inherent uncertainty in this research policy field, well-intentioned ES scientists have no choice but to leave their comfort zone. Governments and other stakeholders in turn should not settle with unidirectional project-based assessments but should engage in a more profound dialogue, while applying stringent quality requirements.

ES research has been very effective in creating awareness and debate concerning the value of biodiversity. However, if we want ES to meet the high expectations of a broadly supported natural resource management framework to be used from local to global scales, while fulfilling the highest sustainability and equity standards, we will have to do much better in solving problems and acknowledging the effects of remaining uncertainty on our conclusions. Despite the pressure for delivery of quick, cheap, clear, and simple results, ES research should strive to develop new methodologies, accept the complexity and uncertainty of results, and structurally address the challenge of transdisciplinarity. Finding the right balance between the three strategies within each ecosystem service assessment is a major challenge. The development of an Intergovernmental Panel on Biodiversity and Ecosystem Services could create a global transdisciplinary community where these considerations are promoted, facilitated, and further developed.

REFERENCES

1. MEA. 2005. *Ecosystems and Human Well-being: Synthesis. Millennium Ecosystem Assessment.* Washington, DC: Island Press.
2. TEEB. 2010. The Economics of Ecosystems and Biodiversity. Mainstreaming the Economics of Nature: A synthesis of the approach, conclusions and recommendations of TEEB.
3. Costanza, R., and Folke, C. 1997. Valuing ecosystem services with efficiency, fairness and sustainability as goals. In *Nature's Services: Societal Dependence on Natural Ecosystems* (pp. 49–70). Washington, DC: Island Press.
4. Eigenbrod, F.P., Armsworth, R.B., Anderson, J.A., Heinemeyer, S., Gillings, D., Roy, B.C., Thomas, D., and Gaston, K.J. 2010. The impact of proxy-based methods on mapping the distribution of ecosystem services. *Journal of Applied Ecology* 47, 377–385.
5. Carpenter, S.R., Bennett, E.M., and Peterson, G.D. 2006. Editorial: special feature on scenarios for ecosystem services. *Ecology and Society* 11(2). Retrieved from http://www.ecologyandsociety.org/vol11/iss2/art32/ES-2005-1609.pdf
6. De Groot, R.S.S., Alkemade, R., Braat, L., Hein, L., and Willemen, L. 2010. Challenges in integrating the concept of ecosystem services and values in landscape planning, management and decision making. *Ecological Complexity* 7(3), 260–272. doi:10.1016/j.ecocom.2009.10.006
7. Seppelt, R., Fath, B., Burkhard, B., Fisher, J. L., Grêt-Regamey, A., Lautenbach, S., Pert, P., et al. 2012. Form follows function? Proposing a blueprint for ecosystem service assessments based on reviews and case studies. *Ecological Indicators* 21, 145–154. doi:10.1016/j.ecolind.2011.09.003
8. Burkhard, B., Petrosillo, I., and Costanza, R. 2010. Ecosystem services—Bridging ecology, economy and social sciences. *Ecological Complexity* 7(3), 257–259. doi:10.1016/j.ecocom.2010.07.001
9. Nicholson, E., Mace, G.M., Armsworth, P.R., Atkinson, G., Buckle, S., Clements, T., Ewers, R.M., et al. 2009. Priority research areas for ecosystem services in a changing world. *Journal of Applied Ecology* 46, 1139–1144. doi:10.1111/j.1365-2664.2009.01716.x
10. Anton, C., Young, J., Harrison, P. A., Musche, M., Bela, G., Feld, C.K., Harrington, R., et al. 2010. Research needs for incorporating the ecosystem service approach into EU biodiversity conservation policy. *Biodiversity and Conservation* 19(10), 2979–2994. doi:10.1007/s10531-010-9853-6
11. Skourtos, M., Kontogianni, A., and Harrison, P.A. 2009. Reviewing the dynamics of economic values and preferences for ecosystem goods and services. *Biodiversity and Conservation* 19(10), 2855–2872. doi:10.1007/s10531-009-9722-3
12. Kosoy, N., and Corbera, E. 2010. Payments for ecosystem services as commodity fetishism. *Ecological Economics* 69(6), 1228–1236. doi:10.1016/j.ecolecon.2009.11.002
13. Gómez-Baggethun, E., De Groot, R., Lomas, P. L., and Montes, C. 2010. The history of ecosystem services in economic theory and practice: From early notions to markets and payment schemes. *Ecological Economics* 69(6), 1209–1218. doi:10.1016/j.ecolecon.2009.11.007
14. Norgaard, R.B. 2010. Ecosystem services: From eye-opening metaphor to complexity blinder. *Ecological Economics* 69(6), 1219–1227. doi:10.1016/j.ecolecon.2009.11.009
15. Vatn, A. 2010. An institutional analysis of payments for environmental services. *Ecological Economics* 69(6), 1245–1252. doi:10.1016/j.ecolecon.2009.11.018

16. Rauschmayer, F., Paavola, J., and Wittmer, H. 2009. European governance of natural resources and participation in a multi-level context: An editorial. *Environmental Policy and Governance* 19(3), 141–147. doi:10.1002/eet.504

17. Skidmore, A.K., Franklin, J., Dawson, T.P., and Pilesjö, P. 2011. Geospatial tools address emerging issues in spatial ecology: A review and commentary on the Special Issue. *International Journal of Geographical Information Science* 25(3), 337–365. doi:10.1080/13658816.2011.554296

18. Pavlovskaya, M. 2006. Theorizing with GIS: a tool for critical geographies? *Environment and Planning A*, 38(11), 2003–2020. doi:10.1068/a37326

19. Currie, W.S. 2011. Units of nature or processes across scales? The ecosystem concept at age 75. *The New Phytologist* 190, 21–34. doi:10.1111/j.1469-8137.2011.03646.x

20. Elmqvist, T., Maltby, E., Barker, T., Mortimer, M., Perrings, C., Aronson, J., Groot, R. De, et al. 2012. Chapter 2, Biodiversity, ecosystems and ecosystem services. In P. Kumar (ed.), *The Economics of Ecosystems and Biodiversity: The Ecological and Economic Foundation* (p. 456). UNEP/Earthprint.

21. Gedney, N., and Cox, P.M. 2003. The sensitivity of global climate model simulations to the representation of soil moisture heterogeneity. *Journal of Hydrometeorology* 4, 1265–1275. doi:http://dx.doi.org/10.1175/1525-7541(2003)004<1265:TSOGCM>2.0.CO;2

22. Merot, P., Squividant, H., Aurousseau, P., Hefting, M., Burt, T, Maitr, V., Kruk, M., Butturini, A., Thenail, C., and Viaud, V. 2003. Testing a climato-topographic index for predicting wetlands distribution along an European climate gradient. *Ecological Modelling* 163(1–2), 51–71.

23. Prentice, I.C., Bondeau, A., Cramer, W., Harrison, S.P., Hickler, T., Lucht, W., Sitch, S., et al. 2007. Dynamic global vegetation modeling: Quantifying terrestrial ecosystem responses to large-scale environmental change. In J.G. Canadell, D.E. Pataki, and L.F. Pitelka (eds.), *Terrestrial Ecosystems in a Changing World* (pp. 175–192). Berlin: Springer Berlin Heidelberg. doi:10.1007/978-3-540-32730-1_15

24. de Groot, R., Fisher, B., Christie, M., Aronson, J., Braat, L., Gowdy, J., Haines-Young, R., et al. (2012). Chapter 1, Integrating the ecological and economic dimensions in biodiversity and ecosystem service valuation. In P. Kumar (ed.), *The Economics of Ecosystems and Biodiversity: The Ecological and Economic Foundation* (p. 456). UNEP/Earthprint.

25. Gowdy, J., Howarth, R.B., Tisdell, C., Hepburn, C., Mäler, K.-G., and Hansjürgens, B. 2012. Chapter 6, Discounting, ethics, and options for maintaining biodiversity and ecosystem integrity. In P. Kumar (ed.), *The Economics of Ecosystems and Biodiversity: Ecological and Economic Foundations,* p. 456. UNEP/Earthprint.

26. Bagstad, K.J., Johnson, G.W., Voigt, B., and Villa, F. 2013. Spatial dynamics of ecosystem services flows: A comprehensive approach to quantifying actual service values. *Ecosystem Services* 4, 117–125.

27. Landuyt, D., Broekx, S., D'hondt, R., Engelen, G., Aertsens, J., and Goethals, P.L.M. 2013. A review of Bayesian belief networks in ecosystem service modelling. *Environmental Modelling and Software* 46, 1–11. doi:10.1016/j.envsoft.2013.03.011

28. Mastrandrea, M.D., Mach, K.J., Plattner, G.-K., Edenhofer, O., Stocker, T.F., Field, C. B., Ebi, K.L., et al. 2011. The IPCC AR5 guidance note on consistent treatment of uncertainties: A common approach across the working groups. *Climatic Change* 108(4), 675–691. doi:10.1007/s10584-011-0178-6

29. Risbey, J.S., and Kandlikar, M. 2007. Expressions of likelihood and confidence in the IPCC uncertainty assessment process. *Climatic Change* 85, 19–31.

30. Budescu, D.V., Broomell, S., and Por, H.-H. 2009. Improving communication of uncertainty in the reports of the intergovernmental panel on climate change. *Psychological Science* 20(3), 299–308. doi:10.1111/j.1467-9280.2009.02284.x.

Chapter 15

Negotiated Complexity in Ecosystem Services Science and Policy Making

Hans Keune[1,2,3,4], and Nicolas Dendoncker[5]

[1]Belgian Biodiversity Platform, [2]Research Institute for Nature and Forest (INBO), [3]Faculty of Applied Economics – University of Antwerp, [4]naXys, Namur Center for Complex Systems – University of Namur, [5]Department of Geography, University of Namur (UNamur). Namur Research Centre on Sustainable Development (NAGRIDD). Namur Centre for Complex Systems (naXys)

1. INTRODUCTION

It's a long way from scientific knowledge to concrete policy action. Along the way many decisions have to be made. A lot of these decisions relate to setting priorities: Given that one can do a lot but not everything, one has to

Ecosystem Services. http://dx.doi.org/10.1016/B978-0-12-419964-4.00015-9

choose. First, we have to define the policy options and ask ourselves questions such as:

- Which ecosystem services are relevant to a decision-making context?
- How do we deal with the incommensurability of policy options?
- Do we invest in biodiversity conservation or in one type of ecosystem service?
- How can we compare ecosystem services-related policy actions with other policy options: Do we invest in nature or employment or food or energy or…?
- How do we deal with differences of opinion among experts and among stake-holders?
- How do we deal with limited information on policy options as there are still many unknowns due to both ecological and social complexity?
- How do we design the decision-making process?
  - Top-down or bottom-up?
  - Complex or simple?
  - Pragmatic (taking into account limited complexity for practical reasons) or realistic (taking into account more complexity as to be more realistic in the approach)?
  - Do we consider this to be only a scientific affair?
  - To what extent do we consider this to be a topic for social debate?
- Which criteria do we choose to qualify our decisions?
  - Do we opt for a single-criterion approach? For example, monetizing every-thing and thereby taking into account only aspects that can be monetized, thus excluding other aspects that might be relevant?
  - Or do we opt for a diversity of criteria? For example, the combination of biophysical criteria (abundance/scarcity, trend, vulnerability), economic criteria (economic contribution to present/future society, costs to replace/rehabilitate the green capital of ecosystem services), social criteria (prefer-ences of different stakeholder groups, equity), public health impact, policy feasibility.

With regard to policy uptake of scientific knowledge on ecosystem services, we consider an integrated decision-making framework to be crucial. Framing com-plexity is a crucial aspect of any ecosystem services approach: How do we deal with ecological and social complexity? The complexity to be taken into account and the approach for dealing with that complexity are part of context-specific negotiation among actors involved in the process of investigation and interpreta-tion, and as such becomes negotiated complexity.

We will first briefly illustrate the complexities involved in ecosystem services and reflect on general complexity approaches and decision making. We then introduce the most important elements of the analytical deliberative multicriteria decision-support framework, which we propose for ecosystem services decision making. Finally, we will reflect on the practical development and use of such a framework in the field of environmental health in Belgium, and on the opportu-nities of implementing it for ecosystem services decision making in Belgium.

2. COMPLEXITY

2.1. The Inherent Complexity of Ecosystem Services

The relationship between the natural environment and humans is highly complex and still poorly understood [2]. In the case of ecosystem services, the complexity is partly caused by the interdisciplinary nature of the issues [3]: Both natural and social sciences have to be involved, and different subject areas need to be integrated. This interdisciplinary challenge is huge at the level of coupled human and natural systems [2], let alone in the young field of ecosystem services. Moreover, complexity causes the potential array of policy options to be diverse and difficult to objectify due to uncertainties, ambiguity, ignorance, and indeterminacy, which challenges the evidence base for ecosystem services management [4]. But the challenge is also transdisciplinary in nature as a new level of complexity [4] comes into play when interpreting knowledge for society, when linking to decision-making processes. We move here from "knowledge about" to deciding "what is important." This approach brings into play not only a diversity of societal sectors, as ecosystem services relate to different aspects of society and thus to various policy fields, but also a diversity of interests and stakes.

2.2. The Nature of Complexity

The magic word *complexity* has been buzzing around in science, policy, and society for quite some time now (e.g., [5, 6]). There seems to be a common feel for a "new way" of doing things, for overcoming the limits of tradition. The "new way" though is seen as two radically different things by two important schools of complexity thinking. Whereas the Santa Fé school [7, 8] merely believes that new scientific strategies in the face of complexity in the end will bring us closer to the modern scientific aim of ever more perfect knowledge and control, the critical complexity school [9–11] points out that limits of knowledge are inherent to complexity, necessitating reduction and critical reflection on the normative basis for any simplification.

Different definitions of complexity have been proposed and adopted in recent years [9, 12–19]. A common feature of many of them, however, is the notion that the complexity of a complex phenomenon cannot be fully described or understood due to the presence of a large number of (often simple) system components that interact in a manner that cannot be explained by the characteristics of the individual components themselves. Moreover, both the description and understanding of complexity are open for debate: "more than one description of a complex system is possible. Different descriptions will decompose the system in different ways. Different descriptions may also have different degrees of complexity" [12, p.257]. Many environmental issues are not only scientifically complex, but also of societal importance and socially complex resulting in "complex problems featuring high uncertainty, conflicting objectives, different forms of data and information, multi interests and perspectives, and the accounting for complex and evolving biophysical and socio-economic systems" [20]. Apart from defining, describing, and interpreting complexity, methodologically dealing with complexity is far from unambiguous.

2.3. Negotiated Complexity

How do we take this complexity into account in decision support? Do we acknowledge complexity in our approach, or do we drastically simplify and reduce it to relatively simple proportions? Decision-support methods have to potentially live up to quite a diversity of expectations, some of which are not easy to combine or reconcile. Moreover, the methodological approach seems open for debate, as neither crystal clear nor undisputed yardsticks for best practices seem to exist [21–25]. The challenge is not only to do justice to the complexity of many decision-making issues and processes, but also to do this as pragmatically as possible.

Deciding on a methodological approach is a topic for negotiation, and as such complexity becomes negotiated complexity. Already in 1989, Rosenhead [22] sketched the need for an alternative methodological paradigm for dealing with issues characterized by complexity, uncertainty and conflict. Almost simultaneously Funtowicz and Ravetz [26–28] presented their critique on normal science, pleading for a postnormal paradigm in cases when facts are uncertain, values in dispute, stakes high and decisions urgent, and do so by referring mainly to environmental issues. Funtowicz and Ravetz build upon the concept of normal science, as coined by Thomas Kuhn [29], and describe it as puzzle solving within a scientific paradigm that is not disputed as such, clearly stipulating how the scientific endeavor should be performed as to solve problems, or more in general defines the truth. The alternative of postnormal science, especially applicable to complex issues, focuses on aspects of problem solving that are often neglected in traditional normal science: uncertainty and values. Funtowicz and Ravetz plead for a wider involvement/participation of actors next to scientific experts and a more explicit account of scientific uncertainties.

Rosenhead [22] characterizes the dominant paradigm in the field of operation research (a field related to decisionsupport methods) as rational comprehensive planning, planning that is organized as a mainly centralized (top-down) expert activity, focusing on objective calculation of the best option for management or policy making. It will do so by identifying a single yardstick for comparing alternatives, collecting extensive datasets that will allow a scientific analysis in which uncertainty is abolished as much as possible, and prescribing convincingly and uniquely what is in the best interest regarding the issues at stake. Rosenhead (1989) points to extensive evidence stating a centralized expert method to have (limited) success with respect to issues of limited complexity, but does not work well regarding complex issues. Rosenhead follows in the footsteps of Lindblom and Cohen [21] who ten years earlier critically discussed the usability of social scientific knowledge for policy making and social problem solving in the face of complexity, and propagated a necessary alliance with ordinary knowledge.

Lindblom and Cohen [21] not only opened the visor to "other" knowledge, but also to "other" knowledge producers who have a role in the process of knowledge production. This tolerance for diversity is echoed in the alternative paradigm propagated by Rosenhead [22]. This means that there is openness for a diversity of solutions (alternatives) and for a diversity of yardsticks for assessing these alternatives that do not necessarily have to be commensurable. Compared to the traditional paradigm, in the alternative paradigm the process is more important than the collection of complete datasets or perfect (as possible) analysis of the data. Uncertainties and differences of opinion or even conflicts are accepted as a fact of complex life, and the active involvement of a diversity of relevant actors in a bottom-up approach is propagated. The importance of the process also means that the analytical part of the decision-support method (data and analysis) should be supportive of the process of interaction and judgment, and should not overload it with too much data and analytical detail. We summarize the traditional and alternative paradigms in Table 15-1 (after Rosenhead [22]):

Similar distinctions as in Table 15-1 are made by several other authors. Vatn [30] sketches more or less the same epistemological divide with respect to environmental appraisal. Vatn views environmental appraisal methods as institutional structures with "rules concerning a) who should participate and in which capacity, b) what is considered data and which form data should take, and c) rules about how a conclusion is reached. p.2207" Vatn distinguishes two distinct underlying assumptions regarding rationality. Individual rationality considers preferences to be individual and given, whereas social rationality stresses the context dependency and dynamic character. Similar to Rosenhead with respect to trade-offs, Vatn stresses the methodological importance of how one views value dimensions: from commensurable to incommensurable. Vatn also distinguishes different types

TABLE 15-1 Overview of key issues of traditional and alternative paradigms

Key issues	Traditional paradigms	Alternative paradigms
Problem formulation	A single objective, optimization; if multiple objectives, then trade-off on one scale	Alternative solutions on separate dimensions without trade-offs
Data demand	Overwhelming importance	Reduced importance
Judgment of data	Science & consensus orientated	Simple and transparent, clarifying terms of conflict
Role of actors	Passive objects	Active subjects
Decision making	Top down, single decision maker	Bottom up, multiple decision makers
Uncertainty	Try to abolish uncertainty	Accept uncertainty

of human interaction, ranging from instrumental/strategic behavior to communicative action. This may have methodological consequences for the involvement of different actors in the process. Finally, he discusses the character of the issues at stake: Does one consider these to be simple and fit for calculation and trade-off, or complex and in need of what he calls a forum type value articulating institution, or in other words, an issue for deliberation? In the field of decision-support methods (DSM), Marakas [24], like Rosenhead and Vatn, also distinguishes different views on rationality. He specifically points out that in the field of DSM the traditional rational actor model in which optimization will tell the decision maker what is the best of all possible solutions has more and more over the years been challenged by the bounded rationality paradigm. According to Marakas, many decisions are qualitative in nature and not fit for quantitative analysis. Moreover, the search for all possible solutions is too complex an endeavor, let alone being able to effectively compare them. Arnott and Pervan [31] distinguish positivist, interpretivist, and critical social science paradigms in their review of DSM literature; the first (positivist) being the more traditional approach; the latter two can be seen as exponents of an alternative paradigm. Arnott and Pervan show how the alternative paradigms seem to gain ground, especially in Europe: Whereas in the United States (1990–2004) more than 95% of all DSM papers can be considered positivistic, in Europe the figure is only some 56%.

3. ANALYTICAL DELIBERATIVE MULTICRITERIA DECISION SUPPORT

We propose to combine two decision-support approaches for dealing with the above-mentioned challenges of dealing with complexity. We hereby specifically focus on a more critical alternative approach to complexity. As such, we follow in the footsteps of criticasters such as Norgaard [32], Spangenberg and Settele [33], Haines-Young [3], and Fish [4] who criticize the paradigm of traditional monetary evaluation, which currently seems dominant in the field of ecosystem services.

3.1. An Analytical Deliberative Approach

The analytical deliberative approach proposed by Stern and Fineberg [34] offers a beautiful hybrid conceptual framework for this combined challenge of scientific analysis, expert elicitation, and social debate: "Analysis uses rigorous, replicable methods, evaluated under the agreed protocols of an expert community . . . to arrive at answers to factual questions. Deliberation is any informal or formal process for communication and collective consideration of issues. p.20" According to Chilvers [35], it is "one of the few evaluative frameworks providing a more symmetrical treatment of 'analytic-deliberative' processes, including how science is conducted and relates to participatory processes. p.161" Stern and Fineberg [34] consider deliberation not only of importance at the end of the pipeline, when analytical scientific results are available, but also at the start, when setting the research agenda or designing decision-making procedures, and at other relevant intermediate steps in the process. We do not need to take this framework as a prescriptive one: "Structuring an effective analytic-deliberative process for informing a risk decision is not a matter for a recipe. Every step involves judgment, and the right choices are situation dependent" [34, p.6].

Stirling [36, 37] points out that with regard to both the analytical and the deliberative aspects of decision-support methods, the political and institutional contexts, and the power structure within, largely influence the way the processes are organized and relate to governance. Public participation in issue framing and decision making is not in itself a guarantee of openness, but very much depends on how such processes are organized. Stirling [37] describes how the increasing acceptance of such participation in scientific and technological assessments, at national and European Union levels, has been to a large extent negated by the still closed, deterministic, and linear approaches to innovation and technological progress. The quality of these processes therefore depends not just on making the process more open, but also on genuine respect for what the process produces and for use of the outcomes. The distinction Stirling makes between opening up and closing down is of relevance to methodological choices and their outcomes in practice and, according to Stirling, in many ways transcends the importance of the distinction between deliberative and analytic approaches. We may add to this that opening up and closure can be seen as part of an iterative process in which discussion and decision making have their place and inform each other (see, e.g., the soft system methodology developed by Checkland [38] referred to in [39]. Closure need not necessarily be regarded as negative (Stirling, [37]); key is the question of who is allowed to be involved or whose views will be taken into account when deciding.

Haines-Young [3] advocates the use of analytical deliberative methods for dealing with nature-human complexity, but judges such approaches to be still underdeveloped, particularly regarding ecosystem services. Fish [4] points at several challenges in this respect, among which overcoming the dominant mainly technocratic approach to ecosystem services is especially vital. Without dismissing the importance of expertise, he pleads for a combination with what he calls interpretive complexity. Spangenberg and Settele [33] advocate a combination of multi-stakeholder approaches with multicriteria analysis. We will make the same connection next.

3.2. Multicriteria Group Decision Support

One of the weaknesses sometimes considered with respect to deliberative or participatory processes is the lack of structuring capacity regarding decision-making processes [40]. Linkov et al. [41] make the same point regarding the field of comparative risk assessment. This structuring capacity often is considered to be one of the main strengths of multicriteria decision analysis (MCDA). MCDA methods have been used extensively over the last decades for environmental issues [42–44]. MCDA is capable of simultaneously embracing, combining, and structuring often incommensurable diversity: diversity of information (such as different types of data, e.g., qualitative and quantitative data, as well as uncertainty), diversity of opinion (also among experts), diversity in actor perspectives (stakes), and diversity in assessment/decision-making criteria. By incommensurability we mean that these aspects do not share likewise measures that make comparison easy. Moreover, as MCDA structures this diversity of information, it is helpful to support deliberation. MCDA helps to pave the way to decision making and communication about the decisions taken, including the argumentations on which the decision was based, particularly by referring to the assessment criteria and different viewpoints of a diversity of actors involved.

According to Gamper and Turcanu [45], MCDA is deemed to overcome the shortcomings of traditional decision-support tools, such as cost-benefit (CBA) or cost-effectiveness analysis (CEA), as it is better fit for dealing with qualitative information and with uncertainties. This ia especially important regarding the support of complex decision problems, such as environmental or sustainability issues. Still Mendoza and Martins [46] emphasize that today many of the MCDA methods should be considered as hard, or consistent with the traditional and rational scientific management approach. Several authors [46, 47] plead for a more integrated approach to MCDA, connecting the analytical approach to a more qualitative soft approach in which social aspects and participatory elements can be included. For example, Proctor and Drechsler [40] applied such an integrated approach in the field of natural resource management, combining a deliberative method with multicriteria evaluation.

4. RELEVANCE FOR BELGIUM

4.1. Practical Application in Belgium: Environmental Health

Has this been done before in Belgium? In the field of environmental health science and policy making in Flanders over the last ten years, an analytical deliberative multicriteria decision-support framework was developed and applied in practice. This was undertaken in the framework of the Flemish Centre of Expertise for Environment and Health (CEH; http://www.milieu-en-gezondheid.be/English), which is funded and steered by the Flemish government. We will briefly introduce this practical application with respect to setting policy priorities [48] and research priorities [49]. We will also briefly discuss some practical lessons.

Between 2001 and 2011, human bio-monitoring research was carried out by the CEH, investigating the very complex relation between environmental pollution and human health by measuring pollutants and health effects in human beings, using biomarkers. The research resulted in an enormous amount of information on a diversity of environmental health issues relevant to policy makers. The main challenge this posed to both policy makers and scientists was how to set priorities, as not all issues could be dealt with by policy measures.

In preparing for the research outcomes, in parallel to the human bio-monitoring, policy makers and scientists discussed the development of an assessment framework. In the beginning, the policy makers asked for a cost-benefit analysis framework. This was heavily refuted by the scientists, as such an assessment would be scientifically unsound, especially regarding health risks. To illustrate this, the example of smoking was offered: From a cost-benefit analysis perspective, smoking should rather be stimulated by the government. Smokers will only get ill at the end of their working careers, most of them will not live long enough to benefit from pension funds, and they will die rather quickly due to long cancer, which will not be a heavy burden for health care. After intense discussion, the (only) monetary yardstick was dismissed and replaced by a multicriteria framework: The research results were to be assessed regarding health, policy, and social aspects [48]. Step by step, moreover, the belief that this would only have to be an expert affair, science being able to unambiguously and objectively decide which policy priorities are best, was challenged. Scientists appeared to be unable to do this, both due to scientific complexity, since specialized expertise is unable to assess issues outside their expert domain, and to policy interpretation entering the sociopolitical domain: Investigating what is happening (problem knowledge) is different from politically defining what ought to be done about it (problem solving). It was thus decided to combine *expert elicitation* (specialists judging aspects that belong to their expertise) with *stakeholder deliberation* (stakeholders discussing what ought to be done based on all information and the expert judgments). Later, the same approach was applied with respect to setting part of the research agenda of a second human bio-monitoring campaign [49].

We briefly sketch the most important steps taken in the practical application. First, in a *deliberative phase,* the decision-support *procedure* is defined, the diversity of decision-making *criteria* is chosen, and the decision *alternatives* that have to be prioritized are designated. Second, in an *analytical phase, desk research* (such as literature and data research) is performed in order to provide the different alternatives with *background information* concerning the different assessment criteria. The environmental and health information relevant to assess public health aspects is collected by natural scientists. The social scientists are responsible for policy-related and social aspects. Next, based on the desk research information, the alternatives are assessed in an *expert elicitation.* Experts on environment and health assess the public health criterion, policy experts assess the policy aspects, and social experts assess the social aspects. These assessments result in both quantitative information (priority rankings of alternatives on different criteria) and qualitative information (arguments, difference of opinion, uncertainties). The outcomes of the expert elicitation are processed in a *multicriteria decision analysis.*

Third, in a *deliberative phase,* the results of both desk research and expert consultation are discussed in a *stakeholder deliberation* that gives advice on the basis of all information: Different from specialized expertise, a societal view deals with the political question of deciding what is important considering all specific aspects together. The diversity of stakeholders that we opted for was inspired by the composition of advisory bodies in Flanders, such as the Flanders Social and Economic Council, the Flanders Health Council, and the Flanders Advisory Council on Environment and Nature. Because (except for the scientific experts) organizations with a focus on the health perspective seemed to be almost absent, environmental health professionals with field experience and contacts with local people, such as general practitioners and the Flemish network of local health and environmental experts, were also invited. Representatives of consumer organizations were also included because of the relevance of a consumer perspective. The procedure aims to result in well-informed *decision making* by the final decision-making body. Finally, in the *external communication* step, transparency was considered to be of the utmost importance, not only at the end when decisions were taken, but also considering important intermediate steps such as defining the procedure and stakeholder deliberation. At key moments in the process, decisions and steps were externally communicated through newsflashes and newsletters, and all steps that were made available on the website were extensively reported.

4.2. Opportunities for Ecosystem Services Decision Making in Belgium

Can an analytical deliberative multicriteria decision-support approach be applied to ecosystem services decision making in Belgium? We believe the ingredients are there. We specifically focus on the multicriteria and deliberative aspects.

Multicriteria

The UK National Ecosystem Assessment (http://uknea.unep-wcmc.org) is a good example of a multicriteria perspective on nature and ecosystem services. The main criteria taken into account for ecosystem services valuation are the economic, health, and social aspects. Can we take this as an example for Belgium? In Belgium information is available on the biophysical aspects of ecosystem services [50, 51]. With regard to the economic aspects of ecosystem services, some work is being done in Belgium: [52]. As for health aspects, some issues are of importance [53], and a diverse set of actors, including science, policy, and stakeholder groups, are interested in contributing [54]. Knowledge capacity is also under development with regard to the social aspects of ecosystem services.

The Analytical Deliberative Approach

As regards analytical deliberative approaches, we want to briefly illustrate the potential in the case of the Regional Master plan development project "De Wijers." The Wijers is located in the northeast of Flanders (Belgium). The project area covers approximately 20,000 ha and is spread out over seven municipalities. The area has diverse attractions: biodiversity, sub-urbanity, industry, highways, and recreational areas. In 2007, the Flemish Land Agency received a mandate from the Flemish minister responsible for Nature, Environment, and Culture to support the development of a Sustainable Master plan for The Wijers. The project focuses on sustainable development based on the approach of systems thinking of ecosystem services. The main objective of the project is to develop a sustainable ecological landscape. In addition to a participatory planning approach, which has to lead to a local supported master plan, the project focuses on cooperation between the different governmental levels (central, regional, and local) and policies (environment, rural development, nature conservation, integrated water policy, mobility and spatial planning, culture, and heritage policy), with the aim of achieving efficient and effective synergies. There are ideas to develop an analytical-deliberative multicriteria decision-support framework for policy making. It is thought that different policy options that may follow from the master plan can be assessed on the basis of a diversity of criteria (ecological, economic, social, policy) and that potentially both experts and stakeholders can be involved in this process.

Challenges Ahead

The idea to develop a deliberative MCA for managing seminatural areas using ES thinking to compare different alternatives seems promising. However, many challenges ahead remain that are somewhat different from those linked to environmental health. In our view, the main challenges relate to the inherently spatial and multiscale nature of ES thinking. Next we briefly outline these challenges in relation to the various phases of the deliberative MCA, as we ahve described for environmental health.

The first challenge lies in defining the scope and boundaries of the study. How do we deal with ES that are produced outside the study areas but affect stakeholders within the study area? What about the opposite, that is, ES that are produced (or destroyed) within the study area but affect stakeholders outside the study area? Should these ES also be studied as part of the analytical phase? Then should stakeholders outside the study area also take part in the study? Where should the limit be placed, and who should decide?

For example, what if, following a deliberative MCA, the chosen alternative leads to a decrease in provisioning services within the study area but, as a result, provisioning services (e.g., food or water) have to be imported to the study area and hence produced elsewhere, thereby changing ES provision elsewhere? More generally, which stakeholders should be involved in the study? Should stakeholders affected by change in ES be included as well as stakeholders who largely affect the provision of ES? This problem relates to distributional issues and to a decision on whose voice counts. How should the first deliberative phase be best designed, given these issues? Who should decide? And who should decide how to decide?

The second challenge relates more closely to the analytical phase. This challenge is linked to assessing ES locally within place-based studies while some changes are global (e.g., climate change), but they could have local impacts that are hard to define locally. This relates to the issue of thresholds and tipping points, which are difficult to predict. Even if data demand should not dominate and even if biophysical data do exist in Belgium, there is still a need for empirical data that will reduce uncertainty and facilitate decision making within deliberative MCA. For example, often expert judgment produces "biophysical-like" data (e.g., data directly linking land uses to ES production) that are used to make up for the lack of field data (see [55, 56]). Caution should be used when valuing ES using these data, and the precautionary principle should be respected (see [57]).

Finally, we would like to remind the reader that the ES approach should by no means be considered the only and best criterion to assess how to manage resources and (seminatural areas. It should rather be complementary to taking biodiversity into account (given the uncertainty still remaining as to how to best link the two concepts). If, however, the deliberative MCA framework is well designed, there is no reason to believe that it could not be applied to the broader field of landscape planning in addition to managing (semi-)natural areas.

5. CONCLUSION

Some will perhaps state that an analytical-deliberative multicriteria decision-support approach will needlessly complicate and slow down the process. An answer to such a critique is that we may expect increased quality and support by a diversity of actors in return. Moreover, the question remains whether we do have an alternative that is both realistic and pragmatic enough. As one Flemish policy representative put it when judging the approach for environment and health policy making in Flanders: "It looks rather complicated, but I cannot think of an alternative approach that better fits our ambitions and the challenges we have chosen."

REFERENCES

1. Keune, H., Springael, J., and De Keyser, W. 2013. Negotiated complexity: Framing multi-criteria decision support in environmental health practice. *American Journal of Operations Rsearch*—ISSN 2160-8830-3(2013), pp. 153–166 http://dx.doi.org/doi:10.4236/ajor.2013.31A015
2. Liu, J., Dietz, T., Carpenter, S.R., Folke, C., Alberti, M., Redman, C. L., et al. 2007. Complexity of coupled human and natural systems. *Science* 317(5844):1513–1516.
3. Haines-Young, R. 2011. Exploring ecosystem services issues across diverse knowledge domain using Bayesian Belief Networks. *Progress in Physical Geography* 35(5), 685–704.
4. Fish, R. 2011. Environmental decision making and an ecosystems approach: Some challenges from the perspective of social science. *Progress in Physical Geography* 35(5), 671–680.
5. Sardar, Z. 2010. Welcome to postnormal times. *Futures* 42, 435–444.
6. Zhu, Z. 2010. Knowledge of the natural and the social: How are they different and what do they have in common? *Knowledge Management Research and Practice* 8,173–188.
7. Kauffmann, S. 1995. *At Home in the Universe.* Oxford: Oxford University Press.
8. Holland, J.H. 1998. *Emergence: from Chaos to Order.* Oxford University Press, New York.
9. Cilliers, P. 1998. *Complexity and Postmodernism: Understanding Complex Systems.* London: Routledge.

10. Morin, E. 2008. *On Complexity*. New York: Perfect Paperbacks.
11. Kunneman, H. 2010. Ethical complexity. In *Complexity, Difference and Identity* (Cilliers, P., and Allen, R., eds.). Dordrecht: Springer Verlag.
12. Cilliers, P. 2005a. Complexity, Deconstruction and Relativism, in: *Theory, Culture & Society* 2005; 22; 255–267.
13. Cilliers, P. 2005b. Knowledge, limits and boundaries. *Futures* 37, 605–613.
14. Funtowicz, S.O., Martinez-Aler, J., Munda, G., and Ravetz, J.R. 1999. *Information Tools for Environmental Policy under Conditions of Complexity*, Environmental Issues Series 9, EEA, Copenhagen.
15. Byrne, D. 1998. *Complexity and the Social Sciences*. London: Routledge.
16. Urry, J. 2005. The complexity turn. *Theory, Culture and Society* 22, 1–14.
17. Richardson, K., Cilliers, P., and Lissack, M. 2001. Complexity science: A "grey" science for the "stuff in between." *Emergence* 3(2), 6–18.
18. Richardson, K. 2005. The hegemony of the physical sciences: An exploration in complexity thinking. *Futures* 37, 615–653.
19. Nowotny, H. 2005. The increase of complexity and its reduction: Emergent interfaces between the natural sciences, humanities and social sciences. *Theory, Culture and Society* 22, 15–31.
20. Wang, J., Jing, Y., Zhang, C., and Zhao, J. 2009. Review on multi-criteria decision analysis aid in sustainable energy decision-making. *Renewable and Sustainable Energy Reviews* 13, 2263–2278.
21. Lindblom, C.E., and Cohen, D.K. 1979. *Usable Knowledge: Social Science and Social Problem Solving*. New Haven; CT: Yale University Press.
22. Rosenhead, J. (ed.). 1989. *Rational Analysis for a Problematic World. Problem Structuring Methods for Complexity, Uncertainty and Conflict*. West Sussex, England: John Wiley & Sons.
23. Weiss, C.H. 1991. Policy research: Data, ideas, or arguments? In Wagner, P., Weiss C.H., Wittrock, B., and Wollmann, H., *Social Sciences and Modern States: National Experiments and Theoretical Crossroads*, pp. 307–332. Cambridge: Cambridge University Press.
24. Marakas, G.M. 1999. *Decision Support Systems in the 21st Century*. NJ: Prentice Hall Pearson Education. New Jersey.
25. Belton, V., and Stewart, T. 2002. *Muliple Criteria Decision Analysis: An Integrated Approach*. Dordrecht: Kluwer Academic.
26. Funtowicz, S.O., and Ravetz, J.R. 1990. *Uncertainty and Quality in Science for Policy*. Dordrecht: Kluwer Academic.
27. Funtowicz, S.O., and Ravetz, J.R. 1991. A new scientific methodology for global environmental issues. In *Ecological Economics: The Science and Management of Sustainability* (Costanza, R., ed.), pp. 137–152. New York: Columbia University Press.
28. Funtowicz, S.O., and Ravetz, J.R. 1994. The worth of a songbird: Ecological economics as a post-normal science. *Ecological Economics* 10(3),197–207.
29. Kuhn, T. 1962, *The Structure of Scientific Revolutions*. Chicago: University of Chicago Press.
30. Vatn, A. 2009. An institutional analysis of methods for environmental appraisal. *Ecological Economics* 68, 2207–2215.
31. Arnott, D., and Pervan, G. 2008. Eight key issues for the decision support systems discipline. *Decision Support Systems* 44, 657–672.
32. Norgaard, R.B. 2010. Ecosystem services: From eye-opening metaphor to complexity blinder. *Ecological Economics* 69(6), 1221–1227.
33. Spangenberg, J.H., and Settele, J. 2010. Precisely incorrect? Monetising the value of ecosystem services. *Ecological Complexity* 7, 327–337.
34. Stern, P., and Fineberg, H. (eds.). 1996. *Understanding Risk: Information Decisions in a Democratic Society*. Washington, DC: National Research Council, National Academy Press.

35. Chilvers, J. 2008. Deliberating competence theoretical and practitioner perspectives on effective participatory appraisal practice. *Science, Technology, and Human Values* 33(2), 155–185.

36. Stirling, A. 2006. Analysis, participation and power: Justification and closure in participatory multi-criteria analysis. *Land Use Policy* 23, 95–107.

37. Stirling, A. 2008. "Opening up" and "closing down": Power, participation, and pluralism in the social appraisal of technology. *Science, Technology and Human Values* 33, 262–294.

38. Checkland, P. 1981. *Systems Thinking, Systems Practice*. Chicester, UK: John Wiley & Sons.

39. Giampietro, M. 2003. *Multi-Scale Integrated Analysis of Agro-ecosystems*. Boca Raton, FL: CRC Press, 472 pp.

40. Proctor, W., and Drechsler, M. 2006. Deliberative multicriteria evaluation. *Environment and Planning C: Government and Policy* 24(2), 169–190

41. Linkov, I., Welle, P., Loney, D., Tkachuk, A., Canis, L., Kim, J.B., Bridges, and T. 2011. Use of multicriteria decision analysis to support weight of evidence evaluation. *Risk Analysis.* 31(8), 1211–1225

42. Kiker, G.A., Bridges, T.S., Varghese, A, Seager, T.P., and Linkov, I. Application of multicriteria decision analysis in environmental decision making. *Integrated Environmental Assessment and Management* 1(2), 95–108.

43. Steele, K., Carmel, Y., Jean Cross, J., and Wilcox, C. 2009. Uses and misuses of multicriteria decision analysis (MCDA) in environmental decision making. *Risk Analysis* 29(1).

44. Huang, I.B., Keisler, J., and Linkov, I. 2011. Multi-criteria decision analysis in environmental sciences: Ten years of applications and trends. *Science of the Total Environment* 409, 3578–3594.

45. Gamper, C.D., and Turcanu, C. 2007. On the governmental use of multi-criteria analysis. *Ecological Economics* 62, 298–307.

46. Mendoza, G.A., and Martins, H. 2006. Multi-criteria decision analysis in natural resource management: A critical review of methods and new modelling paradigms. *Forest Ecology and Management* 230, 1–22.

47. Munda, G. 2008. *Social Multi-Criteria Evaluation for a Sustainable Economy*. Berlin: Springer-Verlag.

48. Keune, H., Morrens, B., Springael, J., Loots, I., Koppen, G., Colles, A., van Campenhout, K., et al. 2009. Policy interpretation of human biomonitoring research results in Belgium: Priorities and complexity, politics and science. *Environmental Policy and Governance* 19(2), 115–129.

49. Keune, H., Morrens, B., Croes, K., Colles, A., Koppen, G., Springael, J., Loots, I, Van Campenhout, K., Chovanova, H., Schoeters, G., Nelen, V., Baeyens, W., and Van Larebeke, N. 2010. Open the research agenda: Participatory selection of hot spots for human biomonitoring research in Belgium. *Environmental Health* 9(33).

50. Chapter 18 Francis Turkelboom et al., this book.

51. Chapter 3 Jacobs et al., Biology, this book.

52. Chapter 2 Liekens et al., this book.

53. Chapter 16 Keune et al., Health, this book.

54. Bauler, T., Brosens, D., Cerulus, T., Chemay, F., Chevalier, C., et al. 2012. Policy Brief: The need for a Community of Practice on Biodiversity—Public Health in Belgium, Belgian Biodiversity Platform, http://www.biodiversity.be/health

55. Burkhard, B., Kroll, F., Müller, F., and Windhorst, W. 2009. Landscapes' capacities to provide ecosystem services—a concept for land-cover based assessments. *Landscape Online* 15, 1–22.

56. Schneiders, A., Van Daele, T., Van Landuyt, W., and Van Reeth, W. 2012. Biodiversity and ecosystem services: Complementary approaches for ecosystem management? *Ecological indicators* 21, 123–133. doi: 10.1016/j.ecolind.2011.06.021.

57. Dendoncker et al., this book.

Chapter 16

The Natural Relation between Biodiversity and Public Health: An Ecosystem Services Perspective*

Hans Keune,[1,2,3,4], Pim Martens,[5] Conor Kretsch,[6] and Anne-hélène Prieur-Richard[1,2,3,4,7]

[1]*Belgian Biodiversity Platform,* [2]*Research Institute for Nature and Forest (INBO),* [3]*Faculty of Applied Economics – University of Antwerp,* [4]*naXys, Namur Center for Complex Systems – University of Namur,* [5]*ICIS, Maastricht University,* [6]*COHAB,* [7]*DIVERSITAS, Reviewed by Francis Turkelboom (INBO)*

Chapter Outline

1. INTRODUCTION

"Human population health should be the central criterion, and is the best long-term indicator, of how we are managing the natural environment." according to McMichael [3]. The plea for human health as a sustainability indicator exemplifies the strategic importance of human health in terms of both ecosystem services and biodiversity conservation. "Sacred Cows and Sympathetic Squirrels" [4] beautifully illustrates the natural relation between biodiversity and human health. The

---

* This chapter builds on the Biodiversity – Public Health text that Hans Keune and Pim Martens prepared for Wittmer et al. (2012) and on the paper Keune H. et al. (2013)

Ecosystem Services. http://dx.doi.org/10.1016/B978-0-12-419964-4.00016-0

cows, sacred in India for religious reasons, also serve as a buffer to malaria mosquitoes as they receive many bites that would otherwise go to humans. The squirrels are known for a similar buffer effect as they help to prevent the spread of Lyme disease among humans. This type of buffer function regarding the spread of infectious diseases within the human population represents one type of important human health-related ecosystem service [1, 5, 6]. In this chapter, we focus on some examples of relationships between health, biodiversity, and ecosystem services (for a broader overview see Figure 16-1), and we briefly discuss the emerging Community of Practice on Biodiversity and Public Health in Belgium.

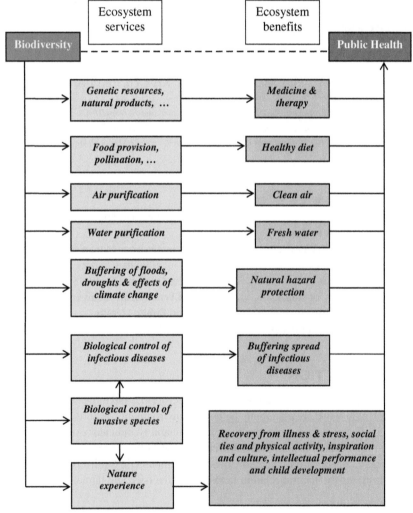

FIGURE 16-1 Overview relation between biodiversity and public health via an ecosystem services perspective *(developed by Hans Keune and Pim Martens for Wittmer et al., 2012)*

2. PUBLIC HEALTH-RELATED ECOSYSTEM SERVICES AND DISSERVICES

2.1. Diseases

One area of intense scrutiny is the relationship between biodiversity and infectious disease, and how ecosystem change and biodiversity loss may affect the ecology of disease organisms and the dynamics of pathogen–host interactions. The emergence and spread of certain pathogens from wildlife to livestock and/or humans, and related social and economic costs, have been well documented. These diseases include HIV, hanta virus, avian influenza, Lyme disease, malaria, dengue fever, leishmaniasis, nipah virus, and ebola [7–11]. Several studies have determined that biodiversity reduces the risk of infectious disease emergence or spread, while its loss or unsustainable exploitation can increase such risks (e.g., [12–14]). The dynamics, however, are complex and probably system-dependent, and high biodiversity does not necessarily reduce disease risk in all situations (e.g., [15, 16]). Nevertheless, land-use change and ecosystem disruption are well-recognized factors influencing the emergence of disease. In a study examining incidences of disease worldwide, Jones et al. [17] established that there has been an increase in disease emergence and an increasing prevalence of zoonotic diseases since 1940. They also established that areas of likely high risk for newly emerging infectious diseases correlated with areas of high biodiversity and with areas of high human population density. Another issue associated with population density is the significant health threat posed by antimicrobial resistance. Among the risk factors for emergence of resistance is pollution from agriculture and from urban areas. Drug-resistant organisms present in the environment can be picked up and carried by wildlife and by feral animals, potentially increasing human health risks in urban areas (e.g., [18, 19]). Biodiversity thus may increase or decrease the risk of spread of infectious diseases among humans and as such may produce both ecosystem services and disservices [10, 13, 20].

2.2. Quality of Food

The relation between biodiversity and food quality is reciprocal: Healthy food depends on biodiversity, and biodiversity depends on how humans organize food production [21]. Agro-biodiversity [22] influences the quality of food by providing a variety of food sources and supporting a diverse diet that is essential for human health [21, 23]. This diversity is under pressure by a decline of farming of traditional food plants, resulting in deficiencies of important nutrients in several regions. Agro-biodiversity also functions as insurance: When an important food source comes under pressure owing to climate change or infections, for example, agro-biodiversity (in opposition to intensive mono-agriculture) may provide alternatives. Furthermore, genetically diverse strains are less vulnerable to disease [24]. Genetic diversity also functions as a biotechnological resource for crop improvement. Intensification of food production by use of pesticides endangers both agro- and wild biodiversity and human health. Still, demand created by worldwide population growth is precipitating an increase in food production. To provide sufficient and healthy food in a sustainable manner presents a huge challenge. The role and effects of biodiversity demand careful consideration. Specific aspects of biodiversity supporting food production are pollinator biodiversity, soil biodiversity, and predators of pest species [23]. The role of pollinators for food production cannot be underestimated, and they are very sensitive to habitat loss, pesticides, and diseases. Soil biodiversity is important, for example, for maintenance of soil fertility, water storage, and carbon storage. Moreover, soil diversity is crucial for soil and water purification by means of pollutant degradation, and as such is also an important human health-related ecosystem service [25].

2.3. Medicine

Biodiversity is an important source for medicine, both traditional and modern, both today and in the future [26–29]. A large part of chemically designed and produced medicines have natural origins, and people in less developed areas depend heavily on medicines directly from nature [26, 28]. Examples are quinine for malaria and pilocarpine for glaucoma. As species are lost at a very rapid rate, so is the potential for future medicine, even though only a small proportion of total available species are analyzed for their medicinal potential. The same is unfortunately true for traditional or indigenous knowledge regarding natural medicines. It is not only plants that potentially contribute to medicine; the potential contribution of animal species should not be underestimated [29]. Osteoporosis prevention and treatment may benefit from further understanding of how denning bears do not seem to suffer from physical nonactivity during the hibernation period, which for other mammals (including humans) results in bone loss. Cone snail peptides, which the animals use to defend themselves or paralyze their prey, can be very effective for chronic pain treatment for humans. Squalamine, found in sharks, is being investigated for its potential antitumor activity. The blood of horseshoe crabs is capable of killing bacteria; study of the functioning of these antimicrobial peptides can be beneficial to designing more effective antibiotic therapies for humans.

2.4. Nature Experience

The natural relation between biodiversity and human health is perhaps best brought to life in the biophilia hypothesis: the innate human tendency to affiliate with nature [25, 30, 31]. This contact with nature has historically biological (genetically programmed) roots, but also represents beneficial health effects for the modern lifestyle: Studies have shown that living near nature enhances health and that contact with nature is beneficial to recovery from illness or stress, it stimulates social ties and physical activity; nature elements close to the work space are beneficial to productivity; and contact with nature stimulates intellectual performance and positively influences children's development (for an overview see [30,31]). The complexity and sensitivity of the relation are also described by the concept of biophobia [31, 32]: The relation between humans and nature is not necessarily only positive; historically, fear of dangers/risks of nature (predators, snakes, poisonous plants) was functional for survival. But nature was also a source for food and water and shelter. Both aspects illuminate human adaptive capabilities [32]. The relation also shows its dual face in the historical development of public health policies. Especially in Western countries in the 20th century, the positive aspects of nature in regards to human health have been mainly discarded. In cities the hygiene principle overshadowed the positive nature experience principle. The idea of coexistence of humans with nature being beneficial to human health slowly seems to be revitalized in current public health thinking. The special role of biodiversity in the nature experience literally is both mysterious and complex: The quality of nature in terms of biodiversity richness enhances the positive influence of nature on human health [31, 33].

2.5. Invasive Species and Diseases

Global travel, transport, and trade drive the introduction of new species to ecosystems. Most introduced species do not cause problems in their new surroundings (and mostly die off), but some of them become invasive and as such become disruptive to ecosystems. Invasive species may result in biodiversity decline, and the success of an invasion may be the result of a lack of different components of biodiversity in the invaded habitat. The impact of invasive biological species on the diversity of the above0mentioned human health-related ecosystem services (and disservices) has gained increasing importance during the past decades [34]. Invasive species may introduce diseases to other habitats, often leading to the dramatic emergence of infectious diseases [10], as well as cause injuries or allergies. Like all species, they may accumulate toxins (e.g., heavy metals) that may end up in human food and may contaminate soil or water. They may also hamper the nature experience, for example, by forming impenetrable stands or water coverage, or by causing the deterioration of the environmental and aesthetic quality of ecosystems [34].

2.6. Disaster Mitigation and Climate Change

Biodiversity and ecosystems can also help mitigate disaster, for example, in lowering the risk of floods [25] and droughts [35], and as such may decrease related public health risks. Climate change plays an important role here and does so also in relation to other of the above-mentioned human health-related ecosystem services (and disservices). Floods may result in water-borne diseases, and droughts can result in malnutrition, with both perhaps resulting in increased food prices [36]. Climate change can have an effect on infectious diseases [8]. For example, climate warming in coastal ecosystems can enhance the spread of cholera. Moreover, some vectors are sensitive to temperature: Owing to climate warming, the spread of mosquitoes and their activity may be enhanced, and so may facilitate the spread of some infectious diseases such as malaria. Extreme weather events such as heavy rainfall, flooding, and droughts may both increase and decrease the activities of vectors. Droughts, for instance, may lead to the formation of small nutrient rich pools favorable to mosquitoes spreading the West Nile virus [8]. The tripartite relation climate change–biodiversity–human health is still poorly understood due to its complexity. Existing studies illuminate the importance of monitoring and the need for better understanding these relationships [36].

3. THE EMERGING COMMUNITY OF PRACTICE ON BIODIVERSITY AND PUBLIC HEALTH IN BELGIUM

Research on the linkages between biodiversity and public health is an emerging issue that nevertheless has not received much concerted attention in Belgium to date. For this reason on November 30, 2011, the first Belgian Biodiversity and Public Health conference was organized by the Belgian Biodiversity Platform: http://www.biodiversity.be/health [37]. The meeting attracted 81 Belgian experts, 68% of whom were scientists (universities and governmental scientific institutes; health, ecological, and social science), 16% represented policy agencies (federal, regions, provinces, cities; health, environmental, nature, and land planning sectors), and the remainder consisted of consultants (policy advice, ecotherapy, education), persons involved in NGOs (nature protection, landscape development, ecological life, and gardening), and media. Apart from introductory keynote speeches, five workshops focused on the following topics: infectious diseases, food, nature experience, spatial tools, and ecosystem services. Discussions during the conference focused on priority scientific and policy challenges and resulted in the identification of several topical issues of priority interest. One outcome of these discussions was to consider the ecosystem services concept as an opportunity to strengthen the linkages between public health and biodiversity. To take this idea forward, preliminary steps were suggested, including (1) development of a catalogue of linkages between biodiversity and public health; (2) development of an overview of existing data and indicators; and (3) reinforcement of communication and collaboration among thematic experts and policy representatives.

A major outcome was a need for further capacity and network building, which was broadly supported by most participants. This will require structural scientific follow-up activities to adequately address the societal challenges related to the biodiversity and public health domain. Therefore, in February 2012, a policy brief was issued in which a variety of experts (science, policy, society) called for support for establishing a Belgian Community of Practice on Biodiversity and Public Health [38]. A Community of Practice is a network made up of individuals and organizations that share an interest and practice, and come together to address a specific challenge as well as to further each other's goals and objectives in a specific topic area [39, 40]. An interesting international example is the Canadian Community of Practice in Ecosystem Approaches to Health [41]. The aim of the Belgian Community of Practice on Biodiversity and Public Health is to build a strong network, to stimulate capacity building, and to produce an overview of the current state of Belgian knowledge capacity regarding the relation between biodiversity and public health. Furthermore, it seeks to facilitate responses by experts to the demands of policy makers and stakeholders regarding biodiversity and public health both in Belgium and at the international level (e.g., the Intergovernmental Platform on Biodiversity and Ecosystem Services [42]).

In 2012, the emerging Belgian Community of Practice on Biodiversity and Public Health decided to organize an inventory of research needs and ideas in order to obtain a clearer view of relevant research topics and the potential for collaboration [43]. The policy-driven research needs, expressed by diverse policy representatives (both national and regional), cover a wide range of topics and policy-relevant issues. There is a general interest in integrated data assessment that couples ecological and public health developments, as well as a general interest in the relations among green space/nature, the living environment, and public health issues. Some specific research topics involve health risks, others health benefits, and still others both. Like the ideas collected from members of the policy community, the research ideas collected from members of national, Wallonian, and Flemish research institutions represent a wide range of expertise, disciplines, and topics. Most of these topics are related to vector-borne diseases. Fewer proposals draw attention to ecosystem health services, for example, in relation to a diversity of habitats, landscapes, and species, urban greening, and the demand for ecosystem services and biodiversity.

REFERENCES

1. Wittmer, H., Berghöfer, A., Keune, H., Martens, P., Förster, J., and Almack, K. 2012. The value of nature for local development. In *The Economics of Ecosystems and Biodiversity in Local and Regional Policy and Management* (Wittmer, H., and Gundimeda, H., eds.), pp. 7–32. London: Routledge.
2. Keune, H., et al. 2013. Science–policy challenges for biodiversity, public health and urbanization: examples from Belgium. *Environmental Research Letters*, special issue Biodiversity, Human Health and Well-Being, http://iopscience.iop.org/1748-9326/8/2/025015/
3. McMichael, A.J. 2009. Human population health: Sentinel criterion of environmental sustainability. *Current Opinion in Environmental Sustainability* 1, 101–106.
4. Dobson, A., et al. 2006. Sacred cows and sympathetic squirrels: The importance of biological diversity to human health. *PLoS Medicine* 3(6), 714–718.
5. European Commission—DG Environment. 2011, October. Biodiversity and health. *Science for Environment Policy Future Briefs* (2). Bristol, UK: University of the West of England. http://ec.europa.eu/environment/integration/research/newsalert/future_briefs.htm

6. COHAB: Co-operation on Health and Biodiversity, http://www.cohabnet.org/
7. Peixoto, I., and Abramson, G. 2006 The effect of biodiversity on the hanta virus. *Epizootic Ecology* 87, 873–879.
8. Molyneux, D.H., et al. 2008. Ecosystem disturbance, biodiversity loss, and human infectious fisease. In *Sustaining Life. How Human Health Depends on Biodiversity* (Chivian, E., and Bernstein, A., eds.), pp. 287–323. New York: Oxford University Press.
9. Marsh Inc. 2008, *The Economic and Social Impact of Emerging Infectious Disease*. New York: Marsh Inc.
10. Thomas M.B., Lafferty, K.D. and Friedman C.S. 2009. Biodiversity and disease. In *Biodiversity Change and Human Health. From Ecosystem Services to Spread of Disease* (Sala, O.E., Meyerson, L.A., and Parmesan, C., eds.), pp. 229–244. Washington, DC: Island Press.
11. Lindgren, E., Andersson, Y., Suk J.E., Sudre, B. and Semenza, J.C. 2012 Monitoring EU emerging infectious disease risk due to climate change. *Science* 336(6080), 418–419.
12. Pongsiri, M., et al. 2009. Biodiversity loss affects global disease ecology. *Bioscience* 59, 945–954.
13. Keesing, F., et al. 2010. Impacts of biodiversity on the emergence and transmission of infectious diseases. *Nature* 468, 647–652.
14. Johnson P.T.J., Preston, D.L., Hoverman, J.T., and Richgels K.L.D. 2013 Biodiversity decreases disease through predictable changes in host community competence. *Nature* 494, 230–233.
15. Randolph, S.E., and Dobson, A.D. 2012. Pangloss revisited: A critique of the dilution effect and the biodiversity-buffers-disease paradigm. *Parasitology* 139, 847–863.
16. Kilpatrick, A.M., and Randolph, S.E. 2012. Drivers, dynamics and control of emerging vector-borne zoonotic infections. *Lancet* 380, 1946–1955.
17. Jones, K.E., Patel, N.G., Levy, M.A., Storeygard, A., Balk, D., Gittleman, J.L., and Daszak, P. 2008. Global trends in emerging infectious diseases. *Nature* 451, 990–993.
18. Radimersky T., Frolkova P., Janoszowska D, Dolejska M, Svec P, Roubalova E, Cikova P, Cizek, A., and Literak, I. 2010 Antibiotic resistance in faecal bacteria (*Escherichia coli, Enterococcus spp.*) in feral pigeons. *Journal of Applied Microbiology* 109, 1687–1695.
19. Allen, H.K., Donato, J., Huimi Wang, H., Cloud-Hansen, K.A., Davies, J., and Handelsman, J. 2010. Call of the wild: Antibiotic resistance genes in natural environments *Nature Reviews Microbiology* 8, 251–259.
20. Turkelboom et al., this book, Chapter 18.
21. Wilby, A., et al. 2009. Biodiversity, good provision, and human health. In *Biodiversity Change and Human Health: From Ecosystem Services to Spread of Disease* (Sala, O.E., Meyerson, L.A., and Parmesan, C., eds.), pp. 13–40. Washington, DC: Island Press.
22. Jackson, L., Bawa, K., Pascual, U., and Perrings, C. (2005). agroBIODIVERSITY: A new science agenda for biodiversity in support of sustainable agroecosystems. DIVERSITAS Report No. 4.
23. Hillel, D., and Rosenzweig, C. 2008. Biodiversity and food production. In *Sustaining Life. How Human Health Depends on Biodiversity* (Chivian, E., and Bernstein, A., eds.), pp. 325–381. New York: Oxford University Press.
24. Zhu, Y.Y., Chen, H.R., Fan, J.H., Wang, Y.Y., Li, Y., Chen, J.B., Fan, J.X., Yang, S.S., Hu,L.P., Leung, H, Mew, T.W., Teng, P.S., Wang, Z.H., and Mundt, C.C. 2000. Genetic diversity and disease control in rice. *Nature* 406(6797), 718–722.
25. Melillo, J., and Sala, O. 2008. Ecosystem services. In *Sustaining Life. How Human Health Depends on Biodiversity* (Chivian, E., and Bernstein, A., eds.), pp. 75–115. New York: Oxford University Press.

26. Newman, D. J. et al. 2008. Medicines from nature. In *Sustaining Life. How Human Health Depends on Biodiversity* (Chivian, E., and Bernstein, A., eds.), pp. 117–161. New York: Oxford University Press.

27. Chivian, E., and Bernstein. A. 2008. *Sustaining Life. How Human Health Depends on Biodiversity.* New York: Oxford University Press.

28. Cox, P.A. 2009. Biodiversity and the search for new medicine. In *Biodiversity Change and Human Health. From Ecosystem Services to Spread of Disease* (Sala, O.E., Meyerson, L.A., and Parmesan, C., eds.), pp. 269–280. Washington, DC: Island Press.

29. Chivian, E., and Bernstein, A. 2008. Threatened groups of organisms valuable to medicine. In *Sustaining Life. How Human Health Depends on Biodiversity* (Chivian, E., and Bernstein, A., eds.), pp. 203–283. New York: Oxford University Press.

30. Kellert, S.R. 2008. Biodiversity, quality of life, and evolutionary psychology. In *Biodiversity Change and Human Health. From Ecosystem Services to Spread of Disease* (Sala, O.E., Meyerson, L.A., and Parmesan, C., eds.), pp. 99–127. Washington, DC: Island Press.

31. Hartig, T., et al. 2011. Health benefits of nature experience: Psychological, social and cultural processes, In *Forests, Trees and Human Health* (Nilsson, K., et al., eds.), pp. 127–168. New York: Springer.

32. Ulrich, R.S. 1993. Biophilia, biophobia, and natural landscapes. In *The Biophilia Hypothesis* (Kellert, S.R., and Wilson, E.O., eds.), pp. 73–137. Washington, DC: Island Press.

33. Kaplan, R., and Kaplan, S. 1989. *The Experience of Nature: A Psychological Perspective,* Cambridge: Cambridge University Press.

34. Pysek, P., and Richardson, D.M. 2010. Invasive species, environmental change and management, and health, *Annual Review of Environmental Resources* 35, 25–55.

35. Chivian, E., and Bernstein, A. 2008. How is biodiversity threatened by human activity? In *Sustaining Life. How Human Health Depends on Biodiversity* (Chivian, E., and Bernstein, A. eds.), pp. 29–73. New York: Oxford University Press.

36. Parmesan, C., and Martens, P. 2009. Climate change, wildlife, and human health. In *Biodiversity Change and Human Health. From Ecosystem Services to Spread of Disease* (Sala, O.E., Meyerson, L.A., and Parmesan,C. eds.), pp. 245–266.

37. Keune, H., et al. 2012) Report of the 2011 Belgian Biodiversity—Public Health Conference, Belgian Biodiversity Platform Brussels www.biodiversity.be/files/1/4/3/1433.pdf

38. Bauler, T., et al. 2012, January-February. Policy Brief: The Need for a Community of Practice on Biodiversity—Public Health in Belgium, Belgian Biodiversity Platform, http://www.biodiversity.be/health

39. Wenger, E., and Snyder, W.M. 2000, January-February. Communities of practice: The organizational frontier. *Harvard Business Review*, 139–145.

40. Meessen, B., Kouanda, S., Musango, L, Richard, F., Ridde, V., and Soucat, A. 2011, August. Communities of practice: The missing link for knowledge management on implementation issues in low-income countries? *Tropical Medicine and International Health* 16(8), 1007–1014. doi:10.1111/j.1365-3156.2011.02794.x

41. COPEH: Canadian Community of Practice in Ecosystem Approaches to Health, http://www.copeh-canada.org/index_en.php

42. IPBES: Intergovernmental Platform on Biodiversity and Ecosystem Services, http://ipbes.net/

43. Keune, H., et al. 2012. Overview results of the inventory of scientific research needs and ideas biodiversity—public health. Belgian Community of Practice Biodiversity—Public Health, Belgian Biodiversity Platform Brussels, http://www.biodiversity.be/1996

Chapter 17

Global Trade Impacts on Biodiversity and Ecosystem Services

Alain Peeters

RHEA Research Centre rue Warichet, Natural Resources, Human Environment and Agronomy (RHEA)

Chapter Outline

1. INTRODUCTION

Global trade is not new. It has existed for centuries and probably for about 3000 years.[1] What has changed are its intensity, its growth rate, and the consequences for human societies, ecosystems and biodiversity in general.

---

1. In Ancient Greece, trade routes existed from Attica to the Indus Valley through the Middle East, the Red Sea, and the Persian Gulf. In Eurasia, the "spice route" developed from the first millennium BD (Before date = before the birth of christ)and the silk route [2] from the first century BD [3]. In the 1400 to 1800 period, the network developed considerably following the discovery of the Americas. It included the "triangular trades" of terrible memory between North America, Africa, and the Caribbean on one hand, and Europe, North America, and the Caribbean on the other hand [4].

Ecosystem Services. http://dx.doi.org/10.1016/B978-0-12-419964-4.00017-2

Prior to the 1970s, global trade was dominated by exchanges between North America, Western Europe and Japan. In addition, South–North flows of raw materials and North–South flows of finished goods were important between developed and developing countries. These conditions were a consequence of the colonial economy and of differences in development levels. From the 1970s, this situation changed. Industrial development took place in many developing countries in Latin America, India, Southeast Asia and East Asia. Many industrial processes which initially took place in developed countries, were relocated to regions characterized by lower production costs, notably lower labour costs. Global trade now includes significant flows of processed goods from developing to developed countries and between developing countries [1].

In the last 20 years, global trade has developed very rapidly among an increasing number of countries, mainly as a result of international agreements that have reduced trade barriers.

Supporters of free trade state that reducing trade barriers is environmentally friendly as a way of using resources more efficiently. They also consider that reducing trade barriers raises incomes and enables countries to spend more on protecting the environment.

Environmentalists argue that, unless accompanied by strict environmental regulations, growth induced by global trade will further deplete and degrade natural resources and destroy biodiversity. They support trade barriers and environmental restrictions in multilateral negotiations as a way of controlling excessive resource depletion and biodiversity decline, particularly in the food product trade.

This chapter introduces the challenges and analyzes some impacts of European standards of life on the ecosystem services of developing countries. It identifies research priorities and suggests some policy options for reducing the impacts of global trade on biodiversity and ecosystem services.

Imports of fossil fuels, metals and steel, chemicals and pharmaceutical products in the EU, although the most important in value (EUROSTAT, 2008), are not considered in this chapter.

The objective here is to present a synthesis of the impacts of a significant number of global trade activities in the areas of food, feed, fiber, and (bio-)fuel on ecosystem services and biodiversity in developing countries. The negative impacts of these trade activities are regularly cited in the media, particularly in specialized media, but their importance is not always precisely quantified. Moreover, they are usually discussed one by one, and there is no synthesis document that analyzes them all together. The relative importance of Europe in this global trade of commodities is also seldom described. This chapter aims to present all this information in a single and well-documented synthesis. Tourism is also considered because it can have a strong positive or negative impact on biodiversity and ecosystems.

2. GLOBAL TRADE AND ITS IMPACTS ON ECOSYSTEMS

2.1. The Importance of Global Trade and EU-27 Imports of Agricultural Commodities

On a global scale, the major food commodity exporters are the United States of America (USA), Brazil, Argentina, Indonesia, and Malaysia. The main products exported by these countries are soybeans and cake (USA: 16.5 Bn US$; Brazil: 16 Bn US$; Argentina: 8 Bn US$), palm oil (Indonesia: 10 Bn US$; Malaysia: 9 Bn US$), sugar (Brazil: 6 Bn US$), and cereals (USA, wheat: 5 Bn US$) (2009 data; FAOSTAT). China is a major importer of soybeans (20 Bn US$) and palm oil (4 Bn US$), and India is also an important importer of palm oil (4 Bn US$) (2009 data; FAOSTAT).

Oil, gas, and other forms of fossil fuels (428 Bn €) are by far the main imports of goods in the European Union (EU). Metals and steel (114 Bn €), chemicals (44 Bn €), and medicinal and pharmaceutical products (38 Bn €) come next. The main categories of agricultural commodity imports (Table 17-1) are vegetables and fruits, fish and other seafood, coffee, tea, cocoa, spices, feeding stuff for livestock, oils, oil seeds, and oleaginous fruits (7.5 Bn €). Cork and wood imports (6 Bn €) are similar to cereal imports (2008 data; Eurostat). Table 17-1 summarizes the most important categories of food and drink (in value) imported by the EU and their relative importance compared with total food and drink imports. This table also shows the rapid increase of imports in the EU.

In terms of volume, the United States and the European Union are the largest world exporters of agricultural commodities such as cereals and oilseeds (Table 17.2), followed by Argentina and Brazil. The EU and China are the largest importers (Table 17.3).

TABLE 17-1 Evolution of EU-27 Imports of Food and Drink, by Main Categories (in Mio €) between 2000 and 2008

	2000	2006	2007	2008
Total imports	54 823	67 922	75 576	80 203
Vegetables and fruit	13 813	18 604	20 495	20 703
Fish and crustaceans	11 716	15 822	16 124	16 082
Coffee, tea, cocoa, and spices	7 956	8 739	9 741	11 110
Feeding stuff for animals	5 569	6 000	7 011	8 732
Cereals and cereal preparations	2 199	2 597	5 218	6 345
Meat and meat preparations	3 099	4 698	5 027	5 194

Source: Gambini, 2009.

2.2. Impacts of Transport

The impacts of transport can be direct, indirect, or cumulative [1]. Global trade necessitates the development of transportation infrastructure and activities. Transportation activities have diverse impacts on the environment. They induce the emission of many pollutants such as GHG in the atmosphere and hydrocarbon in water. Many transport infrastructures have led to the destruction of coastal, wetland, or forest areas. Building routes often require draining land. Routes facilitate access by people to sensitive habitats and species territories. This increases wildlife disturbance and poaching. Transport is a vector for the dissemination of species that can become invasive in new territories.

TABLE 17-2 Exports of Selected Commodities by Large Exporters in Volume

Commodity	Exporters	2003 ('000 tonnes)	Share of world total 2001–2003 (%)
Cereals	United States	82 204	31
Cereals	European Union-15	54 772	22
Oilseeds	United States	29 005	41
Cereals	Argentina	19 584	8
Oilseeds	Brazil	15 978	20
Sugar	Brazil	13 852	26
Oilseeds	Argentina	6 634	9

Source: FAO, 2004.

TABLE 17-3 Imports of Selected Commodities by Large Importers in Volume

Commodity	Importers	2003 ('000 tonnes)	Share of world total (2001–2003) (% of category)
Cereals	European Union (15)	53 619	17
Oilseeds	European Union (15)	27 021	36
Oilseeds	China, Mainland	11 954	19
Meat	European Union (15)	8 683	35

Source: FAO, 2004.

Transport is also continuously increasing due to the rising number of international air passengers worldwide, from 88 million in 1972 to 700 million in 2000. Tourism now accounts for at least 5% of global CO_2 emissions [5].

2.3. General Impacts of Global Trade

Global trade can have both direct and indirect impacts on ecosystems (ex.: biodiversity, GHG emissions, deforestation, water quantity and quality). These impacts on ecosystem and ecosystem services should be estimated by objective methods.

Among the possible methods of estimation, the IPAT equation [6–8] describes impact as a linear result of population, affluence, and technology:

$$\text{Impact} = \text{human Population} * \text{Affluence} * \text{Technology}$$

where affluence = average consumption of each person in the population.

This equation assumes that, since population and affluence are growing rapidly, the only hope for reducing impacts is a progression in technology.

The Environmental Kuznets Curve is another model for estimating impacts on the environment, at a local scale. It is a loose U-shaped relationship between income and environmental quality [9, 10]. Since free international trade is expected to encourage economic development, global trade should finally encourage better environmental protection. Moreover, global trade usually involves a technology transfer from developed to less developed economies; developing countries should also progressively adopt cleaner technologies.

More recent research [11] has shown that the relationship is not always linear as it is in the IPAT equation. On the other hand, Bagliania et al. [12], for instance, were not able to confirm the Environmental Kuznets hypothesis in their study of 2001 ecological footprint data from 141 countries. There is a growing recognition of the complexity and the dynamism of processes, and of the importance of the interactions among social, ecological, and technological systems, leading to nonlinear, cross-scale dynamics and uncertainties [11].

The conclusions of the MEA [13] tried to take this complexity into account and to deliver a balanced and rather positive message: "Increased trade can accelerate degradation of ecosystem services in exporting countries if their policy, regulatory, and management systems are inadequate," "At the same time, international trade enables comparative advantages to be exploited and accelerates the diffusion of more efficient technologies and practices." "For example, the increased demand for forest products … can lead to more rapid degradation of forests in countries with poor systems of regulation and management, but can also stimulate a 'virtuous cycle' if the regulatory framework is sufficiently robust to prevent resource degradation while trade and profits increase."

3. CASE STUDIES OF ECONOMIC ACTIVITIES HAVING MAJOR IMPACTS ON ECOSYSTEMS

On the basis of the literature (e.g., [14]), the economic activities (production, marketing, transport, processing, and use) that are the most regularly cited as having major impacts on ecosystems are as follows:

- Soybean
- Palm oil
- Tropical timber
- Agro-fuels

- Flower and vegetable
- Shrimp and fish farming
- Fishing in oceans
- Tourism

The first two topics receive the most attention in this chapter.

3.1. Soybean

Over a longer period, between 1961 and 2008, feed imports have increased in the 27 countries of the present EU by about 400% (in tonnes) (FAOSTAT). The main increase occurred between 1961 and 1992 (500% of 1961 tonnage), after which a slightly downward trend in total imports was observed. Soy became the main product in feed imports (83% in 2008), showing a continuous increase. Imports of other commodities decreased. Net imports of soybean cakes (cakes imported as such and cakes resulting from the processing of imported soybeans) show an increase from 2.2 to 33.9 million tonnes between 1961 and 2008, with a maximum of 34.8 million tonnes in 2007. Brazil, Argentina, and the United Statesare the main export countries for the EU-27.

Expressed in N and P imported with feed products, the evolution of imports is even more important than in tonnes of products. N imports increased by 600% and P imports by 450%. Expressed in metabolizable energy (ME) and crude protein (CP), annual imports amounted to 490,973 TerraJoules ME and 17 million t CP in 2008. These amounts have been converted to equivalents of the total EU-27 grassland area on the basis of two annual grassland yield hypotheses (5 or 10 t DM/ha), and other assumptions related to the proportion and nutritive values of grazed or harvested (for hay or silage) forages. Feed imports are equivalent to between 7 and 14% of the total EU-27 grassland area on a ME basis, and between 17 and 35% on a CP basis (15 and own calculations). Imports of nitrogen and phosphorus have also negative impacts on the quality of water in the EU, inducing costs for water purification.

In a short-term study (1999–2008), Von Witzke and Noleppa [16] estimated that the EU imported the equivalent of 35 million ha of "virtual land" (land necessary for producing a given tonnage of commodity on the basis of regional yields) in 2007/2008, almost 40% above the value for the years 1999/2000. This area is equivalent to about twice the size of the Utilized Agricultural Area (UAA) of Germany. Feeding stuff imports from soybeans increased by 45% during that period. Soybean represented in 2008 the largest "virtual land" import in the EU, with 19 million ha or 39% of the total virtual land imports.

Soybean and soybean cake imports reveal the protein dependence of the EU for animal feeding. These imported feeds are largely used for pig and poultry feeding but also for supplementation of maize and other cereal-based protein-poor rations for dairy cow feeding and beef cattle fattening.

The expansion of soybean cropping in Argentina and Brazil is leading to deforestation, biodiversity losses (e.g., direct and indirect deforestation of the Amazonian and the Atlantic forests, with about 18.5 Mio ha destroyed in total), conversion of species-rich grasslands of the Pampa, the Campos, and the Cerrado in South America (about −51 Mio ha were destroyed in total), and GHG emissions [17]. The deforestation in the Amazonian Basin is well documented (e.g., [18, 19]) and so is not developed here. Currently, the Atlantic forest only covers 7.5% of its original extent, or 91,000 km^2 in 2010. The Cerrado covered about 1.7 million km^2 in the center of Brazil. It has been massively converted into agriculture in the last 20 years. Since 1970, more than half of the Cerrado's original expanse has been taken for agriculture. At the current rate of loss, the ecosystem could be gone by 2030, according to estimates by Conservation International [20]. The Campos grasslands covered more than 14 million ha in 1970 but, mostly due to conversion to cropping, only some 10.5 million ha remained in 1996 and it is estimated that only 6 million ha is currently left [21].

The Atlantic forest is now highly fragmented; more than 80% of the fragments are smaller than 50 ha each [22]. It contains 20,000 plant species and 1362 vertebrate species, of which 2.5% and 41% are endemic, respectively.

The Cerrado savannah contains 10,000 plant species and 1268 vertebrate species, of which, respectively, 44% and 9% are endemic [23], including emblematic mammal species such as the jaguar (*Panthera onca*), the maned wolf (*Chrysocyon brachyurus*), the giant anteater (*Myrmecophaga tridactyla*), and the giant armadillo (*Priodontes maximus*). It houses 137 threatened species.

The Campos is the habitat for 3000 vascular plants, 385 species of birds, and 90 terrestrial mammals. More than 50 forage species, 16 mammals, and 38 birds, among others, have been classified recently as threatened. Only 2.23% of the Campos surface is officially protected in seven conservation units. It has received little attention compared to other Brazilian biomes, and until now, its threatened status has not been sufficiently recognized [21].

Soy production is the most important economic activity that stimulates investments in transport infrastructures in Brazil. That explain show soy cropping is one of the most important drivers of change leading to deforestation of tropical forests by other economic activities: legal or illegal logging and extensive cattle farming [24, 25]. Crops including soy induce a much more fragmented landscape than pasture-dominated systems [26]. These land-use changes resulted in landscape fragmentation, loss of biodiversity, soil erosion, water pollution, and land degradation of otherwise biodiverse ecosystems.

3.2. Palm Oil

Palm oil is the world's most produced and internationally traded edible oil in 2012. On a total production of 35.6 Mio tonnes, India imports 18.8%, China 16%, the EU 13.8%, Pakistan 5.9%, Bangladesh 2.8%, and the United States 2.8% [27]. About 24% of global production is marketed on local markets in Indonesia, Malaysia, Nigeria, and Thailand. Demand is increasing in all countries. The European Union is a major importer, and imports were recently stimulated by mandating partial substitution of fossil fuels by biofuels for electricity generation and transport.

Palm oil production is concentrated in Indonesia (49% of global exports) and in Malaysia (40% of global exports). Colombia, Nigeria, Thailand, and other countries represent 11% of global production. In Malaysia and Indonesia, closed canopy oil palm plantations occupy 8.3 million ha: 2 Mio ha in Peninsular Malaysia; 2.4 Mio ha in Borneo; and 3.9 Mio ha in Sumatra [28].

The deforestation rate is very high, about 1.5% per year. It occurs everywhere but more rapidly in the lowlands, especially on peat lands (one-tenth of the plantations or ± 880,000 ha) [28, 29].

In Indonesia, palm oil plantations area grew by 11.8% annually from 1997 to 2000, and by 15.8% annually from 2000 to 2007. At a global level, tropical land occupied by palm oil plantations also grew rapidly: 1.55 Mio ha in 1980 and 12.2 Mio ha in 2009 (Figure 17-1). Indonesia plans to double palm oil output in the next ten years. The government encourages deforestation by designating vast tracks of forestland for conversion and issuing long-term leases to a small group of influential families. Selling timber harvested while clearing land generates the capital needed to establish palm oil plantations.

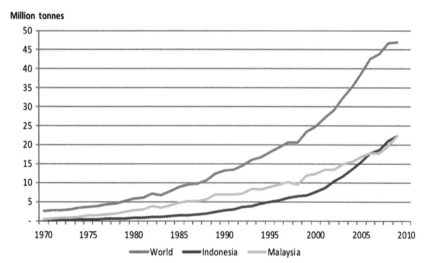

Million tonnes

FIGURE 17-1   Development of palm oil production (including palm kernel oil) (1970–2009)
Source: FAOSTAT and Kamphuis et al., 2010.

Indonesia and Malaysia are two of the main hotspots of biodiversity on Earth. Whereas the original extent of these hotspots was 1,501,063 km^2, the remaining part now covers only 100,571 km^2. Deforestation is thus inducing a very fast decline of biodiversity. There are 15,000 endemic plant species (30% in Borneo). Among 770 bird species, 150 species are endemic and 43 endemic species are threatened. Among 380 mammal species, 170 are endemic and 17 of the 136 genera are endemic, and 60 endemic species are threatened.

Most remaining forest areas are secondary forests, and it is these forests mainly that are now converted into palm oil plantations [29]. A large number of forest animal species are unable to survive within palm monocultures [30–32]; even species living near plantations suffer very negative effects. Regarding mammal species, Clay [33] reports that if 80 species can thrive in primary forests, 30 in disturbed forests, only 11 species can survive in palm oil plantations. Similar reductions of species numbers occur for insects, birds, reptiles, and soil microorganisms. Five mammals exemplify the disaster: the Sumatran Tiger, Sumatran and Bornean Orangutans, Asian Elephant, and Sumatran Rhinoceros. Each of these species is endangered [34]. They once flourished in those areas where rainforests have since been planted with oil palm [35]. Regarding birds, conversion of forest to palm oil plantations results in a reduction in species richness of at least 60%, with insectivores and frugivores suffering greater losses than more omnivorous species [36]. On the basis of a literature review, Kamphuis et al. [37] concluded that oil palm plantations host an average 33% of the number of species found in a primary forest. Plantations have also indirect effects on biodiversity through drainage of peat soils, use of fertilizers and pesticides, loss of connectivity between remaining forest patches, and simplification of food chains [37].

Deforestation for palm oil production is responsible for a significant part of global deforestation. Deforestation is responsible for about 20% of global greenhouse gas emissions. Above-ground biomass carbon of a peat-swamp forest is 179.7 ± 38.2 Mg/ha (±SEM), and that of oil palm is 24.2 ± 8.1 Mg/ ha. In addition, conversion of peat-swamp forest to oil palm would thus result in net carbon loss of 155.5 ± 39.2 Mg/ha. Conversion of peatswamp forest to oil palm would lead to net peat carbon emissions of 5.2 ± 1.1 Mg/ha·year^{-1} [28].

3.3. Tropical Timber

Tropical timber trade is growing rapidly at the global level. In the 2001–2007 period, the trade of wood and wood products almost doubled, reaching 120 billion USD. After 2007, the demand decreased because of the economic and financial crisis in Europe and the United States [37]. The main export countries are presented in Table 17-4 for the four types of timber products. Malaysia is the first exporter followed by two other Asiatic countries, Papua New Guinea and Indonesia. Ivory Coast and Gabon are the two main African players. Brazil is also an important exporter. Table 17-5 shows that Asia Pacific countries produce significantly more timber than African countries and that EU imports are relatively

TABLE 17-4 **Ranking on the Global Importance of Export Country for the Four Types of Tropical Wood Products in 2004**

	1st	2nd	3rd	4th	5th
Logs	Malaysia	Papua NG	Gabon	Myanmar	Rep. of Congo
Sawn	Malaysia	Indonesia	Brazil	Thailand	Cameroon
Veneer	Malaysia	Ivory Coast	Gabon	Brazil	Ghana
Plywood	Malaysia	Indonesia	Brazil	China	Belgium

Source: ITTO, 2005.

TABLE 17-5 **Evolution of EU Imports, Africa, Asia and Global Productions of Tropical Wood Timber (1000 m³) (2006–2010)**

	2006	2007	2008	2009	2010
EU Imports of Tropical Timber					
Logs	1 270	1 232	805	381	398
Sawn	2 490	2 580	2 099	1 295	1 245
Veneer	373	348	303	213	213
Plywood	1 343	1 339	1 329	868	864
Africa Production of Tropical Timber (ITTO Members)					
Logs	18 780	18 150	18 924	18 802	18 752
Sawn	4 720	4 676	4 663	4 615	4 545
Veneer	710	826	918	946	940
Plywood	434	422	452	430	400
Asia - Pacific Production of Tropical Timber (ITTO Members)					
Logs	83 796	87 484	89 322	86 975	83 269
Sawn	19 337	19 202	18 439	17 990	18 680
Veneer	1 473	1 635	1 905	1 714	1 753
Plywood	11 843	11 860	10 169	9 361	9 381
ITTO Total Production of Tropical Timber					
Logs	136 659	141 848	145 585	141 687	138 407
Sawn	43 518	43 402	43 467	42 415	43 280
Veneer	3 467	3 741	4 102	3 914	3 948
Plywood	19 883	19 951	17 847	18 247	18 293

Source: ITTO, 2011.

small compared to the total global production. The decreasing EU imports are partly related to the financial and housing crisis during that period. Malaysia, Indonesia, and Brazil are the main export countries for the European market.

Tropical wood production can have detrimental impacts on deforestation, forest structure, biodiversity, poaching intensity, soil fertility, and GHG emissions. Agricultural activities are the main drivers of these impacts on tropical forests. According to Food and Agriculture Organization (FAO) [38], the leading cause of deforestation is the conversion of forestland to agriculture and urbanization. The impact of wood production depends very much on forest management: clear felling, selective logging (conventional or reduced impact logging—RIL), or plantations. Biodiversity decreases from pristine forest to logged forest and to timber plantations.

Clear cuts result in massive loss of species and are often followed by agricultural expansion or timber plantations. In contrast, well-planned forest management has little environmental impact, especially if RIL techniques are applied. Conventional selective logging is based on a management plan that includes inventories, planning of logging over 25–30 years or more, and recovery time of the resources. A low level of extraction has minimum impact on tree diversity level [39]. A lot of commercial species are light-demanding; there are thus very few seedlings of these species in the understory of a mature forest [40]. Selective timber extraction induces better conditions for their regeneration.

Although primary forests contain the highest number of species, selectively logged forests (conventional selective cuts and RIL) are often hospitable habitats for many species, original species, or species adapted to disturbed habitats (RSR = 88%). In some cases, the populations of original species can even increase. For instance, for six hornbill species, densities are higher in secondary forest sites (30.9 birds/km^2) than in mature forest sites (20.2 birds/km^2) [41]. Elephants prefer clearings and secondary forests [42]. A peak of diversity could even occur after selective logging [43]. However, logged forests are poorer in canopy gleaning species and terrestrial rodents [44]. Secondary forests can host more understory plants, birds, and butterflies than primary forests [45]. Conventional selective logging can, however, create considerable damage, at least in the short term. Untouched forest fragments, representative of the pristine habitat, are very important for conserving biodiversity in managed forests. Timber plantations are as poor in species as oil palm plantations, for instance. Only a few ubiquist species can survive in these monocultures. These plantations often have indirect impacts on biodiversity because they may require soil drainage and pesticide use. The most detrimental indirect impact of all managed forests is the increase of poaching that is facilitated by transport infrastructure built for timber extraction.

Biodiversity levels and threats on Malaysia and Indonesia forests have been described in Section 3.2. In Indonesia, legal log production (in volume) is produced at 46% from natural forest and 54% from plantations. In natural forests, selective logging provides 56% and clear felling 34% of the log production (APHI 2009 in 37).

The net rate of deforestation in the Congo Basin is about 0.16% [46, 47]. Logging intensity is important in this Basin but variable among countries. It ranges from 36 to 65%. The Democratic Republic of Congo appears to be an exception, with the lowest logging intensity (12%) for the largest area of dense moist forest of the region [48]. About 32% of the Congo Basin forests are (temporarily) flooded forests. They contain a huge plant (> 10,000 vascular species) and animal diversity as well as flagship species such as gorillas, chimpanzees, bonobos, and elephants.

3.4. Agro-fuels

First-generation agro-fuels developed rapidly in the last decade, or even earlier in Brazil (since the early 1980s). The main feedstuff sources of biodiesel are soy and palm oil in developing countries and oilseed rape in temperate areas. The main sources of ethanol are sugarcane in developing countries, maize, other cereals, and sugar beet in Europe and the United States. Ethanol based on roots and tubers such as cassava is very low in developing countries. Other crops such as *Jatropha curcas* are developing slowly. The main producers are the United States, Brazil, and the EU. China, Argentina, India, Canada, and Malaysia are secondary producers (Table 17-6) [49]. There is a rapid expansion of agro-fuel feedstuff around the world, in Asia, Oceania, South and Central America, and Africa.

TABLE 17-6 Production (Million Liters) of Main Producers in 2010

Biodiesel		Ethanol	
Country	Million liters	Country	Million liters
EU-27	9 920	USA	48 470
Brazil	2 405	Brazil	26 720
Argentina	2 178	China	7 350
United States	953	EU-27	6 230
Malaysia	886	India	2 073
Australia	640	Canada	1 608
Canada	315	Ukraine	516
India	258	Australia	400
		South Africa	397
		Argentina	378

Source: OECD Stat: http://stats.oecd.org/viewhtml.aspx?QueryId=30104&vh=0000&vf=0&l&il=blank&lang=en

International trade in agro-fuels and agro-fuel feedstuff is modest compared to total global production. It is growing rapidly, though. Long-distance trade is becoming more important. Brazil dominates trade in ethanol, exporting, for instance, to Japan, the EU-27, and the United States. Argentina dominates the trade of biodiesel, followed by Malaysia and the United States (Table 17-7). The EU is a net importer of agro-fuels. Biomass trade is limited because it is more costly. However, Malaysia exports palm kernel shells to the EU. Biofuels produced in tropical regions from sugarcane and palm oil are considerably cheaper than fuels derived from agricultural crops in temperate zones. Biofuel shipping costs are a small proportion of the total value of the fuel itself [50].

There are plans for developing biofuel industries in developing countries on the basis of nonedible feedstock, mainly with jatropha. Competitive large-scale jatropha production does not currently exist, but that could change in the future, which would also require land and resources as use for edible feedstock. Malaysia and Indonesia still have high biofuel production capacities of biodiesel produced on the basis of palm oil. Current production in Malaysia accounts for approximately 45% of the available production capacity (about 1.75 billion liters in 2010). The available capacity of Indonesia is estimated at only about 10% of the current capacity (about 4 billion liters in 2010) [49].

European policies on biomass and biofuel are a strong incentive for developing biomass crops outside Europe. The EU has indeed a limited production capacity. The two main sources of biodiesel for the EU market are soybean oil from Argentina and Brazil, and palm oil from Malaysia and Indonesia. The main source of ethanol is Brazil. The impacts of agro-fuel feedstuff productions on ecosystems are thus similar to those related with soybean imports from Mercosur countries and with palm oil from Southeast Asia. In the 1960s, for

TABLE 17-7 Net Trade (Million Liters) of Main Exporters (Average 2008–2010)

Biodiesel		Ethanol	
Country	Million liters	Country	Million liters
Argentina	1329	Brazil	3502
Malaysia	559	S. Africa	291
USA	548	China	148
Indonesia	98	Thailand	73
		Argentina	63

Source: OECD and FAO. 2011.

instance, soybean cakes were a by-product of soy oil extraction. Nowadays, animal feed tend to become the main product, while oil is used as a source for biodiesel. Agro-fuel production produces an additional pressure on land and ecosystems that is already high because of the human population boom and the increase in living standards in emerging countries. Uncultivated land (forests, peat land, savannahs, and other grasslands) is under threat of conversion to arable land or plantations. That induces massive destruction of natural habitats and biodiversity loss. In Argentina and Brazil, the Pampa, the Campo, the Cerrado, the Atlantic forest, and the Amazonian forest, all species-rich habitats, are being massively converted to arable land, especially for soybean cropping. On cultivated land, polyculture and mixed cropping can be replaced by monoculture of palm, soy, or cane. Monocultures are often cropped on the best land, require intensive use of chemical inputs, and mechanization and are often based on the use of genetically modified organisms (GMOs) (soybean and corn) [51]. Agro-fuel production also has indirect effects by inducing displacement of crops. Sugarcane development in the west of Brazil was not only responsible for the Atlantic forest destruction, but it also pushed other crops to the Amazonian forest, leading to deforestation in this area. Agro-fuel crops necessarily replace something somewhere. If there are no direct effects, there are necessarily indirect effects. On marginal soils, agro-fuels reduce the area available for wildlife.

Environmental impacts are biodiversity loss, habitat destruction, soil fertility degradation, GHG emissions, especially CO_2 emission following peatland drainage, and deforestation and loss of organic carbon in soils, water pollution, and climate modification because of deforestation.

3.5. Flower Production

Kenya has become Europe's first supplier of fresh-cut flowers. Flower production is concentrated around Lake Naivasha. This flower industry began in the 1980s and grew rapidly in the 1990s. Now there are many large farms around the lake. Flowers can be cut in the morning and be in European flower shops by the evening. The average growth of the sector is very high: 20% per annum. This profitable export business has created jobs in the area and has brought wealth to Kenya. About 50,000 jobs have been created on the flower farms, but salaries are low and working conditions difficult, generating social tensions. This economic activity attracted a new population: In 1969 there were 7000 people living around the lake; in 2008 there were 300,000.

This fast economic growth induced environmental impacts, among which deforestation is a major problem. Trees are cut down for firewood and are not replaced. In addition, farms rely on the lake for irrigation, leading to overuse of water and pollution of the lake. Water levels have dropped by three meters. Lake Naivasha has shrunk to half its original size. Water bird and hippo (*Hippopotamus amphibius*) populations are threatened by habitat changes and pollution in the lake. Fish catches are dwindling, putting local fishermen out of business. Water shortage is also a concern for farmers and the Kenyan economy [52].

3.6. Shrimp Farming

Ecuador (14%), India (13%), Argentina (12%), Bangladesh (9%), Thailand (7%), China (6%), Greenland (4%), Indonesia (4%), Vietnam (2%), and Madagascar (2%) are the ten main countries from which the EU imported frozen shrimps in 2009. Frozen shrimps and prawns accounted for 25% of all EU imports of fish and fishery products in 2009 [53].

Shrimp farming began its development in Southeast Asia, in traditional extensive earthen ponds that depended on tidal water exchange for natural feed supply and maintenance of water quality. The system evolved by increasing stocking rates. Intensive systems are supported by feed inputs (pellets derived from fish and cereals) and water management for increasing yields. It can produce 50 times the production from traditional ponds. The development boom of the 1970s and 1980s was followed by busts in the late 1980s and early 1990s. These busts were related mainly to pollution and disease problems, as well as to other ecological effects of shrimp farming (e.g., loss of mangrove ecosystems, nutrient enrichment, and eutrophication of coastal waters, persistence of chemicals and toxicity to nontarget species, development of antibiotic resistance, and introduction of exotic species) [54]. Shrimp farming became particularly important in Thailand, Indonesia, Vietnam, India, South America, Nigeria, and Madagascar. It is often developed in biologically rich mangrove forests and estuaries where it causes pollution and depletes wild fish stocks. Mangrove destruction increases the vulnerability of coastal regions in relation to storm damage and erosion. It destroys the breeding habitats of wild fish, other aquatic species, and birds, including loss of critical spawning and nursery areas of fish and shellfish. Culture ponds for shrimp and fish accounted for the destruction of 20–50% of mangroves worldwide in recent decades [54].

In addition to coastal protection, mangroves provide other ecosystem services such as reduction of shoreline and riverbank erosion, stabilizing sediments and absorption of pollutants. Traditionally, mangroves also provided timber and wild fish. When comparing all ecosystem services provided by mangroves and those provided by shrimp farming, surprising results appear [55]. When only marketed forestry and fishery products are taken into account in valuation efforts, low values (about US$ 100 ha^{-1} year^{-1} in 1990) of mangrove systems are obtained. The traditional subsistence use of fuel wood, medicines, and food increases this value tenfold per household in certain areas of Thailand. When complete systems are considered, mangrove values are comparable with intensive shrimp culture net profits. A larger cost-benefit analysis was calculated on the basis of the whole mangrove ecosystem in Fiji Islands. This study included the interactions of ecological, economic, and institutional factors. It concluded that mangrove reclamation for shrimp farming or rice cropping would not give positive returns. Shrimp farming gives about 30 times the profit of rice farming per ha, but it has adverse effects on food security through (a) loss of surface available for rice cropping by pond conversion or salinization, (b) shifting of culture ponds from milkfish and other domestic food crops to shrimp, and (c) declining coastal fish, crustacean, and mollusc catches associated with mangrove deforestation. The life span of intensive ponds does not exceed 5 to 10 years because of problems linked with self-pollution and disease. In some cases, land is abandoned, and the sterile lands are no longer available for agriculture or aquaculture. These abandoned areas do not appear in worldwide estimates of the land appropriation by shrimp farms. Intensive water use of shrimp culture may draw from freshwater aquifers and reduce the supply of domestic and agricultural water, aside from causing seawater contamination by wastewater, especially of depleted aquifers [54].

In addition to the surface of ponds, other surfaces are required to support production of food inputs, nursery areas, and clean water to semi-intensive shrimp farming [56]. Ecosystem support areas were estimated for six different systems in Colombia. The marine ecosystem support area, for instance, was calculated as the sea-surface area required to produce the fish compressed in the pellets fed to the shrimps. Other support areas include the mangrove postlarval nursery area, areas of mangrove detritus for food, and the part of the lagoon ecosystem providing clean water. Calculations suggest that, to be sustained, a semi-intensive shrimp farm needs a spatial ecosystem support that is 35–190 times larger than the surface area of the farm. All the support areas must maintain good conservation status to ensure sustained production [54].

Similarly, fish farming in cages is dependent on marine ecosystem areas as large as 10,000–50,000 times the area of the cages for producing its food, and 100–200 times for processing parts of its waste. These areas far beyond the site of farms must be integrated in the evaluation of impacts [56].

3.7. Fishing in Oceans

According to the FAO [57] (SOFIA report), over 25% of all the world's fish stocks are either overexploited or depleted. Another 52% is fully exploited and threatened to be overexploited. Thus a total of almost 80% of the world's fisheries are fully to over-exploited, depleted, or in a state of collapse. That induces species losses and ecosystem degradation. Most stocks of large predatory fish stocks have already disappeared. This situation is mainly due to open-access policies to fishery territories and subsidy-driven overcapitalization in fishery boats and other equipment [58].

The EU imported (excluding intra-EU trade) 23.9 Bn US$ from non-EU suppliers in 2010. This makes the EU the largest market in the world, with about 28% of the value of world imports [59]. The six main exporters of fish and fishery products to the EU in 2009 are presented in Table 17-8.

TABLE 17-8 The Six main Exporters of Fish and Fishery Products to the EU in 2009

2009 ranking 2007	Country	Imported fish in value (€ million)	% of total imports of fish
1 (1)	Norway	2861	19%
2 (2)	China	1287	8%
3 (3)	Iceland	916	6%
4 (6)	Vietnam	780	5%
5 (5)	Morocco	777	5%
6 (7)	Thailand	739	5%

Source: European Commission, 2010b.

Overfishing progressively eliminates the largest and the most valuable fish stocks. Fishermen react by fishing down the food web, which usually means fishing smaller, shorter-lived, plankton-eating species (ex.: squid, mackerel, and sardines, and invertebrates such as oysters, mussels and shrimp) instead of larger, longer-lived predators (ex.: tuna, cod, snapper). This trend is particularly marked in northern temperate areas where fisheries are most developed. Below a certain trophic level, which may vary between ecosystems, decreases in trophic levels lead to decreasing catches [58]. The increasing efforts needed to catch commercial value fishes induces marine mammals, turtles, seabirds, and noncommercially viable fish species to be killed as by-catch and discarded (up to 80% of the catch for certain fisheries). The functioning of ocean ecosystems is thus severely affected.

Another consequence of overfishing is that a valuable food source for many human populations is declining, which can have strong social, economical, and dietary implications.

The most famous example of the consequences of overfishing is found in Newfoundland (Canada) where the cod fishing industry suddenly collapsed in 1992. That resulted in almost 40,000 people losing their jobs [60, 61]. Twenty years after the collapse, cod populations have still not been restored.Another example is the collapse of the Peruvian anchoveta fishery in 1972–1973 [62].

3.8. Tourism

The impacts of European societies on remote ecosystems are usually related to imports of ecosystem goods to Europe. Regarding tourism, ecosystem services are consumed locally by European citizens moving toward these remote ecosystems in quest of the sun but also of other ecosystem goods.

Tourism can produce many types of adverse environmental effects. Some impacts are linked with the building of infrastructure (e.g., roads, airports, hotels, restaurants, golf courses, and marinas). The tourism industry generally overuses water resources for hotels, swimming pools, golf courses, and gardens. This can result in water shortages and water pollution. Tourism can create great pressure on other local resources such as energy, food, soil, and biodiversity that locally may already be in short supply or threatened. Tourism can cause the same forms of pollution as any other industry: air emissions, noise, solid waste and littering, release of sewage, oil and chemicals, and visual pollution. It can degrade sensitive ecosystems that offer attractive landscapes such as sandy beaches, lakes, riversides, and mountain tops and slopes. In particular, coastal wetlands are often drained and filled for construction of tourism facilities and infrastructure. Infrastructures are often extended at the expense of coral reefs, mangroves, and hinterland forests, leading to coast erosion, destruction of habitats, and depletion of fish stocks. Coral reefs often suffer from trampling by tourists and divers who may break coral structures, ship groundings, pollution from sewage, overfishing, and fishing with poisons and explosives that destroy coral habitat. Among the 109 countries with coral reefs, in 90 of them coral reefs are being damaged. The negative impacts of tourism can gradually destroy the environmental resources on which it depends [5, 63].

On the positive side, tourism has the potential to create beneficial effects on environmental protection and conservation. It can raise awareness of tourists by providing them with environmental information, notably on the consequences of their actions. It can contribute to the conservation of natural areas and serve as a tool to finance their protection. Revenue from park-entrance fees and similar sources can be used to pay for the protection and management of protected areas. User fees, income taxes, taxes on sales or rental of recreation equipment, and license fees for activities such as hunting and fishing can provide funds needed to manage natural resources. These funds can be used for paying park ranger salaries and park maintenance, for example. They can also be used for improving incomes of surrounding populations that may more easily accept the hunting and poaching ban if they receive support for farming and animal husbandry. Tourism can create jobs and provide alternatives to illegal timber extraction, hunting, and poaching, or slash-and-burn agriculture that may have great environmental impacts. Regulatory measures for tourism activities also help to offset negative impacts [64, 65]. Tourism has often had a positive effect on wildlife conservation efforts in national parks and wildlife parks, notably in Africa but also in South America, Asia, Australia, and the South Pacific. In the Great Lakes region of Africa, mountain gorillas provide revenue from ape-related tourism. A gorilla tracking permit, which costs US$ 250, has been established. Three gorilla groups of about 38 individuals were habituated to human presence. In total, they can generate over US$ 3 million in revenue per year, making each individual worth nearly US$ 90,000 a year to Uganda (UNEP Great Apes Survival Project and Discovery Initiatives: http://www.unep.org/grasp/).

Wildlife conservation has thus been successful in a great number of cases of protected areas, when it comes to conservation and the appeal to global tourism (with the most symbolic animals, who in the case of the Serengeti were sometimes imported and did not necessarily represent the marketed "pristine environment"). This can also have a cost of ES provision to the local people, who are–often violently–denied access to their lands. Successes are at least to be considered with more attention to trade-offs. Production of value by and for people living outside of protected areas is not always satisfactory or sufficient. The categorizations of protected areas within national park can also have subtle but profound local social effects on the representation of nature and environment by local people excluded from these areas or required to change their practices and their thinking system [66, 67]. Adams and Hutton [68] argue that key political issues have to be taken into account in contemporary international conservation policy: the rights of indigenous people and the relationship between conservation of biodiversity and reduction of poverty.

4. CONCLUSIONS AND RECOMMENDATIONS

The case studies described in this chapter show the many threats but also some of the opportunities of global trade for ecosystem services and biodiversity. It is not likely that global trade will decrease in the medium term. The challenge thus consists in defining the trade-off between opportunities and threats.

Since economic growth requires the use of increased quantities of fossil energy and natural resources, including land, natural, and seminatural habitats, global trade often has adverse environmental consequences. The pressure on the environment is all the more important in less developed countries, which lack regulations concerning the protection of this environment.

European citizens are sometimes aware of global environmental problems linked with global trade and would like to improve the impact of their way of life on tropical ecosystems. That would open perspectives for a labeling system certifying low impacts on the environment. However, European consumers are still mainly looking for cheap food, flowers, clothes, and other commodities. Labeled products are likely to be more expensive. The integration of the value of ecosystem services in commodity prices and a sustainable management of ecosystems and biodiversity has a cost. These two contrasted attitudes of European consumers are thus not compatible and limit market opportunities of labeled products. However, there is some room for these products, as shown by the relative success of "fair-trade" labels.

The problem can be tackled in two ways: from the point of view of citizens of less developed countries and from the perspective of the EU.

Poor people are more concerned about how to survive in the short term than about environmental protection. The demand for environmental protection increases as living standards improve, at least in developed countries. This effect is described by the Environmental Kuznets Curve [9, 10]. The optimistic conclusions of this statement and this curve can be applied to pollution, but they do not take into account the fact that, during the environmental degradation phase, high-nature value habitats and biodiversity can be irreversibly destroyed.

EU citizens often do not perceive that their consumption patterns may contribute to ecosystem degradation in remote countries. For instance, they do not understand that the consumption of shrimps causes mangrove destruction and may result in decreased fish reproduction, which can affect local subsistence economies and economies of other countries [69]. An ecological footprint labeling could help to improve information. Even more importantly, ecosystem value degradation should be included in prices for sending a strong signal to consumers [70]. Consumer health can also be a strong argument that can orient their consumption decisions. The case studies on soybean and palm oil imports in the EU shows a clear link between ecosystem destruction and biodiversity loss on one hand and risk for human health on the other hand.

4.1. The Case of Soy

The case of soy is discussed here briefly to illustrate policies that could be undertaken to decrease the negative impacts of global feedstuff trade on biodiversity and ecosystem services.

Protein independence and soybean import reduction in the EU are desirable for strategic, environmental, and economic (including farmer's income) reasons. They could be achieved by a combination of several means in the EU:

- decrease of meat consumption (in the EU; total per-capita protein consumption is about 70% higher than recommended, and average intake of saturated fatty acids is about 40% higher);
- food waste reductions and recycling;
- partial shift from pork and poultry meat to sheep and beef meat[2];
- supports for larger use of grasslands;
- development of annual and perennial legume crops (pulse and forage) in the EU;
- substitution of maize and soybean by grass feeding for ruminants.

Reduction of food wastes can also have an impact on the EU land structure by increasing the services provided by grassland ecosystems.

Westhoek et al. [71] identified three strategies for the EU livestock sector: (1) increasing resource efficiency, (2) consuming less or different animal products, and (3) producing with fewer local impacts. Present policies are encouraging for the first strategy but with regard to the second strategy, policies are practically nonexistent, and with respect to the third strategy, policies are usually secondary to free-market policies. The CAP could be an efficient tool for orienting the livestock sector in the right direction.

---

2. Imported feedstuff is mostly used by the pork and poultry industry.

In Brazil, original initiatives have been taken. A soy moratorium (2006 to 2011) on the purchase of soy from illegally logged forests in the Amazon between the Brazilian government, international soy traders, and Greenpeace has been negotiated. Its objective is to avoid selling soy that is cultivated on land in the Amazon Biome deforested after July 2006. ABIOVE, the Association of the Brazilian Vegetable Oil Industry, and ANEC, the Association of Brazilian Cereal and Oilseeds Exporters, claimed that the measures aimed at preventing lodging in the tropical forest areas were successful. Based on field observations, NGOs stated that deforestation was reduced by 59% compared to the previous year. The moratorium allows the Brazilian Ministry of Environment together with NGOs and the soy industry to map protected areas precisely, to establish an adequate monitoring system, and to elaborate rules on land-use rights in the Amazon. The end of the moratorium will reveal the government's willingness and aptitude to impose rules in the Amazon region.

The Task Force Sustainable Soy has been established. In 2008, the Task Force counted 19 members: private companies including Unilever, FEDIOL (EU Vegetable Oil and Proteinmeal Industry), and FEFAC (European Feed Manufacturers' Federation). This Task Force frequently meets and consults NGOs (for example, the Soy Coalition), producer organizations, EU Ministry representatives, and banks. In addition, the international Round Table on Responsible Soy (RTRS) is a platform that defines pragmatic and suitable solutions for enhancing sustainable soybean production. The Task Force supports the RTRS positions [25].

These initiatives encourage more sustainable solutions, but they cannot stop deforestation and do not address the heart of the problem. They can thus influence the trend but they cannot fundamentally change or reverse it.

4.2. Research Priorities

Research can help provide a better understanding of the following topics and mechanisms:

Description and understanding of fluxes
- description (understanding) of commodity (mass of material and chemical compounds) and energy fluxes (ex.: [72, 73]);
- translation of fluxes of commodities from agriculture, forestry, and fisheries into a virtual flow of land, freshwater, and marine ecosystems (ex.: [14, 74]);
- translation of these fluxes into CO_2 (ex.: [75]), energy, and nutrient fluxes.

Managing trade-offs and synergies (after [55])
- identification of empirical patterns of trade-offs and synergies among sets of ecosystem services;
- identification of the most effective ways to mitigate trade-offs or enhance the synergism of ecosystem services;

description of the intensity of the relationships between ecosystem services;

determination of the evolution of strength of the relationships over time, management, and across scales.

Understanding regulating services and regime shifts (after [55])

determination of shifts in ecosystem services that can occur during regime shifts;

assessment of the speed of these shifts;

description of the variability of ecosystem service provision change with declines in regulating services;

determination of the point before regime shifts in the provision of ecosystem services.

Description and understanding of the consequences of global trade

assessment of the impacts on ES and biodiversity in developing countries and in Europe;

assessment of the impacts on human health.

Design of alternative solutions and tools that can be used to promote alternatives

identification of drivers and actors of economic activities;

testing scenarios of alternative solutions (modeling);

development of pilot projects of more sustainable systems in developing countries and in Europe;

design of accounting tools (ex.: Life Cycle Assessment and Green National Accounts) dealing with international aspects of ecosystem services is essential for the integration of ES criteria in society functioning (ex.: [14, 74]).

This list of research topics is not meant to be exhaustive.

4.3. Policy Aspects

Answers to the biodiversity issues generated by global trade are multiple. The questions of global governance and the judicious use of technical trade barriers are crucial, however. Tariffs on agricultural commodities have been progressively reduced in the last 50 years. These tariff reductions are sometimes very important. For example, there are no import duties for most protein crops and all oilseed crops and their meals in the EU [76].

Competition of different types of agriculture on the global market differs from that in many other economic sectors. EU farmers have to comply with sanitary, environmental, and animal welfare rules, but agriculture is heavily subsidized in the EU and in other Organization for Economic Cooperation and Development (OECD) countries in the Northern Hemisphere. In emergent economy countries, like Brazil, Indonesia, and Malaysia, subsidies do not exist for farmers, but salaries can be very low, land is cheap and available, and the destruction of natural capital (ex.: forest, soil fertility) gives a competitive advantage for exporting several agricultural commodities. This competitive price is obtained partly because of the destruction of natural resources in the case of soybean and oil palm, for instance, and it is also often the case for shrimp and fish farming, or fishing in oceans. Thus, these economic activities are destructive by nature and are stimulated by global trade. Corrective measures have to be taken.

World Trade Organization (WTO) trade rules are aimed at liberalizing trade in products without reference to the processes by which those products are made. Exceptions to this principle are provided in General Agreement on Tariffs and Trade (GATT) Article 20 [77]. Technical trade barriers based on social, sanitary, food safety, and environmental criteria can be considered and are widely used by developed countries [78].

In, GATT members negotiated Agreements on Technical Barriers to Trade (TBT) as part of the Uruguay Round Agreement [79, 80]. The agreements specify that technical measures should not be used as disguised trade barriers or in an arbitrary or discriminatory manner. They should only be applied to the extent necessary and must be based on scientific principles and on risk assessment [79, 80]. The Agreement on Technical Barriers to Trade tries to ensure that regulations, standards, testing, and certification procedures do not create unnecessary obstacles, while also providing members with the right to implement measures to achieve legitimate policy objectives, such as the protection of human health and safety, or the protection of the environment (http://www.wto.org/english/tratop_e/tbt_e/tbt_e.htm). Countries are encouraged to use international standards developed by international scientific organizations [80].

Technical trade barriers are defined as regulations and standards governing the sale of products into national markets. Their first objective is the correction of market inefficiencies stemming from externalities associated with the production, distribution, and consumption of these products. They can significantly impede trade in some circumstances [78], but they differ from many other trade barriers because they can be economically efficient [81]. They may be adopted in instances when regulatory authorities judge that markets fail to provide optimal amounts of unowned or commonly owned environmental resources [78]. They are increasingly important in the international trade of agricultural products.

Environmental governance is well developed. There are currently over 500 Multilateral Environmental Agreements (MEAs), including 45 of global geographical scope with at least 72 signatory countries [82] and about 155 on biodiversity [83]. These agreements apply internationally or regionally, and they concern a variety of environmental issues, including the atmosphere, living matter, marine life, desertification, ecosystem protection, and control of persistent organic pollutants. A regional agreement concerns the deforestation in Borneo, for instance. The implementation of MEAs is extremely difficult in developing countries for several reasons (lack of funding, of human resources and of political willingness).

Since the urgency and extent of the problems go beyond the capacity of existing institutions, the possibility arises of creating an international organization that centralizes these issues—a World Environment Organization (WEO) that could implement an International Environmental Governance (IEG) [84]. Another option would consist of reforming the United Nations Environment Program (UNEP) so that it could play this major political role, enlarging its functional structure and clarifying its operational mandate.

It is urgent that the management of conflicts between the international trade regime and the international environmental regimes be improved [85]. In the short term, a better collaboration between the MEAs and the WTO would be desirable. In the longer term, the dialogue and the coordination between a future WEO and the WTO would be a better solution. States and large economic blocks, like the EU, could also use well-documented environmental trade barriers for restricting the imports of some products and require environmental-friendly production processes if these products have to be imported in the EU. Some products with a very detrimental impact on the environment could even be banned.

Among policy options for decreasing the adverse effects of global trade on ES and biodiversity, the following can be cited:

- international agreements providing a mechanism for improving the global environment. Environmental protection requirements should be a feature of future international trade agreements;
- information and negotiation in developing countries with governments and local industries and business companies (ex.: the Roundtable on Sustainable Palm Oil (RSPO) = Association of growers, industrial users and NGOs; the Task Force Sustainable Soy and the international Round Table on Responsible Soy (RTRS));
- adoption of a moratorium when justified (ex.: Soy Moratorium in the Brazilian Amazon);
- development of protected areas on large territories in developing countries (ex.: the WWF Program of 220,000 km^2 of protected areas in Borneo);
- environmental (ex.: agro-fuels, Directive 2009/28/CE) and consumer's health (ex.: palm oil) regulations and action plans (ex.: FLEGT initiative = Forest Law Enforcement, Governance and Trade) in the EU;
- import penalties and bans for high impact on environment products [14];
- integration of all costs including ecosystem service destruction in product prices (ex.: [56]);
- development of eco-labeling (ex.: fish [56]; Marine Stewardship Council); tourism [64]; wood: Forest Stewardship Council (FSC));
- cessation of perverse subsidies (ex.: support to industrial fishery) in the EU;
- development of sustainable procurement in public and private sectors;
- development of EU and international programs for the payment of ecosystem services (PES) in less developed countries (ex.: REDD+ = Reducing

emissions from deforestation and forest degradation programs and policies, plus pro-forest activities);
- development and support of alternative techniques (ex.: forest, reduced impact logging (RIL); farming, integrated agriculture, or agroecology);
- development of alternative solutions (ex.: grasslands in general and legumes in the EU/soybean, butter and olive oil/palm oil, temperate wood, and certified tropical wood/illegal tropical timber).

Many policy tools are thus available. In many regions, the degradation of the status of biodiversity and habitat quality requires urgent action and fast implementation of a combination of these tools.

ACKNOWLEDGMENT

This research has been funded by the Belgian Science Policy Office (BELSPO) within a cluster project called Belgian Ecosystem Services (BEES). The text was inspired by a conference organized in Brussels on October 24, 2011, with the participation of the following speakers: Eric Arets (ALTERRA. Wageningen University and Research Centre), Dieter Cuypers (VITO), Nicolas Dendoncker (University of Namur), Pierre Devillers and Roseline Beudels (RBINS), Jean-Louis Doucet (University of Liège), Sander Jacobs (University of Antwerp), Monique Munting and Mark van Oorschot (Netherlands Environmental Assessment Agency). The author thanks the reviewers for their useful comments: Edwin Zaccaï (ULB) and Frederic Huybrechts (University of Antwerp).

REFERENCES

1. Rodrigue, J.-P., Comtois, C., and Slack, B. 2009. *The Geography of Transport Systems*, 2nd ed. New York: Routledge, 352 pp.
2. Keay, J. 2006. *The Spice Route: A History.* Berkeley: University of California Press.
3. Elisseeff, V. 2001. *The Silk Roads: Highways of Culture and Commerce.* UNESCO Publishing/ Berghahn Books.
4. Pétré-Grenouilleau, O. 2004. *Les Traites négrières, essai d'histoire globale.* Paris: Gallimard, 468 p.
5. Gössling, S. 2002. Global environmental consequences of tourism. *Global Environmental Change* 12(4), 283–302.
6. Chertow, M.R. 2001. The IPAT Equation and its variants. Changing views of technology and environmental impact. *Journal of Industrial Ecology* 4(4), 13–29.
7. Commoner, B. 1972. *The Environmental Cost of Economic Growth in Population, Resources and the Environment.* Washington, DC: Government Printing Office, pp. 339–363.
8. Ehrlich, P.R., and Holdren, J.P. 1971. Impact of population growth. *Science* 171, 1212–1217.
9. Shafik, N. 1994. Economic development and environmental quality: An econometric analysis. *Oxford Economic Papers* 46, 757–773.
10. Suri, V., and Chapman, D. 1998. Economic growth, trade and energy: Implications for the Environmental Kuznets Curve. *Ecological Economics* 25(2), 147–160.
11. Leach, M., Scoones, I., and Stirling, A. 2010. *Dynamic Sustainabilities: Technology, Environment, Social Justice.* London: Earthscan.

12. Bagliania, M., Bravo, G., and Dalmazzone, S. 2008. A consumption-based approach to environmental Kuznets curves using the ecological footprint indicator. *Ecological Economics* 65(3), 650–661.

13. Millennium Ecosystem Assessment. 2005. *Ecosystems and Human Well-being: Synthesis.* Washington, DC: Island Press, 137 p.

14. Koellner, Th. 2011. Ecosystem services and global trade of natural resources. *Routledge Explorations in Environmental Economics*, 292 p.

15. Swolfs, S. 2011. Evolution of animal feed import in the EU: Possible ecological impacts. Universiteit Antwerpen, Master in de Milieuwetenschap, Academiejaar 2010–2011, 92 p.

16. Von Witzke, H., and Noleppa, S. 2010. EU Agricultural Production and Trade: Can More Efficiency Prevent Increasing "Land-Grabbing" Outside of Europe? Agripol, OPERA—Research Centre, 36 p.

17. Fearnside, P.M. 2001. Soybean cultivation as a threat to the environment in Brazil. Environmental Conservation 28, 23–38.

18. Fernside, P.M. 2007. Deforestation in Brazilian Amazonia: History, rates, and consequences. *Conservation Biology* 19, 680–688.

19. Malingreau, J.-P., and Tucker, C.J. 1988. Large-scale deforestation in the Southeastern Amazon Basin of Brazil. *Ambio* 17(1), 49–55.

20. Steinfeld, H., Gerber, P., Wassenaar, T., Castel, V., Rosales, M. and de Haan, C. 2006. Livestock's long shadow, environmental issues and options. FAO and LEAD, 390 p.

21. Carvalho, P.C.F., and Batello, C. 2009. Access to land, livestock production and ecosystem conservation in the Brazilian Campos biome: The natural grasslands dilemma. *Livestock Science* 120, 158–162.

22. Ribeiro, M.S., Metzger, J.P., Martensen, A.C., Ponzoni, F.J. and Hirota, M.M. 2009. The Brazilian Atlantic forest: How much is left, and how is the remaining forest distributed? Implications for conservation. *Biological Conservation* 142, 1141–1153.

23. Myers, N., Mittermeier, R.A., Mittermeier, C.G., da Fonseca, G.A.B., and Kent, J. 2000. Biodiversity hotspots for conservation priorities. *Nature* 403, 853–858.

24. Geist, H.J., and Lambin, E.F. 2001. What drives tropical deforestation? A meta-analysis of proximate and underlying causes of deforestation based on subnational case study evidence. LUCC Report Series 4. LUCC International Project Office, CIACO, 116 p.

25. Van Berkum, S., and Bindraban, P.S. 2008. Towards sustainable soy: An assessment of opportunities and risks for soybean production based on a case study Brazil. Report 2008–080. LEI Wageningen UR, The Hague, The Netherlands, 99 p.

26. Carvalho, F.M.V., De Marco, Júnior P., and Ferreira, L.G. 2009. The Cerrado into-pieces: Habitat fragmentation as a function of landscape use in the savannas of central Brazil. *Biological Conservation* 142, 1392–1403.

27. The AOCS Lipid Library. 2012. Oils and fats in the market place. Commodity oils and fats. Palm oil. Downloadable from: http://lipidlibrary.aocs.org/market/palmoil.htm

28. Koh, L.P., Miettinen, J., Liew, S.C., and Ghazoul, J. 2011. Remotely sensed evidence of tropical peatland conversion to oil palm. Proceedings of the National Academy of Sciences USA 108, 5127–5132.

29. Baghwat, S.A., and Willis, K.J. 2008. RSPO principles and criteria for sustainable palm production. *Conservation Biology* 22, 1368–1370.

30. Danielsen, F., Beukema, H., Burgess, N.D., Parish, F., Bruhl, C.A., Donald, P.F., Murdiyarso, D., Phalan, B., Reijnders, L., Struebig, M., Fitzherbert, E.B. 2009. Biofuel plantations on forested lands: Double jeopardy for biodiversity and climate. *Conservation Biology*, 348–358.

31. Fitzherbert, E.B., Struebig, M., Morel, A., Danielsen, F., Bruhl, C.A., Donald, P.F., and Phalan, B. 2008. How will oil palm expansion affect biodiversity? *Trends in Ecology and Evolution* 23, 538–545.

32. Koh, L.P., and Wilcove, D.S. 2008. Is oil palm agriculture really destroying tropical biodiversity? *Conservation Letters* 1, 60–64.

33. Clay, J. 2004. *World Agriculture and the Environment: A Commodity-by*-Commodity *Guide to Impacts and Practices.* Washington, DC: Island Press.

34. Robertson, J.M.Y., and Van Schaik, C.P. 2011. Causal factors underlying the dramatic decline of the Sumatran orang-utan. Oryx 35, 26–38.

35. Brown, E., and Jacobson, M.F. 2005. CRUEL OIL. How Palm Oil Harms Health, Rainforest & Wildlife. Center for Science in the Public Interest (CSPI), 40 p.

36. Devillers, P., Devillers-Terschuren, J., and Beudels-Jamar, R. (2004. Habitats of South-East Asia. Report to ASEAN. Royal Belgian Institute of Natural Sciences.

37. Kamphuis, B., Arets, E., Verwer, C., van den Berg, J., van Berkum, S., and Harms, B. 2010. Dutch trade and biodiversity. Biodiversity and socio-economic impacts of Dutch trade in soya, palm oil and timber. Alterra Report 2155, 146 p.

38. FAO. 2011. *State of the World's Forests 2011*. Rome: FAO, 164 p.

39. Hall, J.S., Harris, D.J., Medjibe, V., and Ashton, P.M.S. 2003. The effects of selective logging on forest structure and tree species composition in a Central African forest: implications for management of conservation areas. *Forest Ecology and Management* 183: 249–264.

40. Doucet, J.-L. 2003. L'alliance délicate de la gestion forestière et de la biodiversité dans les forêts du centre du Gabon. PhD Thesis, Faculté universitaire des Sciences agronomiques de Gembloux, 323 p.

41. Whitney, K.D., Fogiel, M.K., Lamperti, A.M., Holbrook, K.M., Stauffer, D.M., Hardesty, B.D., Parker, V.T., and Smith, T.B. 1998. Seed dispersal by *Ceratogymna* hornbills in the Dja Reserve, Cameroon. *Journal of Tropical Ecology* 14, 351–371.

42. Struhsaker, T.T., Lwanga, J.S., and Kasenene, J.M. 1996. Elephants, Selective Logging and Forest Regeneration in the Kibale Forest, Uganda. *Journal of Tropical Ecology* 12(1), 45–64.

43. Johns, A.G. 1997. *Timber Production and Biodiversity Conservation in Tropical Rain Forests.* New York: Cambridge University Press.

44. Cleary, D.F.R., Boyle, T.J.B., Setyawati, T., Anggraeni, C.D., Van Loon, E.E., and Menken S.B.J. 2007. Bird species and traits associated with logged and unlogged forest in Borneo. *Ecological Applications* 17, 1184–1197.

45. Schulze, C.H., Waltert, M., Kessler, P.J.A., Pitopang, R., Shahabuddin, D., Veddeler, M., Muhlenberg, M., Gradstein, S.R. Leuschner, C., Steffan-Dewenter, I., and Tscharntke, T. 2004. Biodiversity indicator groups of tropical land-use systems: Comparing plants, birds and insects. *Ecological Applications*, 1321–1333.

46. Duveiller, G., Defourny, P., Desclée B., and Mayaux, P. 2008. Deforestation in Central Africa: Estimates at regional, national and landscape levels by advanced processing of systematically-distributed Landsat extracts. *Remote Sensing of Environment* 112, 1969–1981.

47. Hansen, M.C., Stehman, S.V., Potapov, P.V., Loveland, T.R., Townshend, J.R.G., DeFries, R.S., Pittman, K.W., Arunarwati, B., Stolle, F., Steininger, M.K., Carroll, M., and Dimiceli, C. 2008. Humid tropical forest clearing from 2000 to 2005 quantified by using multitemporal and multiresolution remotely sensed data. *Proceedings of the National Academy of Sciences USA* 105(27), 9439–9444.

48. De Wasseige, C., Devers, D., de Marcken, P., Eba'a, A.R., Nasi, R. and Mayaux, P. 2009. Les forêts du Bassin du Congo: Etat des forêts 2008. Office des publications de l'Union Européenne. Luxemburg, 425 p.

49. OECD and FAO. 2011. OECD-FAO Agricultural Outlook 2011–2020. Chapter 3: Biofuels. OECD and FAO, 77–93.

50. IEA (International Energy Agency). 2007. World Energy Outlook 2006. OECD and IEA, 596 p.

51. Munting, M. 2010. Impact de l'expansion des cultures pour biocarburants dans les pays en voie de développement (Impact of biofuel crop expansion in developing countries). CETRI, 98 p.

52. Food & Water Watch and The Council of Canadians. 2008. Lake Naivasha Withering Under the Assault of International Flower Vendors. Food & Water Watch–The Council of Canadians: 5 pp. Downloadable from: http://www.canadians.org/water/documents/NaivashaReport08.pdf

53. European Commission. 2010a. Trade, Fisheries, Products, Shrimps. Downloadable from: http://ec.europa.eu/trade/creating-opportunities/economic-sectors/fisheries/products/

54. Primavera, J.H. 1997. Socio-economic impacts of shrimp culture. *Aquaculture Research* 28, 815–827.

55. Bennett, E.M., Peterson G.D., and Gordon L.J. 2009. Understanding relationships among multiple ecosystem services. *Ecology Letters* 12(12), 1394–1404.

56. Folke, C., Kautsky, N., Berg, H., Jansson, A., and Troell, M. 1998. The ecological footprint concept for sustainable seafood production: A review. *Ecological Applications*, 8(1) Supplement, S63–S71.

57. FAO. 2010. The State of World Fisheries and Aquaculture 2010. Rome: FAO, 197 p.

58. Pauly, D., Christensen, V., Dalsgaard, J., Froese, R., and Torres, F., Jr. 1998. Fishing Down Marine Food Webs. *Science, New Series*, 279(5352), 860–863.

59. Pulvenis de Séligny, J.-F., Grainger, R., Gumy, A., and Wijkström, U. (eds.). 2010. The State of World Fisheries and Aquaculture 2010. FAO, Fisheries and Aquaculture Department, 197 pp.

60. Hannesson, R. 1996. *Fisheries (Mis)management: The Case of the North Atlantic Cod.* Oxford, England: Fishing News Books.

61. Mason, F. 2002. The Newfoundland cod stock collapse: A review and analysis of social factors. *Electronic Green Journal* 1(17), 21 p. Downloadable from: http://escholarship.org/uc/item/19p7z78s

62. FAO. 1997. Review of the state of world fishery resources: Marine fisheries. FAO Fisheries Circular No. 920 FIRM/C920. Downloadable from: http://www.fao.org/DOCREP/003/W4248E/w4248e30.htm

63. GDRC (Global Development Research Center). 2001. Environmental impacts of tourism. Downloadable from: http://www.gdrc.org/uem/eco-tour/envi/index.html

64. Font, X., and Buckley, R. 2001. Tourism ecolabelling: Certification and promotion of sustainable management. CABI, 359 p.

65. Buckley, R. 2009. Ecotourism: Principles and practices. CABI: 368 p.

66. Brockington, D. 2002. Fortress Conservation: The Preservation of the Mkomazi Game Reserve, Tanzania. James Currey.

67. West, P., Igoe, J., and Brockington, D. 2006. Parks and peoples: The social impact of protected areas. *Annual Review of Anthropology* 35, 251–277.

68. Adams, W.M., and Hutton, J. 2007. People, parks and poverty: Political ecology and biodiversity conservation. *Conservation and Society* 5, 147–183.

69. Primavera, J.H. 1993. A critical review of shrimp pond culture. *Reviews in Fisheries Science* 1, 151–201.

70. Ekins, P., Folke, C., and Costanza, R. 1994. Trade, environment and development: the issues in perspective. *Ecological Economics* 9, 1–12.

71. Westhoek, H., Rood, T., van den Berg, M., Janse, J., Nijdam, D., Reudink, M., and Stehfest, E. 2011. The Protein Puzzle. The consumption and production of meat, dairy and fish in the European Union. The Hague, PBL Netherlands Environmental Assessment Agency, 218 p.

72. Martinez-Alier, J. 1990. *Ecological Economics: Energy, Environment and Society.* Blackwell.
73. Martinez-Alier, J., and Ropke, I. (eds.). 2008. *Recent Developments in Economics*, Ecological 2 vols. Cheltenham, UK: Edward Elgar.
74. Würtenberger, L., Koellner, T., and Binder, C.R. 2006. Virtual land use and agricultural trade: Estimating environmental and socio-economic impacts. *Ecological Economics* 57, 679–697.
75. Davis, S.J., and Caldeira, K. 2010. Consumption-based accounting of CO_2 emissions. *Proceedings of the National Academy of Science of the United States of America* Edition, 6 p.
76. European Commission (Directorate-General for Agriculture and Rural Development, Unit C5). 2011. Oilseeds and protein crops in the EU: http://ec.europa.eu/agriculture/cereals/factsheet-oilseeds-protein-crops_en.pdf)
77. Hoekman, B.M., and Kostecki, M.M. 1995. *The Political Economy of the World* Trading *System: From GATT to WTO.* Oxford: University Press.
78. Roberts, D., Josling, T.E., and Orden, D. 1999. A framework for analyzing technical trade barriers in agricultural markets. Market and Trade Economics Division, Economic Research Service, U.S. Department of Agriculture, Technical Bulletin No. 1876: 44 p.
79. World Trade Organization. 1994a. Agreement on Technical Barriers to Trade. Geneva: World Trade Organization.
80. World Trade Organization. 1994b. Agreement on the Application of Sanitary and Phytosanitary Measures. Geneva: World Trade Organization.
81. Weyerbrock, S., and Xia, T. 2000. Technical trade barriers in US/Europe agricultural trade. *Agribusiness* 16, 2: 235–251.
82. Inomata, T. 2008. Management Review of Environmental Governance within the United Nations System. United Nations, Joint Inspection Unit, Geneva.
83. Kanie, N. 2007. Governance with multilateral environmental agreements: A healthy or ill-equipped fragmentation. In *Global Environmental Governance: Perspectives on the Current Debate Centre for UN Reform.*
84. Maskus, K.E. 2000. Regulatory Standards in the WTO. Peterson Institute for International Economics, Working Papers, 9 p.
85. Eckersley, R. 2004. The Big Chill: The WTO and multilateral environmental agreements. *Global Environmental Politics* 4(2), 24–50.
86. Gambini, G. 2009. EU-27 consistent world leader in trade of food and drink. *External trade. EUROSTAT Statistics in Focus* 78/2009, 7 p.
87. FAO. 2004. The State of Agricultural Commodity Markets 2004. FAO: 52 p.
88. ITTO (International Tropical Timber Organization). 2005. Annual review and assessment of the world timber situation 2004. ITTO, 243 p. Downloadable from: http://www.itto.int/annual_review/
89. ITTO (International Tropical Timber Organization). 2011. Annual review and assessment of the world timber situation 2010. ITTO, 186 p. Downloadable from: http://www.itto.int/annual_review/
90. European Commission. 2010b. Trade, Fisheries. Downloadable from: http://ec.europa.eu/trade/creating-opportunities/economic-sectors/fisheries/index_en.htm

Part IV

Ecosystem Services: Tools & Practices

Shutterstock.com / © *kropic1*

Chapter 18

CICES Going Local

Ecosystem Services Classification Adapted for a Highly Populated Country

Francis Turkelboom[1], Perrine Raquez[2], Marc Dufrêne[3], Leander Raes[4], Ilse Simoens[1], Sander Jacobs[5,6], Maarten Stevens[1], Rik De Vreese[7], Jeroen A.E. Panis[8], Martin Hermy[9], Marijke Thoonen[1], Inge Liekens[10], Corentin Fontaine[2], Nicolas Dendoncker[11], Katrien van der Biest[12], Jim Casaer[1], Hilde Heyrman[13], Linda Meiresonne[1] and Hans Keune[14,15,16,17]

[1]*Research Institute for Nature and Forest (INBO),* [2]*University of Namur (UNamur),* [3]*ULG-GxABT,* [4]*UG,* [5]*Research Institute for Nature and Forest (INBO),* [6]*University of Antwerp. Department of Biology, Ecosystem Management Research Group (ECOBE),* [7]*VUB,* [8]*Agency for Nature and Forests, Government of Flanders,* [9]*KULeuven,* [10]*Flemish Research and Technology Organisation (VITO),* [11]*Department of Geography, University of Namur (UNamur). Namur Research Centre on Sustainable Development (NAGRIDD). Namur Centre for Complex Systems (naXys),* [12]*University of Antwerp, Ecosystem Management Research Group (ECOBE),* [13]*VLM,* [14]*Belgian Biodiversity Platform,* [15]*Research Institute for Nature and Forest (INBO),* [16]*Faculty of Applied Economics – University of Antwerp,* [17]*naXys, Namur Center for Complex Systems – University of Namur*

Chapter Outline

Ecosystem Services. http://dx.doi.org/10.1016/B978-0-12-419964-4.00018-4

1. WHY WE NEED A COMMON CLASSIFICATION SYSTEM FOR ECOSYSTEM SERVICES IN BELGIUM?

Although the concept of ecosystem services (ES) has been popularized widely since publication of the Millennium Assessment (MA) in 2005 [1], different classification schemes have been proposed by several authors, such as Costanza et al. [2], Daily [3], de Groot et al. [4], Wallace [5], and TEEB [6]. Costanza [7] argued that due to the dynamic complexity of ecosystem processes, the inherent characteristics of ecosystem services, and the diverse decision contexts, different types of classification schemes should be considered. He concludes: "Any attempt to come up with a single or 'universal' classification system should be approached with caution."

Although it is recognized that a diversity of approaches is probably necessary, the use of multiple classifications makes comparison and integration between studies and assessments more difficult. With the fast-growing number of ES assessment and valuation studies around the world, the need to design a common base that enables comparison between ES assessments at different places has become more urgent [8]. This common base should be specific enough to be operational, while remaining relevant to a multitude of objectives for which frameworks and implementation plans may be developed [9].

This need has become especially acute since the new European Biodiversity strategy requires all EU member states to map and assess the state of the ecosystems and their services in their national territory by 2014 (Target 2, Action 5). For that reason, a working group on Mapping and Assessment of Ecosystems and their Services (MAES) has been set up to support European member states in undertaking the necessary work. The MAES working group decided to apply Common International Classification for Ecosystem Services (CICES) v4.3, which will be used throughout Europe [10].

CICES was initiated by the European Environment Agency (EEA) and is coordinated by the University of Nottingham [11–13]. One advantage of the CICES approach is that it allows adjustment to local conditions. In highly populated and developed areas, such as Belgium (337 inhabitants/km²), open space is rapidly declining and fragmenting, and the natural water cycle is getting disturbed (e.g., peak flows due to compaction, nutrient loads). In 2009, built-up areas (e.g., residential housing and transport infrastructure) covered 20% of the Belgian surface, while forest and wooded land covered only 23%. The high population density and the recent land-use changes have caused several environmental pressures, such as flooding risk, drought, air pollution, eutrophication, and loss of biodiversity. These pressures have had a negative effect on health and well-being, and are increasing the cost of environmental management measures. Consequently, the demand for specific services that can be provided by nature is increasing, while claims from different sectors often overlap or are contradicting. To adapt and fine-tune the latest CICES classification to the specific Belgian conditions, it was decided to design a Belgian version of CICES (CICES-Be).

2. CICES-BE: GOAL AND CONSULTATION APPROACH

The purpose of CICES-Be is to provide a standardized, but flexible, ES classification system that can accommodate different kinds of use in Belgium, but that can be further adapted in the future. It must be usable for the upcoming regional ecosystem services assessments for Wallonia and Flanders (respectively, for 2013 and 2014), valuation studies, payments for ecosystem services (PES) schemes, local planning exercises based on ES, and others. The aim is also for a robust list of ES that can be used as a basis for studies at different spatial scales. For example, if an ES assessment is conducted on a local scale, the CICES-Be classification can be further refined by adding another sublevel with more specific ES. For the national scale, the classification can be limited to a few broad classes (e.g., division or group level).

The initiative for CICES-Be was taken by the Research Institute for Nature and Forests (INBO) and the Université de Namur. The starting point was CICES v3 [12]. Where discrepancies with the Belgium context were found, modifications were made. Where important ES for Belgium were missing, new ES were added. In order to improve the classification from different perspectives and to increase support for the final product, the resulting CICES-Be v1 was then sent to Belgian experts who showed interest in this topic. Through iterative feedback loops, CICES-Be was further improved until consensus was reached with CICES-Be v6. The consultation lasted one year, from May 2012 until April 2013. In total, 19 experts from 11 organizations contributed to CICES-Be. The contributing experts are based at research centers, administrations, and policy-support units, have diverse disciplinary backgrounds, and come from both the Flemish and Walloon regions. The results of this Belgian consultation process were also used as an input to the international e-consultation process to improve the international CICES classification (http://cices.eu/).

3. ES DEFINITIONS AND ES CASCADE

Before we could embark on the development of CICES-Be, however, we first needed a common understanding about the framework and definitions.

3.1. What are Ecosystems Services?

The concept of ecosystem services is inherently anthropocentric. Human beings are value-expressing agents who translate basic ecological structures and processes into value-laden entities [4]. One can visualize this with a simple thought experiment: in an Earth-like planet with no humans, there could be a wide array of ecosystem structures and processes, but there would be no services [14].

CICES defines **ecosystem services** as "the contributions that ecosystems make to human well-being," and that arise from the interaction of biotic and abiotic processes. Ecosystem services refer to the *final* outputs or products from ecological systems, which are the items directly consumed or used by people

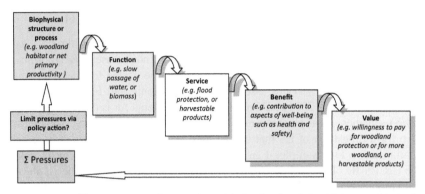

FIGURE 18-1   The ecosystem service cascade model, showing the relationship between biophysical structures and processes and benefits and values for human well-being [18].

[12]. In other words, ecosystem services are actually conceptualizations of the *useful things* ecosystems *provide* for people. As for consistency with the MA, the term *services* is generally taken to include both goods and services.

3.2. The Ecosystem Services Cascade

The definition makes it clear that ES cannot stand by themselves, but that there is something of a production chain linking ecological and biophysical structures and processes on the one hand and elements of human well-being on the other, and that there is potentially a series of intermediate stages between them. To disentangle the pathway from ecosystems and biodiversity to human well-being, a conceptual framework was proposed: the **ES cascade** structure (Figure 18-1; 12). The advantage of this construct is that it clearly demonstrates to decision makers and ecosystem service users that functional ecosystem structures and processes are required before services and benefits can be provided. In addition, the cascade adequately shows that, in order to maintain the sustainable flow of services, it requires the protection of and investment in the supporting ecosystems and biodiversity. The cascade also helps to frame a number of important questions about relationships between people and nature, such as: What are the critical levels or stocks of natural capital[1] needed to sustain the flow of ecosystem services?; Can natural capital be restored once damaged?; What are the limits to the supply of ecosystem services in different situations?; How do we value the contributions that ecosystem services provide to human well-being? The judgment made about the seriousness of these issues or pressures partly shapes policy action (= the feedback arrow in the diagram) [12].

Although the cascade model is a useful conceptual device for understanding the links between ecosystems and people, it is of course a simplification of

---

1. Natural capital is defined as the stock of natural ecosystems that yields a flow of valuable ecosystem goods or services into the future [2].

the real world [8]. For example, it should be realized that ecosystem processes and services do not always show a one-to-one correspondence: sometimes a single ecosystem service is the product of two or more processes, whereas a single process can contribute to more than one service [4]. For example, the function *wave regulation* provides services, such as flood prevention, drinking water, and recreation potential. Also, the benefits of a certain service can be manifold: for example, the provision of food has multiple benefits, such as health, employment, pleasure, and even cultural identity [15, 16]. These multiple linkages between both processes and structures on the one hand, and services and benefits on the other make the decision-making process complex [5]. The cascade model also does not really clarify the fact that ecosystems are usually not capable of generating all potential services simultaneously [8].

To make practical use of the ES cascade, all the steps need to be defined clearly:

- The actual use of goods or services provides **benefits** to humans, such as nutrition, health, and pleasure. Benefits are defined as "the gains in welfare and well-being generated by ecosystem services" [17].

- **Value** is defined as the measurement of the benefit, which can be expressed in monetary or nonmonetary terms. Metrics from various scientific disciplines can be used (e.g., economics, sociology, ecology). In economics, value is always associated with trade-offs, that is, something has (economic) value only if we are willing to give up something else to get it or enjoy it. Benefits and values are separated because the way we value these benefits is subjective: Different groups may value these gains in different ways at different times and at different places. Thus, different values can be attached to a particular benefit. When we try to measure an overall value, these different appreciations should be included [14]. Benefits are usually generated by ecosystem services in combination with human inputs, such as labor, institutions, knowledge, or equipment (e.g., hydroelectric power is dependent on water regulation services of nature, but also needs human engineering and construction materials). So attributing a value entirely to ecosystems would be misleading. Any attempt to value nature's services would have to try to disentangle the contribution that natural and human-made capital make to the benefit being considered [18].

- For many years, the terms *ecosystem function* and *ecosystem service* have been used interchangeably by some authors, creating a confusion that still exists today. **Ecosystem function** is defined as the "capacity or capability of the ecosystem to do something that is potentially useful to people" [2–4, 19, 20]. Or more specifically: "a subset of the interactions between ecosystem structure and processes that underpin the capacity of an ecosystem to provide goods and services" [6]. The capacity to deliver a service exists independently of whether anyone wants or needs that service. That capacity only becomes a service when some beneficiary can be identified. For example: The presence of ecological structures like woodlands or wetlands in a catchment area may have the capacity (function) of slowing the passage of surface water. This function of the ecosystem becomes a service, when it modifies the intensity of flooding in downstream residential areas [8, 17].

- The building blocks of ecosystem functions are the interactions between structure and processes. **Ecosystem structure** is "the biophysical architecture of an ecosystem." The composition of species making up this architecture may vary. **Ecosystem process** is defined as "any change or reaction which occurs within ecosystems" [1]. Processes may be physical (e.g., infiltration of water, sediment movement), chemical (e.g., reduction, oxidation), or biological (e.g., photosynthesis, denitrification), whereby biodiversity is more or less involved in all of them [17]. Although there are still quite a lot of knowledge gaps about the relationship between biodiversity and ecosystem services, scientific understanding has improved over the last decade and existing knowledge has been reviewed in a few recent papers (e.g., [21–23]).

While these definitions help us further, the application of these definitions is situation-dependent. Whether or not something is called a service depends often on the perspective of the beneficiary [15, 24, 25]. For example, if someone is interested in the benefit of timber, then primary productivity is a service, but for someone who is interested in drinking water, primary production can be considered an ecosystem process.

3.3. Do Ecosystems also Produce *Disservices*?

By definition, ES refers only to the goods and services produced by biodiversity and ecosystems benefiting human well-being. However, not all impacts of nature on human well-being are positive [26, 27]. Ecosystems may also (or are perceived to) provide disservices. In urban settings, Lyytimaki and Sipila [28] argued that it may be counterproductive to frame ecosystem services only in a positive way, without paying adequate attention to the various nuisances and disservices that ecosystems inevitably produce. Consequently, they argue that green spaces in urban settings should be managed not only to generate more services and biodiversity, but also to produce fewer disservices.

As no widely agreed definition of ecosystem disservices exists, we propose the following definition: "functions of ecosystems that are (or are perceived) as negative for human well-being"[2] [28]. Ecosystem disservices can be subdivided into four categories:

- Species negatively affecting human health: Some type of biodiversity is directly deleterious for human health—for example, wetlands providing habitat for

---

2. Some literature uses the term *disservices* to indicate the negative effects of ecosystem degradation caused directly by human activities. For example in the context of agriculture, the term *ecological disservices* is typically understood as disturbed or missing services as the consequence of loss of biodiversity by agricultural practice, such as nutrient runoff and erosion, loss of wildlife habitat, greenhouse gas emissions, and pesticide poisoning of humans and nontarget species. As this definition of disservices covers quite a varying content, it is suggested that this second interpretation of the definition of disservices be included.

malarial mosquitoes, pathogen populations, and toxic plants. As biodiversity is a necessary component of healthy, well-functioning ecosystems, conversion of natural habitats to managed or disturbed habitats can increase the prevalence of disease. In this way, habitats can become worse for humans in terms of their disservices [29, 30].

- Species causing production damage: An example is damage to crops and live-stock by pests and wild animals [31, 32].
- Discomfort caused by nature: Biodiversity elements can cause distress to human welfare. Examples are species generating nuisance [33], natural areas in urban setting that generate a feeling of fear at night [34], presence of large carnivores that cause a feeling of insecurity, and insects that cause discomfort.
- Natural disasters: Natural phenomena, such as damages caused by floods and natural occurring wildfires.

Assessing disservices can, however, be complicated: First, the same ecosystem function can be perceived as a service or disservice depending on the context or the person. The balance between disservice and service can be subtle and therefore requires a concerted effort to understand the involved species in detail [35]. Second, a certain ecosystem function that generates a positive ecosystem service can negatively affect another ES. For example, the existence of a roe deer population in a certain area can contribute to opportunities for hunting and recreation (nature experience and wildlife photography), but they can be negative for the regeneration of a tree species, thereby negatively impacting timber production; natural areas in cities are positive for recreation and quality of life, but can cause slippery roads in autumn or feelings of insecurity at night; water regulation provided by a vegetated landscape might be valued by someone who is dependent on a steady water supply, but for someone interested in using the water for boating, this vegetation can be a burden. Finally, ecosystem disservices can be perceived as a result of changes in biodiversity, or because of changes in human perceptions alone. On the other hand, adverse effects for human health can be caused by ecosystem services that are not noticed at all or are not perceived as negative. Differentiating perceived disservices from actual disservices can be challenging [28].

The issue of disservices is to a large extent a matter of positive or negative appreciation by humans. Depending on the situation and stakeholders, an ES can provide either a benefit or a liability. When it is important to look at the whole picture (for example, for a management plan of a specific region), disservices should be included as well. However, as both positive and negative impacts are part of the same continuum, they can be linked to the list of services below. We therefore have chosen not to make a separate category for disservices within CICES-Be.

4. AN ES CLASSIFICATION SYSTEM FOR BELGIUM: CICES-BE

4.1. Key Principles of CICES

The proposal for CICES was based on the requirement that any new classification has to be consistent with accepted typologies of ecosystem goods and services currently being used in the international literature, and that it should be compatible with the design of the System of Integrated Environmental and Economic Accounting (SEEA) methods and UN standard classifications (ISIC4, CPC, COICOP). In constructing CICES, three main principles were applied [11, 12]:

- Hierarchical structure: In the present one-dimensional ES listings, each time a new service is identified, the list has to be updated. Therefore, a hierarchical structure was proposed into which new and specific elements can be fitted without disrupting the general structure of the classification. A hierarchical classification also enables summaries of services' outputs at different levels of generality, a feature that is difficult to accomplish with a simple listing. At the highest level, the three usual "service themes" are listed: provisioning, regulating and maintenance, and cultural ES (called Sections). Below the Sections level, different service groups are nested (i.e., Division, Group, and Class). The labels of the classes used in CICES have been selected to be as generic as possible, so that other more specific or detailed categories can progressively be defined, according to the interests of the user or country, or the concerned scale.

- Final outputs only: CICES refers specifically to the "final" outputs or products from ecosystems. Following common usage in the ES literature, the classification recognizes these outputs to be provisioning, regulating and maintenance, and cultural services, but it does not cover the so-called supporting services originally defined in the MA. As the supporting services are only indirectly consumed or used, they are treated as part of the underlying structures, processes, and functions that characterize ecosystems. The distinction between final and intermediate products was also proposed to avoid the problem of double-counting when undertaking monetary valuation. Valuation should only be applied to the item directly consumed or used by a beneficiary because the value of the ecological structures and processes that contribute to it is already wrapped up in this estimate [24, 25, 36]. It was therefore proposed that supporting services are best dealt with in other ways in environmental accounts [11–13]. In reality, this division between final and intermediate outputs is not always clear. Some of the ES can be intermediate as well as final services, depending on the user of the service. For example, pollination is a final service for the fruit grower (as it is an essential production factor for the producer) and a supporting service for the fruit consumer. But as this is a generic classification system, this type of ES is included as long as at least one stakeholder can be identified that directly benefits from a certain ES.

- Finally, a key point of CICES is that it is a classification of services and not of benefits [13].

4.2. Role of Supporting Services and Abiotic Resources in CICES

The fact that supporting services are not included in CICES should not be taken to mean they are unimportant. Any given ES depends on a range of interacting and overlapping ecosystem functions, and one supporting service may simultaneously facilitate the delivery of many final outputs. Typical examples of supporting services are nutrient cycling, photosynthesis, water cycling, and maintenance of the gene pool. As the category supporting services comprises every function and structure that is somehow involved in sustaining service flow, providing resilience, energy, and substrate; then it will probably include nearly "all biophysical complexity. A consequence is that any attempt to seriously define the set of supporting services is likely to oversimplify the role of nature. So, every list will be necessarily incomplete and illustrative, and any valuation will be incomplete [12]. A second implication is that each plan or intervention that changes land use and its related supporting services will have profound implications for delivery of related ecosystem services. In other words, lists of desirable ES should not be goals by themselves, but a starting point to reflect on the underlying processes and functions and on how to achieve sustainable ecosystem management.

The inclusion or exclusion of **abiotic materials** (e.g., minerals, salt) and **renewable abiotic energies** (e.g., wind, hydro, solar, waves, tides, thermal energy) was quite a controversial issue within the CICES and Belgium ES communities. The most important points that were raised during the Belgian discussion are summarized below:

- The first perspective is related to the definitions. If ecosystems are defined as the interactions between living organisms and their abiotic environment, then it is argued that ecosystem services have to be traceable back to some living process, that is, be dependent on biodiversity [25]. Others argue that the ecosystem consists of biotic and abiotic processes. Many included ecosystem services, such as flood control, hydrology-related services, but also water and air purification de facto depending (partly) on abiotic structures and processes. The latter is used as an argument for including abiotic-based services in the ES classification.
- The second perspective is related to the renewability of the resource. Most authors agree that nonrenewable materials that are mined, such as fossil fuels, gold, and uranium, should not be included. For renewable natural resources, the opinions are divided. Some argue that the level of renewability could be a distinguishing feature for inclusion in CICES. This requires a consensus about the renewal period. If this period is set, for instance, at 100 years, this means that ES would include the extraction of sand in dynamic rivers and salt mines, but not the mining of fossil fuels [4]. However, defining a renewal period is always controversial. Therefore, some have suggested basing the argument on extraction rate versus delivery rate. For example, in the case of petroleum, it is the speed of extraction that makes its use unsustainable. If oil would only be extracted at the rate at which it can be replaced, it could be considered a renewable resource. This approach is consistent with the idea of sustainable resource use: Only those goods and services are included that can be used on a sustainable basis.

- The third perspective is related to abundance. Wind, solar, tidal, and other energies are abundant and nondepletable. If the ES framework is aimed to be a tool that assists society in making decisions about scarce or limited natural resources and their services, it could be argued that it is not very useful to include them in the context of an ES analysis.
- Fourth, there is the aspect of attribution. As the origin of wind and solar energy cannot be attributed to a certain ecosystem type, it is proposed to exclude them [4]. Others argue including them, based on the fact that the amount of generated energy depends on topography, orientation, and local climate.
- A final argument for inclusion is that abiotic resources play an essential role in the transition to sustainability.

For CICES v4.3, it was decided to leave the "pure" abiotic resources out of the classification system of ES [13], and for the time being, CICES-Be will follow CICES. Nevertheless, when abiotic resources and energies play an important role in the issues at stake, it makes a lot of sense to include them in mapping and planning exercises.

4.3. Modifications of CICES for the Belgian Context

When the final international CICES v4.3 was published in January 2013, it was decided to harmonize CICES-Be v5 as much as possible with CICES v4.3. The purpose of this exercise was twofold: on the one hand, to keep the classification adapted to Belgian conditions; on the other hand, to keep the system compatible with the international one, at least at the section and division level. This resulted in CICES-Be v6 with 8 divisions, 18 groups, and 41 classes. Where we felt it was relevant for Belgium, additional subclasses were defined (34 in total). All the elements of CICES-Be that differ with CICES are marked In blue colour in Table 18-1. The major differences between CICES-Be and CICES are the following:

Additional ES in CICES-Be: Where important ES for Belgium were missing, new ES were added, such as: prevention and control of fire, control of invasive species, control of nature-borne human diseases, moderation of certain diseases by exposure to nature, and some specific cultural services (see below).

Modified ES in CICES-Be:

- Biomass production for nutrition in CICES-Be is split up according their origin: terrestrial, freshwater, or marine. This is done because these ES can be associated with very distinct professional and recreational activities.
- The ES division *mediation of waste, toxics and other nuisances* in CICES is split up according to media (biota versus ecosystems) and processes (e.g., bioremediation, dilution, filtration, sequestration). For CICES-Be, we did not find this division to be practical, as in reality many of these processes interact. Therefore, it was decided to subdivide them based on the type of service they provide (soil and water quality regulation, air quality regulation, shielding). Consequently, the group "water conditions" in CICES was omitted in

TABLE 18-1 ES Classification for Belgium CICES-Be v6

Section	Division	Group	Class	Subclass for Belgium	Examples of Service Providing Units	Benefits (non exhaustive) Availability of:
Provisioning	Nutrition	Biomass	Terrestrial plants, fungi, and animals for food	Commercial crops	Cereals, vegetables, fruits	Food
				Kitchen garden crops	Vegetables, fruit	
				Land-based commercial livestock	Free-range dairy and meat cows, chickens	
				Hobby animals for food	Sheep, goat, chicken, rabbit, bees	
				Edible wild animals, plants, and fungi	Game, wild honey, mushrooms, berries, nuts, wild plants (e.g., young nettle branches)	
			Freshwater plants and animals for food	Freshwater fish and shellfish	Freshwater fish (trout, eel)	
				Cultivated freshwater fish	Carp	
				Edible water plants	Water cress	
			Marine algae and animals for food	Sea fish and shellfish	Marine fish (sea bass)	
				Cultivated seafood and shellfish	Mussel culture	
				Edible plants from salt and brackish waters	Macro and microalgae, saltwort	
		Potable water	Surface water for drinking		Rivers, lakes, reservoirs, collected precipitation	Drinking water for domestic use
			Groundwater for drinking		Springs, (nonfossil) aquifers	

Continued

TABLE 18-1 ES Classification for Belgium CICES-Be v6—cont'd

Section	Division	Group	Class	Subclass for Belgium	Examples of Service Providing Units	Benefits (non exhaustive) Availability of:
Materials		Biomass	Fibere and other materials from plants, algae, and animals for direct use or processing	Ornamental plants and animals	Bulbs, cut flowers, decorative plants, shells, feathers, pearls	Ornamental plants & animal products
				Plant fibers and materials	Timber trees, flax, straw, herbs, resins,	Timber, paper, natural medicines, dyes, clothes
				Animal fibers and materials	Animal parts (skin, bones)	Soap, leather, gelatine, wool
			Materials from plants, algae, and animals for agricultural and aquaculture use	Organic matter for fertilization and/or soil improvement	Manure, litter, bark, algae, "plaggen"	Fertilizer for crop production, improved soil structure
				Fodder and forage	Maize, grasses	Food for animal raising
			Genetic materials from all biota		Genetic material (DNA) from wild plants, algae and animals	Medicines, breeding programs
		Nonpotable water	Surface water for nondrinking purposes		Rivers, lakes, reservoirs, collected precipitation	Water for irrigation, industrial production, cooling
			Groundwater for nondrinking purposes		Springs, (nonfossil) aquifers,	

Section	Division	Group	Class	Sub-class for Belgium	Examples of Service Providing Units	Benefits (non exhaustive)
Energy	Biomass-based energy sources	Plant-based energy resources	Energy crops and plant residues	Yellow mustard, wheat, beetroot, straw, grass and herb residues form nature and roadside management		Energy
			Energy trees and woody residues	Fuel wood (e.g., poplar, willow trees), woody residues form nature management		
		Animal-based energy resources		Dung, fat, oils, biogas		
Regulation and maintenance	Mediation of waste, toxics and other nuisances	Soil and water quality regulation	Bioremediation of polluted soils (phyto-accumulation/degradation/stabilization)		Plants & micro-organisms	Less polluted soils
			Water purification and oxygenation		Wetlands, lagoons, molluscs	Improved water quality
			Nutrient regulation		Buffer strips, soils, water bodies, estuaries, coastal zones	Stable nutrient levels
		Air quality regulation	Capturing (fine) dust, chemicals and smells		Trees, shrubs, forests	Improved air quality
		Shielding	Mitigation of noise & visual impacts		Vegetative buffers, landscape structures	Quieter environment

Continued

TABLE 18-1 ES Classification for Belgium CICES-Be v6—cont'd

Section	Division	Group	Class	Sub-class for Belgium	Examples of Service Providing Units	Benefits (non exhaustive)
	Mediation of flows	Mass flow	Mass stabilization and control of erosion	Gravity flow protection (e.g. landslides, creep)	Land coverage, roots of large trees	Land stability
				Protection against water and wind erosion	Cover crops, buffer strips, vegetation along the hydrological network, woodlands	Mudflow protection, less dredging costs, less impact of wind erosion
			Buffering and attenuation of mass flows		Rivers, lakes, sea	Transport and storage of sediment
		Liquid flow	Hydrological cycle and water flow maintenance		Permanent vegetation, land coverage	Secure navigation, drought prevention, protection against salt intrusion, hydro-power
			Flood protection	Natural flood protection & sediment regulation	Natural flood plains, wetlands	Flood safety, less dredging costs, navigation
				Coastal protection to waves, currents energy & sea level rise	Dunes, marshlands, sea grass	Coastal safety

Maintenance of physical, chemical, biological conditions	Lifecycle maintenance, habitat and gene pool protection	Pollination	Bees, butterflies	(Better) fruit setting
		Seed dispersal	Birds, insects and mammals	Improved tree propagation
		Maintaining nursery populations and habitats	Wetlands suitable for spawning grounds	Bigger commercial fish and shellfish population
		Prevention and control of fire	Fire resistant vegetation buffers, wetlands, wet heath'	Fire safety
		Control of (alien and/or local) invasive species	Competing plants and animal species	Reduced impact of undesirable invasive species
	Pest and disease control	Pest control	Beetle banks, hedgerows, vegetation strips, heterogeneous landscapes, agroforestry	Better health of agricultural plants and animals
		Disease control		
		Control of nature-borne human diseases	Diversity of plants and animals result in dilution of competition with vectors	Lower risk for nature-borne human diseases
		Moderation of certain diseases by exposure to nature	Trees, pollen, plants, animals, micro-organisms	Less susceptible to allergies, better resistance to infections
	Soil formation & composition	Weathering processes, decomposition and fixing processes	Green mulches, N-fixing plants, soil organisms	Fertile soils

Continued

TABLE 18-1 ES Classification for Belgium CICES-Be v6—cont'd

Section	Division	Group	Class	Sub-class for Belgium	Examples of Service Providing Units	Benefits (non exhaustive)
		Atmospheric composition and climate regulation	Global climate regulation by reduction of greenhouse gas concentrations		Vegetation, soils, sediments, oceans	More stable global climate
			Micro and regional climate regulation	Regional climate regulation (e.g. maintenance of regional precipitation patterns & temperature)	Forests	More stable regional climate
				Rural micro-climatic regulation	Windbreaks, shelter belts, shading trees, droves	Buffered micro-climate, air ventilation
				Urban micro-climatic regulation	Shading trees, parks, green roofs	

Section	Division	Group	Class	Sub-class for Belgium	Examples of Service Providing Units	Benefits for Wellbeing
Cultural	Physical and intellectual interactions with biota, ecosystems, and land-& seascapes	Natural environment suitable for outdoor activities	Area for non-excludable outdoor activities	Green environment suitable for daily outdoor activities	Neighbourhood green, shading trees, park, natural play area, green schoolyard drove, cemetery, fallow land, dike, trail	Physical, social and mental wellbeing, motoric and creative development of children

Landscape for outdoor recreation	Forest, beach, agricultural landscape, river, areas with wild food, pick-nick spot in nature, sport facility	Walking, jogging, cycling, horse riding in forest, mountain biking, surfing, canoeing, skiing, motorized activities, pick-nick, collecting natural products	Physical, social and mental well-being
Natural landscapes and species for nature experience & education	Area of outstanding natural beauty (e.g. nature reserve, natural spring, lake, river, rare species, natural smell & noises), attractive and charismatic species, area and species with educational value	Eco-tourism, bird watching, nature conservation activities, nature photographing and filming, landscape painting, spiritual activities, eco-therapy, nature education	Physical, social, mental, spiritual well-being, inspiration, cognitive development, spiritual development, nature awareness
Landscape and biodiversity suitable for research	Ecological patterns, pollen, tree rings, genetic patterns	Understanding of natural processes, technological applications, biomimicry	Better understanding of our dependency and relationship to nature

Continued

TABLE 18-1 ES Classification for Belgium CICES-Be v6—cont'd

Section	Division	Group	Class	Sub-class for Belgium	Examples of Service Providing Units	Benefits (non exhaustive)	Benefits for Wellbeing
			Area for excludable outdoor activities	Area for land-consuming recreation	Private land: Private garden, pasture for hobby animals. Areas with entrance fees: Camping site, zoo, botanical garden, safari park, golf course, horse riding school, licensed fishing areas	Relax and playing in gardens, golf, camping, riding horse, relaxation in theme park, non-consumptive angling	Physical, social, mental well-being, motoric and creative development of children
				Area for land-consuming productive activities	Farm land, pasture, kitchen garden, leased land for hunting, licensed fishing areas	Outdoor work for farming, forestry, firewood collection, vegetable growing for home consumption, hunting, consumptive angling	Physical, social and mental well-being, nature awareness

Spiritual, symbolic and other interactions with biota, ecosystems, and land-/seascapes	Natural surroundings around build-up areas			
	Natural surroundings around buildings for living, working and studying	Green/blue views from residences, schools, offices, elderly homes	Positively influence on living, working and indoor learning (better concentration, more creative, less stress) Higher prices of real estate	Physical, social, mental well-being
	Natural surroundings around institutions for recovery and therapy	Green/blue views from hospitals, psychiatric institutes, revalidation centres	Recovering from mental or physical illness positively influenced by the green environment,	Improved mental and/or physical health
Spiritual and/or emblematic	Landscapes and species with cultural and symbolic values	Typical cultural landscape (e.g. heath, pine forests, hedgerows), symbolic/emblematic species (e.g. stork, sky lark, wild boar)	Cultural heritage, folklore, flagship species for promoting regional identity Hunting, fishing, photographing and observing emblematic species	Sense of place/identity Sense of possession of skills

Note: Text in blue font indicates where CICES-Be differs from CICES v4.3.

CICES-Be because they are considered to be part of "soil and water quality regulation" in CICES-Be.

- Under the group soil formation and composition, the classes *weathering processes* and *decomposition and fixing processes* were merged in CICES-Be, as these processes are closely related to each other.
- The services under group *gaseous/air flows* in CICES are very much related to the microclimate and were therefore included under the class micro and regional climate regulation in CICES-Be.The cultural services section is conceptualized quite differently under CICES-Be:

Cultural services are primarily regarded as the "environmental settings, locations or situations that give rise to changes in the physical or mental states of people, and whose character are fundamentally dependent on living processes." Over millennia these environmental settings have been co-produced by the constant interactions between humans and nature [13, 37].

Following this logic, all cultural service classes in CICES-Be refer to a biophysical setting that provides cultural services (e.g., landscapes, individual species, and whole ecosystems). The direct benefits we derive from these cultural services are recreation, nature exploration, living in a nice environment, nature education, and others. These activities provide consequential benefits, such as physical, social, and mental well-being, and motoric and creative development for children. These benefits for well-being are mentioned in the last column of CICES-Be. This is in contrast to CICES, where benefits (e.g., use, education, entertainment, and symbolic) are categorized as ES themselves. CICES also lists bequest value (importance for future generations) and existence value (right of existence) as cultural services. They are not, however, included as ES in CICES-Be, as they are considered part of a valuation analysis.

The CICES Division Physical and Intellectual Interactions is subdivided in two groups within CICES-Be: natural environment suitable for outdoor activities and natural surroundings of built-up areas:

- For the group natural environment suitable for outdoor activities, we made a distinction between two service classes based on the concept of *excludability*. To be excludable means that "one person/party (can) keep another person/party from using a certain good or service" [14]. For CICES-Be, two classes are distinguished:

  1. *Area for nonexcludable outdoor activities*: These are public areas that everyone can use. Examples are green environment suitable for daily outdoor activities (e.g., daily stroll, cycling to work), landscape for outdoor recreation (e.g., jogging, mushroom picking), natural landscapes and species for nature experience and education (e.g., bird watching, landscape painting, and spiritual activities), and landscape and biodiversity suitable for research.

  2. *Area for excludable outdoor activities*: These are the areas where one group can exclude another group. We distinguish this as a separate

class, as some categories of this class are rapidly expanding in Belgium, and as excludability controls to a large extent how many people can benefit from them. The level of excludability can, however, vary, ranging from nonaccessible land (e.g., private gardens) to areas with restricted access (e.g., land accessible to only club members or paying visitors). We distinguish two subclasses: land that is occupied to make a certain type of recreation possible (such as private gardens or grazing land for hobby horses) and land that is used for productive activities (such as farming and kitchen-garden). The benefit of the latest type is the satisfaction and mental well-being one gets from outdoor work; the agricultural products are classified under the provisioning services.

- *Natural surroundings around built-up areas*: This is the passive use of natural settings and does not require any outdoor activity—for example, the view on green environment from residences, offices, and therapeutic institutions. This service is not included in CICES, but for Belgium it was chosen to give this a separate group in the cultural ES section. The reason is that owing to the high population pressure in Belgium, this service is becoming a more and more scarce—and therefore highly valued—resource.
- Finally, there is the Division/Group/Class that focuses on the cultural and symbolic values of landscapes and species. For this ES, it is not essential to visit these places, but the mere fact that these landscapes and species exist in people's mind is sufficient to generate a benefit for them.

5. CONCLUSION

The advantage of an inventory of ecosystem goods and services (and disservices) is that it shows in a systematic way the contributions of ecosystems to human well-being. This can assist in sensitizing policy makers, administrations, and the general public to the significance of ecosystems and enable giving suitable weight to environmental considerations within political decision making [38]. On the scientific side, the process of drawing up the classification among experts boosted discussion on definitions and conceptual assumptions regarding ecosystem services and on their application in a Belgian context..

The inventory list of CICES-Be aims to provide a complete overview of all the potential ecosystem goods and services that can be relevant in the Belgian context (summarized in Table 18-2). The hierarchical approach makes it possible to adapt the classification to more general uses (e.g., mapping on scale Belgium) or to more specific uses (e.g., sustainable management planning at the level of a municipality or a park). It is important to note that a list of (desirable) ES is not a goal by itself, but is rather a starting point to reflect on the underlying functions, processes, and structures, and on how to achieve sustainable ecosystem management.

TABLE 18-2 **Summary of CICES-Be v6**

Section	Division	Group
Provisioning	Nutrition	Biomass
		Potable water
	Materials	Biomass
		Nonpotable water
	Energy	Biomass-based energy sources
Regulation and maintenance	Mediation of waste, toxics, and other nuisances	Soil and water-quality regulation
		Air-quality regulation
		Shielding
	Mediation of flows	Mass flow
		Liquid flow
	Maintenance of physical, chemical, and biological conditions	Lifecycle maintenance, habitat, and gene pool protection
		Pest and disease control
		Soil formation and composition
		Atmospheric composition and climate regulation
Cultural	Physical and intellectual interactions with biota, ecosystems, and land- and seascapes	Natural environment suitable for outdoor activities
	Spiritual, symbolic, and other interactions with biota, ecosystems, and land-/seascapes	Natural surroundings of built-up areas
		Spiritual and/or emblematic

On the one hand, the link to the internationally accepted CICES classification is a great advantage for future international reporting and comparisons. On the other hand, the operationalization of CICES-Be will require further work, such as the development of proper ES indicators, integration of ES and their indicators into environmental reports, consideration of ES in specific sector reports and in debates about societal "hot" issues. It is expected that by applying the CICES-Be inventory in practical cases, additional improvements to the classification scheme will be made in the future.

REFERENCES

1. MA [Millennium Ecosystem Assessment]. 2005. *Ecosystems and Human Well-being: Synthesis*. Washington, DC: Island Press.
2. Costanza, R., D'Arge, R., De Groot, R., Farber, S., Grasso, M., Hannon, B., et al. 1997. The value of the world's ecosystem services and natural capital. *Nature* 387, 253–260.
3. Daily, G.C. 1997. Introduction: What are ecosystem services? In Daily GC (ed.), *Nature's Services: Societal Dependence on Natural Ecosystems*. Washington, DC: Island Press, pp. 1–10.
4. De Groot, R.S., Wilson, M.A., and Boumans, R.M.J. 2002. A typology for the classification, description and valuation of ecosystem functions, goods and services. *Ecological Economics* 41(3), 393–408.
5. Wallace, K.J. 2007. Classification of ecosystem services: Problems and solutions. *Biological Conservation* 139(3–4), 235–246.
6. TEEB. 2010. *The Economics of Ecosystems and Biodiversity: Ecological and Economic Foundations*. London: Earthscan.
7. Costanza, R. 2008. Ecosystem services: Multiple classification systems are needed. *Biological Conservation* 141(2), 350–352.
8. Haines-Young. R., and Potschin, M. 2009. Methodologies for defining ecosystem services. CEM report No.14.
9. Nahlik, A.M., Kentula, M.E., Fennessy, M.S., and Lander, D.H. 2011. Where is the consensus? A proposed foundation for moving ecosystem service concepts into practice. *Ecological Economics* 77, 27–35.
10. Maes, J., Teller, A., Erhard, M., Liquete, C., Braat, L., Berry, P., Egoh, B., Puydarrieux, P., Fiorina, F, Santos, F., Paracchini, M.L., Keune, H., Wittmer, H., Hauck, J., Fiala, I., Verburg, P.H., Condé, S., Schägner, J.P., San Miguel, J., Estreguil, C., Ostermann, O., Barredo, J.I., Pereira, H.M., Stott, A., Laporte, V., Meiner, A., Olah, B., Royo Gelabert, E., Spyropoulou, R., Petersen, J.E., Maguire, C., Zal, N., Achilleos, E., Rubin, A., Ledoux, L., Brown, C., Raes, C., Jacobs, S., Vandewalle, M., Connor, D., and Bidoglio, G. 2013. Mapping and Assessment of Ecosystems and Their Services. An analytical framework for ecosystem assessments under action 5 of the EU biodiversity strategy to 2020. Publications Office of the European Union, Luxembourg.
11. Haines-Young, R., and Potschin, M. 2010a. Proposal for a common international classification of ecosystem goods and services (CICES) for integrated environmental and economic accounting. European Environment Agency.
12. Potschin, M., and Haines-Young, R.H. Common International Classification of Ecosystem Services (CICES): 2011 Update. Paper prepared for discussion at the expert meeting on ecosystem accounts organised by the UNSD, the EEA and the World Bank, London, December 2011. Centre for Environmental Management School of Geography, University of Nottingham/ European Environment Agency.
13. Haines-Young, R., and Potschin, M. 2013. CICES V4.3—Revised report prepared following consultation on CICES Version 4, August–December 2012. EEA Framework Contract No EEA/IEA/09/003.
14. Fisher B., Turner R. K., and Morling, P. 2009. Defining and classifying ecosystem services for decision making. *Ecological Economics* 68(3), 643–653.
15. Fisher, B., Turner, R.K., Zylstra, M., Brouwer, R., de Groot, R., Farber, S., Ferraro, P., Green, R., Hadley, D., Harlow, J., Jefferiss, P., Kirkby, C., Morling, P., Mowatt, S., Naidoo, R., Paavola, J., Strassburg, B., Yu, D. and Balmford, A. 2008. Ecosystem services and economic theory: Integration for policy-relevant research. *Ecological Applications* 18(8), 2050–2067.
16. Chan, K.M.A., Satterfield, T., and Goldstein, J. 2012. Rethinking ecosystem services to better address and navigate cultural values. *Ecological Economics* 74, 8–18.

17. Haines-Young, R., and Potschin, M. 2010a. The links between biodiversity, ecosystem services and human well-being. In Raffaelli, D., and C. Frid (eds.), *Ecosystem Ecology:A New Synthesis*. BES Ecological Reviews Series. Cambridge: Cambridge University Press, pp.110–139.

18. Potschin, M., and Haines-Young, R. 2011. Ecosystem services: Exploring a geographical perspective. *Progress in Physical Geography* 35(5), 575–594.

19. De Groot, R.S. (1992). *Functions of Nature: Evaluation of Nature in Environmental Planning, Management and Decision Making*. Groningen: Wolters-Noordhoff.

20. Brown, T.C., Bergstrom, J.C., and Loomis, J.B. 2007. Defining, valuing, and providing ecosystem goods and services. *Natural Resources Journal* 47(2), 329–376.

21. Hooper, D.U., Chapin, F. S., Ewel, J.J., Hector, A., Inchausti, P., Lavorel, S., Lawton, J.H., Lodge, D.M., Loreau, M., Naeem, S., Schmid, B., Setälä, H., Symstad, A. J., Vandermeer, J. and Wardle, D.A. 2005. Effects of biodiversity on ecosystem functioning: A consensus of current knowledge. *Ecological Monographs* 75, 3–35.

22. Mace, G.M., Norris, K., and Fitter, A.H. 2011. Biodiversity and ecosystem services: A multi-layered relationship. *Trends in Ecology and Evolution* 27(1), 19–26.

23. Cardinale, B.J., Duffy, J.E., Gonzalez, A., Hooper, D.U., Perrings, C., Venail, P., Narwani, A., Mace, G.M., Tilman, D., Wardle, D.A., Kinzig, A.P., Daily, G.C., Loreau, M., Grace, J.B., Larigauderie, A., Srivastava, D.S., and Naeem, S. 2012. Biodiversity loss and its impact on humanity. *Nature* 486, 59–67.

24. Boyd, J., and Banzhaf, S. 2007. What are ecosystem services? The need for standardized environmental accounting units. *Ecological Economics* 63, 616–626.

25. Fisher, B., and Turner, K. 2008. Ecosystem services: Classification for valuation. Biological *Conservation* 141, 1167–1169.

26. Relph, E. 1985. Geographical experiences and being-in-the-world: The phenomenological origins of geography. In Seamon, D., and Mugerauer, R. (eds.), *Dwelling, Place and Environment*, pp. 15–31. New York: Columbia University Press.

27. Manzo, L. 2005. For better or worse: Exploring multiple dimensions of place meaning. *Journal of Environmental Psychology* 25, 67–86.

28. Lyytima, J.J., and Sipila, M. (2009). Hopping on one leg–The challenge of ecosystem disservices for urban green management. *Urban Forestry and Urban Greening* 8, 309–315.

29. Schmidt, K.A., and Ostfeld, R.S. 2001. Biodiversity and the dilution effect in disease ecology. *Ecology* 82, 609–619.

30. Vanwambeke, S.O., Lambin, E.F., Eichhorn, M.P., Flasse, S.P., Harbach, R.E., Oskam, L., Somboon, P., Van Beers, S., Van Benthem, B.H.B., Walton, C., and Butlin, R.K. 2007. Impact of land-use change on dengue and malaria in northern Thailand. *Ecohealth* 4, 37–51.

31. De Boer, W.F., and Baquete, D.S. 1998. Natural Resource use, crop damage and attitudes of rural people in the vicinity of the Maputo Elephant Reserve, Mozambique. *Environmental Conservation* 25(3), 208–218.

32. Rao, K.S, Maikhuri, R.K., Nautiyal, S., and Saxena, K.G. 2002. Crop damage and livestock depredation by wildlife: A case study from nanda Devi Biosphere Reserve, India. *Journal of Environmental Management* 66(3), 317–327.

33. DeStefano, S., and Deblinger, R.D. 2005. Wildlife as valuable natural resources vs intolerable pests: a suburban wildlife management mode. *Urban Ecosystems* 8, 131–137.

34. Koskela, H., and Pain, R. 2000. Revisiting fear and place: Women's fear of attack and the built environment. *Geoforum* 31, 269–280.

35. Dunn, R.R. 2010. Global mapping of ecosystem disservices: The unspoken reality that nature sometimes kills us. *BIOTROPICA* 42(5), 555–557.

36. Wallace, K.J. (2008). Ecosystem services: Multiple classifications or confusion? *Biological Conservation* 141, 353–354.

37. Church, A., Burgess, J., Ravenscroft, N., Bird, W., Blackstock, K., Brady, E., Crang, M., Fish R., Gruffudd, P., Mourato, S., Pretty, J., Tolia-Kelly, D., Turner, K., and M. Winter. 2011. Cultural Services. In The UK National Ecosystem Assessment Technical Report. UK National Ecosystem Assessment, UNEP-WCMC, Cambridge, 633–692.

38. Staub, C., Ott, W., Heusi, F., Klingler, G., Jenny, A., Häcki, M., and Hauser, A. 2011. Indicators for Ecosystem Goods and Services: Framework, methodology and recommendations for a welfare-related environmental reporting. Federal Office for the Environment, Bern. *Environmental Studies*, no. 1102, 17 S.

Chapter 19

The Ecosystem Services Valuation Tool and its Future Developments*

Inge Liekens[1], Steven Broekx[1], Nele Smeets[1], Jan Staes[2], Katrien Van der Biest[2], Marije Schaafsma[3], Leo De Nocker[1], Patrick Meire[2] and Tanya Cerulus[4]

[1]*Flemish Research and Technology Organisation (VITO)*, [2]*University of Antwerp, Ecosystem Management Research Group (ECOBE)*, [3]*University of East Anglia, Centre for Social and Economic Research on the Global Environment, England, United Kingdom*, [4]*Environment, Nature and Energy Department, Environmental, Nature and Energy Policy Division, Government of Flanders.*

Chapter Outline

1. INTRODUCTION

Humankind benefits from a multitude of resources and processes that are supplied by natural ecosystems, collectively referred to as ecosystem services and goods [1]. Degradation of the world's ecosystems during the past 50 years due to urban expansion, agricultural intensification, and industrialization has led to a serious decline in ecosystem service delivery [2]. The key challenge of policy making today is to prevent or reduce this incessant degradation of ecosystems and their services while meeting the increasing demands of society.

Flanders is an example of a region facing ecosystem degradation and the loss of ecosystem services and replacement of these services by costly technical measures and infrastructure in aspects as flood prevention, drought

---

* Based on *Environmental Impact Assessment Review*, Broekx S., et al., "Environmental Impact Assessment Review," 65-74 © 2013 with permission of Elsevier.

Ecosystem Services. http://dx.doi.org/10.1016/B978-0-12-419964-4.00019-6

management, climate change policy, and health care [3]. The Flemish Region is encountering enormous challenges to improve the quality of the environment in order to comply with EU environmental standards and conserve the natural capital to guarantee the health and quality of life of its inhabitants. Current measures and actions are not adequate to reach these standards and already generate very high costs for environmental management. The efforts that are needed to maintain both environmental quality and the different functions are pushing the limits of acceptable cost efficiency and societal support.

Since the Millennium Ecosystem Assessment [2], the services natural ecosystems deliver are being more and more recognized (e.g., [4, 5]). This is supported by a rapidly growing amount of literature and models on ecosystem service classification, quantification, and valuation.

Although methodologies for classification, quantification, and valuation are improving, applications of the ecosystem services concept are mainly restricted to illustrating the importance of preserving or restoring ecosystems in regional to global ecosystem service mapping or ecosystem services accounting. Important examples of large-scale ecosystem service assessments include the National Ecosystem Assessment in the UK [6], the Natural Capital/INVEST project [7], and the Valuing the Arc initiative [8]. Its use in day-to-day decision-making processes remains limited, however, especially at the planning level [9].

Both limited interest among geographers and excessive complexity of currently available models are depicted in the literature as major reasons for this lack [4, 5, 10]. Frequently occurring mismatches between the spatial scale of research and the spatial scale of applications can be another reason for limited applicability of current research in spatial planning [11]. Nevertheless, spatial planning decisions would benefit from systematic considerations of their effects on ecosystem services [12]. Estimating the impacts of policy on a wide range of ecosystem services can also serve as an element in the development of more cost-effective policy implementation, establishing win-win situations across different environmental domains such as water, air, and climate change. To date, most tools for environmental impact assessment (e.g., cost-benefit analysis, strategic environmental assessment, and life cycle analysis) do not include impacts on ecosystems (alterations in vegetation and biodiversity). In this chapter, we present the "nature value explorer" (*natuurwaardeverkenner* in Dutch), a web application specifically built to explore the quantity and value of ecosystem services in day-to-day decision making in Flanders. Belgium, as part of a, for instance, cost-benefit analysis (CBA). The nature value explorer combines spatially sensitive and site-specific inputs with generic quantification and valuation functions, allowing effective and straightforward identification of service providing areas to support spatial planning in Flanders. The application is developed to estimate the impact of land-use and land-cover change on ecosystem services. It does not address degradation of habitat quality.

As the end-user perspective is a crucial first step in the design of practical tools, we start from an inventory of user requirements for quantification and valuation of ecosystem services in Flemish policy making in Section 2.

These requirements are used to define the design characteristics of the web application, described in Section 3. Section 3 also describes the applied methodology for quantification and valuation. The use and an example are found in Section 4. The conclusion appears in Section 5.

2. USER REQUIREMENTS

User requirements and potential policy applications were derived from 26 individual end-user consultations. The end-users are a mix of organizations involved in policy preparation, policy execution, policy evaluation, and civil society organizations. The list covers all actors with a prominent role in the management of the open space in Flanders (recreation, agriculture, nature, and water management).

An important conclusion that can be drawn from all consultations is that the general interest from potential end-users in Flanders in the ecosystem services concept is very large. Not only typical nature conservation administrations or civil society organizations express an interest but also end-users focused on spatial planning, agriculture, land, and water management consider a more in-depth knowledge of ecosystem services as added value for policy making.

The expected advantages of applying ecosystem service-based approaches confirm the typical advantages listed in [13]. Demonstrating the importance of nature and biodiversity and arguing for the protection of existing nature or for additional nature development were often mentioned. A clear common need exists in an operational decision support tool/instrument for both balancing different land-use types and developing specific nature areas as for global cost efficiency of measures over different policy domains. For use in cost-benefit analysis (CBA) and other decision-support tools, different end-users were eager to have easy-to-calculate indicator data or supply and demand maps for quantifying as well as for monetizing ecosystem services, taking into account spatial aspects.

Important requirements listed specifically for the tool were user-friendliness, transparency, flexibility and scientific reliability. User-friendliness is especially important for nonspecialist users. The tool needs to make clear what a specific service exactly means, how this service can be quantified and valued, and where required input data can be found. Reflecting the complexity of the ecosystem processes and accuracy on the other hand are very important for specialist users. Unfortunately, these properties do not match easily, and trade-offs between accuracy and applicability are unavoidable.

3. METHODOLOGY

An ecosystem goods and services approach was followed to develop the tool. First, a list was compiled identifying the important ecosystems in Flanders. Second, the international classification of ecosystem services (CICES) was used to list all possible ecosystem services [4]. Then quantification and value functions were developed based on different methods and studies for use within value transfer exercises. For the provisioning services, crop production and wood production were considered. For cultural services, we consider the recreation, amenity, and nonuse value. Regulating services include nutrient retention and climate regulation (sequestration in soils and biomass), air quality regulation, and noise mitigation.

Table 19-1 presents an overview of the ecosystem goods and services included in the tool.

The web application does not allow for detailed spatially explicit ecosystem service quantification and grid-based computations. Instead, a flexible system of service providing units (SPU) for which end-users can define specific properties (e.g., soil characteristics, vegetation type), and thus the potential to vary the spatial detail was advocated. If budgets are more extensive and the availability

TABLE 19-1 **Quantified and Valued Ecosystem Services in the Tool**

	Service	Quantification Method / Important Variables	Valuation Method
Cultural services	Recreation, amenity, and nonuse value	Choice experiment with attributes such as size, accessibility, nature type, surrounding environment	
Provisioning services	Crop production	Standard gross margin	
	Wood production	Potentially produced volume and harvest factor	
Regulating services	Water quality regulation: Denitrification	Seitzinger: residence time Pinay: soil moisture and texture	Avoided cost method for N
	Climate change: C sequestration in soils	Meersman: soil drainage, vegetation type and soil texture	Avoided cost method for C
	Water quality regulation: N, P sequestration in soils	C/N/P ratios	Avoided cost method for N and P
	Climate change: C-sequestration in forest biomass	Wood increment and species-specific carbon density	Avoided cost method C
	Air quality: removal of PM_{10}	Removal factors in Oosterbaan et al. 2006	Avoided damage costs
	Noise mitigation	Huisman 1990: noise level, width forest	Hedonic pricing

of data is not an issue, users can decide to define for each scenario a large amount of SPUs on a relatively small scale (up to 1 ha).

The tool can be consulted on the Internet via http://www.natuurwaarde-verkenner.be. End-users are able to create and save scenarios, share scenarios with other registered users, and consult public scenarios. Interactive discussions are stimulated through a discussion forum. User-friendliness is increased by adding information boxes explaining each service and its required input data, a section with frequently asked questions and an information page containing background documents and publications related to the nature value explorer.

3.1. Cultural Services

Cultural services include use values related to recreation, amenity and education, and nonuse values related to bequest values and existence values. For all of these individual services, specific quantification methods and valuation techniques can be used or stated preference techniques that are able to capture all cultural services in single willingness to pay estimates. A stated preference study (choice experiment) surveying people's willingness to pay for nature restoration was performed to capture all cultural services in a single value function. This experiment was described in detail in [14]. The idea behind the development was that the value of a nature area is not captured by one characteristic but depends on a number of characteristics of the area studied; on the characteristics of the beneficiaries, that is, the people who attach a value to this area; and on spatial characteristics such as size and distance [15]. In a choice experiment (CE), respondents are presented with a number of alternatives from which they are asked to choose. The alternatives can be a good or a service, characterized in terms of different attributes, but also policy alternatives or land-use change scenarios [16]. Each alternative is defined in terms of the same attributes, including a price, but with different values (attribute levels). Examples include varying levels in biodiversity (high-low), accessibility (accessible or not), and size of the area (between 1 and 200 ha).

Respondents usually are shown two or three alternatives on a choice card and an option that allows them to choose none of the two, also referred to as the opt-out. In the case of land-use changes, this opt-out can also represent the current situation or no land-use change. As respondents express their preferences by making choices between different alternatives, they trade off the different attributes and levels. A statistical function can then be estimated that links choice probabilities to the characteristics of the alternatives. The trade-off between price and other attributes is especially relevant, as this reflects how much a respondent is willing to pay (WTP) for a particular change in this attribute. This allows determining marginal values for changes in the attributes and combinations of attributes.

In the choice experiment for the nature value explorer, respondents are asked to choose between different land-use changes related to the creation of different types of nature area, with different spatial and nonspatial characteristics and impacts on their current tax levels. Agricultural land use, with no particular nature or landscape value, is the reference situation in the rural areas where these land-use changes can take place.

Based on the information obtained in focus groups with laypeople and expert interviews, seven attributes were included in the CE: nature type, including marshes, natural grasslands, forests, open water and swamps, heath land, inland dunes, and pioneer vegetation; species richness; and spatial attributes, including size of the area, accessibility, surrounding land use, and distance to the respondents' residence. Finally, the monetary attribute is a mandatory annual tax to be paid by all Flemish households to a fund exclusively used for the creation and conservation of nature areas in Flanders. The data were obtained from an Internet survey conducted through a marketing bureau panel from which respondents were randomly chosen in three different provinces of Flanders. A total of 3000 residents filled out the survey. After removing incompletes (no choice section) and protest bidders (6%), approximately 2300 respondents (approximately 10,000 observations) were included in the analysis. The analysis of the sociodemographic information of the respondents suggests that the sample is mostly representative for the Flemish population (see [13]).

The resulting value function is used to value a nature area, according to some selected biophysical characteristics (nature type, size, surrounding land use, access, and species richness) as well as household-related characteristics (income, mean age, member of nature organizations, distance to the created nature area). This function for additional nature development in Flanders expressed in annual € per household can be written as:

WTP = 122 * pioneer vegetation + 93 * mudflat and marsh + 92 * natural grass land + 157 * forest + 133 * open water, reed and swamp + 133 * heath land and inland dunes + 0,05 * size in ha + 28 * species + 34 * availability of walking trails − 0,63 * distance in km + 8 * natural surroundings + 8 * residential surroundings − 15 * industrial surroundings − 0,36 * high number of species * age + 0,01 * monthly net income - 37 * % women+ 108 *% membership.

The respondents are willing to pay more for easily accessible nature, but it is not dominant over the other attributes, so people also attach a high value to nonaccessible nature. The nature type is important. Forests are valued most, followed by open water, swamps, and heath land. Pioneer vegetation, marshes, and grasslands are valued least.

As distance to the respondent's residence was an attribute in the survey, we automatically generated a distance decay function [15], which adjusts individual WTP downward as respondents live further away from the proposed land-use change. The number of households with positive WTP can thus be based on our empirical findings, rather than on arbitrary assumptions about the relevant spatial size of the economic market. By combining the value function with GIS data on number of households per spatial unit in the surroundings (up to 50 km), sociodemographic data, and distances to the created area, the total amenity and nonuse value of the land-use change can be calculated.

As the choice experiment asked for the utility for one area with specific charac-teristiscs between 10 ha and 200 ha and substitution effects are not yet included in the function, we followed a pragmatic approach in downsizing the parameters of the value function on the basis of the size parameter so that the function can also be used for upscaling to a more regional level or if multiple areas will be created.

3.2. Provisioning Services

Crops such as grains, vegetables, and fruits are the cultivated plants or agricul-tural products harvested by people for human or animal consumption as food. Agricultural services may under some schemes not be considered as ecosystem services but are referred to as environmental services. In this assessment, they are considered as ecosystem services. The main argument is that including pro-visioning services derived from agriculture or agro-ecosystems is essential in a trade-off analysis. Furthermore, agricultural systems comply in a strict sense with the definition of an ecosystem [17].

For the valuation of this ecosystem service, we use a pricing approach and try to estimate the market prices for animals and crops. The estimated value of the biodiversity resource based on market price is equal to the quantity of sold resource x (market price – costs related to production). We take the standard gross margin as indicator. The meat and dairy standard gross margins are linked to fodder production and so are linked to land use.

Wood can be used for different purposes ranging from construction material to packaging, raw material, and energy. We use the Moonen et al. [18] study to esti-mate the potentially produced volume, and we multiply this with a harvest factor to determine the actually produced volume. The produced volumes are generated only for the eight main commercial species (Beech, Oak, Poplar, European Larch, Scots Pine, Corsican Pine, Spruce, Douglas fir, deciduous and coniferous) per soil type. The soil type is based on the Belgian classification system (texture, drain-age, and profile), http://geo-vlaanderen.gisvlaanderen.be/geo-vlaanderen/bodem-kaart/. Market prices per lot are used to estimate the value per m³.

3.3. Regulating Services

Quantification

Valuing the change in quantities of different regulating services is a complex, but crucial, element in the valuation of impacts on ecosystems (or the creation of new ecosystems). We often lack tools and models to assess the changes in physical, biochemical, and ecological processes on the delivery of ecosystem services. The nature value explorer offers less detail than some specific models that focus in detail on a single service or area, but offers a more detailed and accurate assessment than fixed €/ha values per vegetation type. The latter are not preferred as for spe-cific services, the vegetation type is not the major factor influencing the magnitude of the service and is insufficient to capture the spatial variation in the delivery of ecosystem services. At the same time, extensive, process-based model calculations were considered too complicated and too computationally intensive to include in a web application to explore the impact on ecosystem services. Instead, quantification functions were developed that on the one hand take into account the main driving factors of the underlying ecological processes such as soil texture, groundwater level, and vegetation type and on the other hand require little computation time. The quantification functions build on regional datasets (existing land-use/land-cover and soil map classifications) and studies to increase the accuracy and transparency.

The quantification of denitrification processes in wetland ecosystems is based on [19]. Removal efficiency depends mainly on the residence time of the water in the ecosystems. For terrestrial ecosystems we used [20] to deduct potential denitrification. Removal efficiency depends on soil moisture and soil texture.

Carbon sequestration in soils is based on estimates from [21]. Meersman et al. performed a multiple regression approach to assess the spatial distribution of Soil Organic Carbon (SOC) and its dependency on soil characteristics in Flanders, Belgium. Meersman et al determined a potential maximal carbon content for a given soil drainage, vegetation type, and soil texture. Changes in soil drainage and/or vegetation will change the potential maximal carbon content. The annual carbon sequestration potential is a percentage of the difference in potential and actual carbon content. This approach is process based and incorporates changes in potential storage and the associated temporal dynamics. Literature estimates of net ecosystem exchange are broad ranging, as they capture a moment and do not incorporate long-term dynamics and driving variables such as soil properties, climate, and soil hydrology. The N and P content of soils is indirectly derived from the carbon content. Based on analyses performed in Flanders, the C/N ratio varies between 10 and 30, depending on the nature type. Based on [22], we set the average N/P ratio at 15.

Carbon sequestration in forest biomass is linked to the method for wood production. The increment in biomass per ha per year is converted in the annual carbon sequestration per ha per year using the species-specific carbon density [23].

It is well documented that trees and vegetation can serve as effective sinks for air pollutants and PM10 (particulate matter < 10 μg) and thus contribute to air quality improvement and related public health benefits [24, 25]. As PM10 is the most important pollutant, accounting for 60% of health impacts from environmental pollution [26], the focus of this analysis is on PM10. Because there are no data available for Flanders, the estimates in the web application are based on removal factors for individual trees and shrubs from [27] and [28]. The removal factors (expressed in kg/ha) are in the same range (+/– 50 %) of those used by [24] or [25] for grasslands. Trees and vegetation also have an impact on other air pollutants, but there is more uncertainty about the removal factors and on the valuation of sinks.

Nature areas can contribute to the mitigation of noise from, for example, traffic. The effect of the soil and especially the vegetation is often underestimated in models for noise simulation [29, 30]. The service is only important when people are affected. Noise mitigation for soft soils and forests is derived from [29] and [31]. Huisman [29] measured the decrease in decibel (dB(A)) based on the frequency of the source, the soil characteristics, the meteorological effects, and the noise penetration in the forest. He found an average decrease of 6–16 dB(A)for 100- to 300-m-wide forests.

Monetary Valuation

We use a combination of avoided abatement costs (for nutrients and carbon sequestration), damage costs (for air pollution), and hedonic pricing (for noise mitigation). The avoided abatement cost method is used to value nutrient removal because, due to the natural denitrification that an ecosystem delivers, costly abatement measures to obtain environmental goals can be avoided. The value of an additional kg nitrogen removed by an ecosystem can be derived from the marginal cost curve of nitrogen removal. This cost curve was calculated in preparation for the Flemish river basin management plan to reach a good water status according to the European Water Framework Directive [32,33]. The costs of the measure with the highest marginal cost included in the program of measures to reach water quality objectives are 74€/kg N and 800€/kg P. Most measures have an impact on both N and P, and it is therefore impossible to individually link avoided costs to separate pollutants. To avoid double-counting, we estimate the value of nutrient retention for both pollutants but only apply the maximum value. The valuation of nutrients applied here is significantly higher than figures in literature, which vary between 2 and 20 €/kg for N. [34,35] and 70 €/kg for P [36]. This reflects on the one hand the fact that nutrients are a large problem in Flanders and on the other hand that already a lot of relatively cheap measures (e.g., advanced treatment in wastewater treatment plants) are taken and less cost-effective measures are necessary to reach environmental objectives.

The benefits of carbon sequestration are not directly related to the place of sequestration, but rather are experienced at a global level, through the impact on climate change. To assess the value of carbon sequestration by ecosystems, theoretically two approaches can be followed: [1] marginal damage costs; and [2] avoided abatement costs. As impacts are global, the selected data are based on studies at the global level. The results of these studies range very broadly from an external cost close to zero to 160 $1995 /tonne CO_2 [37–39]. It is difficult to pick a meaningful average for these studies inasmuch as it varies a lot in function of statistical models, inclusion of gray literature, and older studies. Against this background, we have chosen a central value of 50 €/tonne CO_2-eq., which is higher than recommended values for emissions in 2010 and close to recommended values for 2020. This value corresponds to 183 €/tonne of C. As this central estimate does not, however, reflect the value of CO_2 sequestration in the long run, it is recommended that a higher value of 200 €/ton CO_2-eq (737 €/tonne C) be used for sensitivity analysis, especially for projects in which C-sequestration in the longer run may be an important part of total benefits.

Air quality improvement for PM_{10} has important benefits for public health, especially related to cardiovascular and respiratory impacts [41]. These impacts are typically valued using indicators related to avoided costs for health care and medicine, loss of productivity at the workplace and at home, and willingness to pay to avoid suffering and loss of life expectancy. The data are based on results from air quality models for Flanders, dose-response functions, and valuation data from European research projects [40, 41]. We further account for the size and origin of the particles . This results in a value of 54€/kg PM_{10}.

To assess noise mitigation, we used a noise sensitivity depreciation index based on the results of two large studies using hedonic pricing [42, 43]. Market value of properties decrease with 0.4% per dB(A) at lower noise levels (40 dB(A)) and 1.9% at higher noise levels (60 dB(A)).

4. USING THE INFORMATION

The use of all the numbers and functions in the tool are explained and illustrated in a separate manual. The manual bundles the methods and functions to quantify and value the ecosystem services, and gives the assumptions made. To help users better understand calculation procedures, each function is illustrated.

This manual and tool are not static. It was not possible to derive quantification functions for all ecosystem services. The quantification and valuation functions that are presented are built on the current state of knowledge and on data availability, but can be improved in the future when new scientific insights emerge and /or better data is available. The manual and reports can be found on the website of the nature value explorer.

Below we give an hypothetical example of a case calculated with the nature value explorer. The case is the creation of a 190 ha wetland on a agricultural land (fields and meadows). The following output (Figure 19-1) is given after filling out the requested information on soil type, drainage, and groundwater levels. More detailed information on the calculations can be exported.

Kwantificering van ecosysteemdiensten

	Gebied	Eenheid
Ecosysteemdienst	**opstal**	
Belevings- en overdrachtswaarde:	1 052 582	huishoudens in 50km
Nitraatverwijdering via biologische denitrificatie:	0	kg N/jaar
C opslag in de bodem:	230	ton C/jaar
N opslag in de bodem:	-4 360	kg N/jaar
P opslag in bodem:	-291	kg P/jaar
C opslag in de strooisellaag en biomassa van bossen:	0	ton C/jaar
N opslag in de strooisellaag en biomassa van bossen:	0	kg N/jaar
P opslag in de strooisellaag en biomassa van bossen:	0	kg P/jaar
Verbetering luchtkwaliteit(vnl. fijn stof):	-265	kg PM/jaar

Waarde van ecosysteemdiensten in euro/jaar

	Gebied	Totaal
Ecosysteemdienst	**opstal**	
Belevings- en overdrachtswaarde:		9 827 901
Nitraatverwijdering via biologische denitrificatie:	0	0
C opslag in de bodem:	42 061	42 061
N opslag in de bodem:	-322 668	-322 668
P opslag in bodem:	-232 553	-232 553
C opslag in de strooisellaag en biomassa van bossen:	0	0
N opslag in de strooisellaag en biomassa van bossen:	0	0
P opslag in de strooisellaag en biomassa van bossen:	0	0
Verbetering luchtkwaliteit(vnl. fijn stof):	-7 939	-7 939
Totaal:	-198 431	9 629 470

Let op: De natuurwaardeverkenner berekent niet de intrinsieke/absolute waarde van natuur. Deze is oneindig aangezien de mens van de natuur afhankelijk is voor zijn overleving. De natuurwaardeverkenner berekent de sociaaleconomische waarde van een (negatieve of positieve) verandering in de bijdrage die de door u bestudeerde natuurgebieden leveren aan de menselijke welvaart. Deze sociaal-economische waarde moet binnen een ruimer beslissingskader geplaatst worden naast andere niet-economische doelstellingen zoals instandhoudingsdoelstellingen.

Local intranet

FIGURE 19-1 Output of the explorer for a hypothetical wetland restoration project.

5. CONCLUSION

The need for a valuation tool such as is described in this chapter is illustrated by the success of the webtool. Since its launch in September 2010, approximately 120 users have registered and 200 scenarios have been simulated. The tool was originally set up for use in cost-benefit analysis for large infrastructure projects with an impact on nature. Administrations also explicitly refer to the tool when setting up new cost-benefit analyses related to infrastructure projects. In addition to this purpose, environmental NGOs see the tool as a means to demonstrate the value of nature areas and to motivate investments in nature development/restoration. Other organizations see the tool as a support for payments for ecosystem services. First cases for which the tool was applied are for policy appraisal on infrastructure decisions (transportation infrastructure) to support the development of effective flood risk management plans, to advocate the protection of existing natural areas, and to support the design of green, built-up areas.

The application illustrates the possibilities and limitations of a simple, ready to use assessment tool to provide scientifically based information for decision making and interaction with stakeholders. Even if some of the scientific underpinnings are subject to debate among scientists and the use of simplified models introduces additional uncertainty, it allows nonspecialists to get an impression of the relative importance of different ecosystem services.

This tool is a first step. The methods and data are open for further elaboration, refinement, and updates. Improvements require a transparent and open framework that has the advantage of frequent end-user feedback. This will help to trace methodological errors, define the focus for further development, and refine the tool to better fit end-users' needs. During ensuing years, the tool will be improved to make further applications possible and will be expanded with more ecosystem services, when new scientific information becomes available. Increasing end-user interaction is also a key feature contributing simultaneously to user-friendliness and scientific reliability. The objective is to establish a learning system allowing end-users to define case-specific input values, pose questions, exchange results and experiences on a discussion forum, and improve calculation procedures. Overall, the web application is aimed at providing a platform where the stakeholders of ecosystem services can exchange knowledge and further enhance practical methodologies. These knowledge exchanges are paramount to develop a multidisciplinary research domain as ecosystem services assessment.

REFERENCES

1. Daily, G.C. 1997. Valuing and safeguarding Earth's life support systems. In *Nature's Services: Societal Dependence on Natural Ecosystems* (Postel S, Bawa K, Kaufman L, Peterson, C.H., Carpenter, S., Tillman, D., et al., eds.), pp. 365–374. Washington, DC: Island Press.
2. Millennium Ecosystem Assessment. 2005. *Ecosystems and Human Well-being: Synthesis.* Washington, DC: Island Press.

3. MIRA Indicatorrapport. 2010. Marleen Van Steertegem (eindred.), Milieurapport Vlaanderen, Vlaamse Milieumaatschappij (in Dutch).

4. Haines-Young, R., and Potschin, M. 2011. Common International Classification of Ecosystem Services (CICES). London: European Environment Agency. http://unstats. un. org/unsd/envaccounting/seeaLES/egm/Issue8a.pdf

5. Seppelt, R., Fath, B., Burkhard, B., Fisher, J.L., Grêt-Regamey, A., and Lautenbach, S. 2011. Form follows function? Proposing a blueprint for ecosystem service assessments based on reviews and case studies. *Ecological Indicators* 21,145–154.

6. Bateman, I.J. 2011. Valuing changes in ecosystem services: Scenario analyses. The UK National Ecosystem Assessment: Technical Report. Cambridge: UNEP-WCMC.

7. Kareiva, P., Tallis, H., Ricketts, T.H., Daily, G.C., and Polasky, S. 2011. *Natural Capital: Theory and Practice of Mapping Ecosystem Services.* Oxford: Oxford University Press.

8. Fisher, B., and Turner, R.K. 2008. Ecosystem services: Classification for valuation. *Biological Conservation* 141, 1167–1169.

9. Daily, G.C., and Matson, P.A. 2008. Ecosystem services: From theory to implementation. *Proceedings of the National Academy of Sciences* 105, 9455–9456.

10. Koschke, L., Fürst, C, Frank S., and Makeschin, F. 2012. A multi-criteria approach for an integrated land-cover-based assessment of ecosystem services provision to support landscape planning. *Ecological Indicators* 21, 54–66.

11. Meinke, H., Nelson, R., Kokic, P., Stone, R., Selvaraju, R., and Baethgen, W. 2006. Actionable climate knowledge: From analysis to synthesis. *Climate Research* 33,101–110.

12. Geneletti, D. 2011. Reasons and options for integrating ecosystem services in strategic environmental assessment of spatial planning. *International Journal of Biodiversity Science, Ecosystem Services and Management* 7(3),143–149.

13. TEEB. 2010. The Economics of Ecosystems and Biodiversity: Ecological and Economic Foundations.

14. Liekens, I., Schaafsma, M., De Nocker, L., Broekx, S., Staes, J., Aertsens, J., and Brouwers, R. 2013. Developing a value function for nature development and land use policy in Flanders, Belgium. *Land Use Policy* 30(1), 549–559.

15. Bateman, I.J., Georgiou, S., and Lake, I. 2006. The aggregation of environmental benefit values: Welfare measures, distance decay and BTB. *BioScience* 56(4), 311–325.

16. Louviere, J.J., Hensher, D.A., and Swait, J.F. 2000. *Stated Choice Methods: Analysis and Applications.* Cambridge: Cambridge University Press.

17. Maes, J., et al. 2011. A spatial assessment of ecosystem services in Europe: Methods, case studies and policy analysis—phase 1. PEER Report No 3. Ispra: Partnership for European Environmental Research.

18. Moonen, P., Kint, V., Deckmyn, G., and Muys, B, 2011. Wetenschappelijke onderbouwing van een lange termijnplan houtproductie voor Bosland. Eindrapport opdracht LNE/ANB/LIM-2009/19

19. Seitzinger, S., Harrison, J. A., Bohlke, J. K., Bouwman, A. F., Lowrance, R., Peterson, B., et al. 2006. Denitrification across landscapes and waterscapes: A synthesis. *Ecological applications* 16(6), 2064–2090.

20. Pinay, G., Gumiero, B., Tabacchi, E., Gimenez, O., Tabacchi-Planty, A. M., Hefting, M. M., et al. 2007. Patterns of denitrification rates in European alluvial soils under various hydrological regimes. *Freshwater Biology* 52(2), 252–266.

21. Meersman, J., De Ridder, F., Canters, F., Debaets, S., and Van Molle, M. 2008. A multiple regression approach to assess the spatial distribution of Soil Organic Carbon (SOC) at the regional scale (Flanders, Belgium). *Geoderma* 143(1–2), 1–13.

22. Koerselman, W., and Meuleman, F.M. 1996. The vegetation N:P ratio: A new tool to detect the nature of nutrient limitation. *Journal of Applied Ecology* 33(6), 1441–1450.

23. Van de Walle, et al. 2005. Growing stock-based assessment of the carbon stock in the Belgian forest biomass. *Annals of Forest Science* 62, 1–12.

24. Nowak, D. J., Crane, D. E., and Stevens, J. C. 2006. Air pollution removal by urban trees and shrubs in the United States. *Urban Forestry and Urban Greening*, 4(3–4), 115–123.

25. Tiwary, A., Sinnett, D., Peachey, C., Chalabi, Z., Vardoulakis, S., Fletcher, T., Leonardi, G. Grundy, C., Azapagic, A., and Hutchings, T.R. 2009. An integrated tool to assess the role of new planting in PM 10 capture and the human health benefits: A case study in London. *Environmental Pollution* 157, 2645–2653.

26. MIRA. 2008. Brouwers, J., De Nocker, L., Schoeters, K., Moorkens, I., and Jespers, K. Milieurapport Vlaanderen: Achtergronddocument: Thema klimaatverandering. Vlaamse Milieumaatschappij.

27. Oosterbaan, A., and Vries, E.A.D. 2006. *Kleine landschapselementen als invangers van fijn stof en ammoniak* (p. 58).

28. Oosterbaan A., and Kiers, M. 2011. Landelijke kaart "potentiële fijnstofinvang door groene vegetaties" (Alterra Wageningen UR), in Melman, T.C.P. en C.M. van der H. *Ecosysteemdiensten in Nederland: verkenning betekenis en perspectieven. Achtergrondrapport bij Natuurverkenning 2011.* Wageningen.

29. Huisman, W. 1990. Sound Propagation Over Vegetation-covered Ground (Open Library). Retrieved April 22, 2011, from http://openlibrary.org/books/OL12852710M/Sound_Propagation_Over_Vegetation-covered_Ground.

30. Goossen, C.M. and F. Langers (2003). Geluidbelasting in het centraal Veluws Natuurgebied: een quick scan van de geluidbelasting in het Centraal Veluws Natuurgebied in zijn geheel en in afzonderlijke delen die belangrijk zijn voor recreatie. Wageningen, Alterra, Research Instituut voor de Groene Ruimte, Alterra-rapport 798. (p. 56).

31. DeFrance, J., Barriere, N. and Premat, E. 2002. Forest as a meterological screen for traffic noise. In *Proceedings of the 9th International Congress on Sound and Vibration.*

32. Cools, J., Broekx, S., Vandenberghe, V., Sels, H., Meynaerts, E., Vercaemst, P., et al. Orlando, Cd-rom. 2011. Coupling a hydrological water quality model and an economic optimization model to set up a cost-effective emission reduction scenario for nitrogen. *Environmental Modelling & Software* 26(1), 44–51.

33. Broekx, S., Meynaerts, E., Wustenberghs, H.D., Heygere, T., and De Nocker, L. 2011. Setting up a cost effective programme of measures to improve surface water status in the Flemish region of Belgium with the Environmental Costing Model. In Pulido-Velazquez, M., Heinz, I., Lund, J., Andreu, J., Ward, F., and Harou, J. (eds.), Hydro-economic models for water management: Applications to the EU Water Framework Directive.

34. Gren, I.M. 1995. Costs and benefits of restoring wetlands: Two Swedish case studies. *Ecological Engineering* 4(2), 162–153.

35. Jenkins, W.A., Murray, B. C., Kramer, R., and Faulkner, S. P. 2010. Valuing ecosystem services from wetlands restoration in the Mississippi Alluvial Valley. *Ecological Economics* 69(5), 1051–1061.

36. Borjesson, P. 1999. Environmental effects of energy crop cultivation in Sweden II : Economic valuation. *Biomass and Bioenergy* 16, 155–170.

37. Tol, R. 2005. The marginal damage costs of carbon dioxide emissions: An assessment of the uncertainties. *Energy Policy* 33(16), 2064–2074.

38. Stern, N. 2006. *Stern Review: The Economics of Climate Change.* Cambridge: Cambridge University Press.

39. Kuik, O., Brander, L., and Tol, R.S.J. 2009. Marginal abatement costs of greenhouse gas emissions: A meta-analysis. *Energy Policy* 37(4), 1395–1403.
40. Michiels, H., Mayeres, I., Int Panis, L., De Nocker, L., Deutsch, F., and Lefebvre W. 2012. PM2.5 and NOx from traffic: Human health impacts, external costs and policy implications from the Belgian perspective. *Transportation Research Part D: Transport and Environment* 17(8), 569–577.
41. Rabl, A., and Holland, M. 2008. Environmental Assessment Framework for Policy Applications: Life Cycle Assessment, External Costs and Multi-criteria Analysis. *Journal of Environmental Planning and Management* 51(1), 81–105.
42. Theebe, M.A.J. 2004. Planes, trains, and automobiles: The impact of traffic noise on house prices. *The Journal of Real Estate Finance and Economics* 28(2–3), 209–234.
43. Udo, J., Janssen, L.H.J.M., and Kruitwagen, S. 2006. Stilte heeft zijn prijs. *Economisch Statistische Berichten* 91(4477).

Chapter 20

EBI—An Index for Delivery of Ecosystem Service Bundles[*]

Katrien Van der Biest[1], Rob D'Hondt[2], Sander Jacobs[4,5], Dries Landuyt[2,3], Jan Staes[1], Peter Goethals[2] and Patrick Meire[1]

[1]*University of Antwerp, Ecosystem Management Research Group (ECOBE),* [2]*University of Ghent, Laboratory of Environmental Toxicology and Aquatic Ecology-AECO,* [3]*Flemish Institute for Technological Research (VITO),* [4]*Research Institute for Nature and Forest (INBO),* [5]*University of Antwerp. Department of Biology, Ecosystem Management Research Group (ECOBE)*

Chapter Outline

1. INTRODUCTION

A community, situated in an upstream river valley, wants to develop an economically efficient and socially fair long-term sustainability land-use strategy. This is not an easy task. Many services, of which only a minority is currently valued and considered in land-use decisions, have to be included in this strategy. Farmers and foresters prefer to increase productivity; global warming predictions and legislation argue for carbon storage; space for housing is needed; recreational potential of the area has to be increased. To consider a broad range of ecosystem services and benefits, the complexity of the ecological functions underpinning them ,as well as the social

---

[*] Based on *Ecological Indicators*, Van der Biest K., et al., "EBI: An index for delivery of ecosystem service bundles." © 2013 with permission of Elsevier.

Ecosystem Services. http://dx.doi.org/10.1016/B978-0-12-419964-4.00020-2

complexity of beneficiaries, landowners' and other stakeholders' needs have to be captured. Societal decisions on the importance of several services have to be combined with reliable information about the landscape's potential to deliver these services.

This chapter presents a prototype of a tool that is being developed to capture the ecological complexity of delivering ecosystem service bundles. Rather than providing a prescription of the ideal land use, it aims to inform decision makers by demonstrating the consequences of land-use decisions in a given landscape context. The tool is demonstrated here in a pilot case; to apply it in real-life contexts, the model needs further enlargement, improvement, and testing in socioeconomic valuation and deliberative decision methods.

Integrating the ecosystem service concept into land-use planning requires tools that allow rapid and transparent assessment of ecosystem services. The demand for simple indicators has stimulated the emergence of land-use-based proxy methods. Although these methods have been very powerful in creating policy awareness on different levels, they are insufficient when it comes to land-use and policy planning for ecosystem service delivery. Most ecosystem services originate from complex interactions between abiotic and biotic elements, which are not straightforwardly integrated into land-use categories. Discarding the complex ecological reality poses serious risks for adverse effects of policies, as important mistakes can be made if local decisions are based solely on simple proxies.

The approach presented here constitutes the first steps toward developing an index that will allow ecosystem service assessment and optimization in a scientifically rigorous way, capturing the biophysical complexity in a user-friendly and transparent tool.

1.1. Why Do We Need This Index?

The tool was developed in response to the urgent need to enhance land-use decision making. Degradation of ecosystems as a result of land-use optimization toward mainly provisioning services is causing socioeconomic costs and puts future well-being at risk. As a pilot case, the River Nete Basin (Flanders, Belgium) was selected. The valley is densely populated and intensively used, but important surfaces consist of seminatural freshwater wetland areas. Freshwater systems are worldwide one of the most threatened ecosystems, despite the important services they deliver: They provision services such as fish, shellfish, and reed; regulate services, such as water purification, climate regulation, and flood control; and supply cultural services such as recreation. Global change and increasing human demands for water will increase the pressure on these systems. As most of their services are not directly valued, they are bound to be further degraded in favor of directly valuable provisioning services such as food production, wood production, or just built-up surfaces. Trade-offs between all services should be considered in decisions. For example, if a parcel is converted into agricultural land, it is important to know how much less carbon can be stored in soils, how much less nitrate can be prevented to leak to ground-water reserves, how much less habitat is available for protected species, and so on. Which are the parcels where an agricultural land use would have the highest efficiency and the least undesirable impact? Such questions can only be addressed by looking at the biophysical characteristics of the ecosystem that determine whether or not an area is potentially suitable for the delivery of ecosystem services. The Ecosystem service Bundle Index (EBI) aims to consider relationships between services and thus allow assessment of management decisions on multiple ecosystem services. Furthermore, it is developed to be integrated with multiple disciplines, but remain user friendly and transparent. This is a first step toward such a model.

2. DEVELOPMENT OF THE INDEX

2.1. Model Basis

The EBI builds on a Bayesian belief network, which is a statistical tool that allows integration of several data sources and data types. They can combine both empirical quantitative data and expert-based qualitative information. In this respect, use of Bayesian models is especially promising in ecosystem service research where quantitative data is often lacking owing to the heterogeneity of processes that drive service delivery, links, and trade-offs between services and their associated values [1, 2]. The model results were converted into maps that allow evaluation and testing of the model output and that can be used as a graphical focus of discussion/negotiation/participatory approaches in management development.

2.2. Study Aea

The Grote Nete catchment was used as a case study to develop and test the pilot tool (Figure 20-1). It is a typical lowland landscape with numerous brooks and small rivers, infiltration areas, and dune relicts. The soil type varies from sand

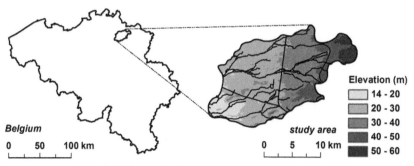

FIGURE 20-1   Location, topography, and major hydrology of the study area.

with sandy loam and loamy sand in the floodplains to loamy and clayey soils in the southernmost part. Land coverconsists mainly of agriculture (22% pasture and 15% cropland), paved area (28%), forest (17%), and wetland (4%). The area is characterized by a broad range of hydrological and biophysical conditions affecting the potential for delivering different types of services.

2.3. Selected Ecosystem Services

The EBI prototype was developed using three interacting ecosystem services exhibiting trade-offs and competitive land claims in the catchment: provisioning services such as food production and wood production, and regulating service climate regulation as long-term storage of soil organic carbon. Forests, extensively managed grasslands and wetlands, for example, are capable of storing carbon, but wetlands are unable to provide substantial food or wood. On the other hand, intensively managed croplands and forests exhibit suboptimal capability to store large amounts of carbon. Developing the model with a limited amount of services facilitated testing of the basic functionality. Yet, to make the model applicable on the management level the whole bundle of ecosystem services needs to be included.

2.4. Data Input

The pilot model is driven by a combination of existing qualitative ([3] for agricultural provision and [4] for wood production) and quantitative ([5, 6, 7] for soil organic carbon) research data as well as expert judgment. Four input variables are needed to run the model: soil texture, hydrology, soil profile development, and land use. Each of these variables is represented by a raster with a certain pixel resolution—in this pilot case 100 x 100m. The model is run for each pixel individually.

2.5. Model Structure

The backbone of the model is the elementary distinction between potential supply of ecosystem services (based on the natural capacity of an ecosystem) and the actual delivery, depending on land-use and biophysical conditions. It is essential to take this distinction into account when optimizing land use and maximizing the extent to which the potential for service delivery is being realized. For each pixel, the model calculates three different indicators (Figure 20-2): [1] for each service separately, the potential service delivery indicator, depending entirely on biophysical characteristics; [2] for each service separately, the actual service delivery indicator, depending on the potential delivery and land use; and [3] the EBI, which is a weighted sum of the actual provision indicators of the three services.

Input from stakeholders is not required solely for calculation of a certain land-use scenario. As the demand for ecosystem services is locally (and temporally) variable, the composition of the bundle of services is a societal decision too. Therefore, the final index can be severely influenced by weighting the sum of the provisioning indicators. To illustrate this concept, three examples of different weighting schemes are given in Figure 20-3. If a purely food-production-oriented approach is taken by assigning a double weight to agricultural production compared to the two other services, the optimal resulting land use will be cropland or grassland over the whole surface (Figure 20-3, left). Similar results are obtained when prioritizing wood production (Figure 20-3, middle) or carbon sequestration (Figure 20-3, right).

Deciding on these weights (in fact, on the relative societal importance of the services in the bundle) is a political question with a high socioeconomic impact and should optimally be based on ethics of fairness and sustainability. The bundle of services depends on societal demand for the individual services and the delivery potential of the system. Inventorying this demand is very complex in its own right, and resulting service delivery maps generated by a certain

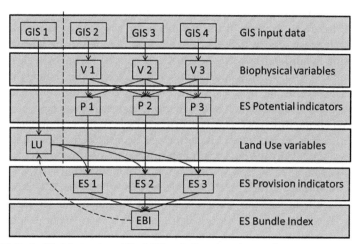

FIGURE 20-2   **Model structure.** The full lines indicate how the actual ecosystem service bundle index (EBI) is determined based on spatial data of the input variables. The dashed arrow illustrates how the optimal land uses are selected based on the maximal attainable EBI value under certain biophysical conditions.

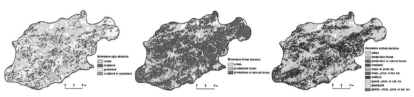

FIGURE 20-3   Example of model output for most optimal land use using three different weighting schemes: high weight for food production (left), wood production (middle), and climate regulation through soil organic carbon sequestration (right).

weighting might influence this weighting decision. Also, combining data on the demand and potential supply of ecosystem services and, for instance, legislative or property claims (NATURA2000, "confirmed agricultural expansion zones") in a spatial explicit way could be very instructive and useful in obtaining an efficient, sustainable, and fair land-use policy. In this respect, maps of the model output are likely to be used during policy negotiations, stakeholder workshops, deliberative approaches, and so on, requiring appropriate socioscientific expertise.

For this pilot case, weighting was performed so that the total surface area in which a service is indicated as most important (= highest delivery) would be more or less equal for the three services.

3. MODEL APPLICATION

One of the model's applications is to develop scenarios and compare the results of the EBI score for a given land-use scenario with the EBI score for the actual land use. One could, for example, construct a most optimal land-use scenario in which for every location (pixel) the model suggests the most optimal land use(s) (= highest service delivery) for the given biophysical characteristics. Calculating the difference between actual and optimal EBI allows identifying opportunities for optimizing ecosystem service delivery in a spatially explicit way. In other words, the model identifies areas where land-use conversion may yield higher service production rates and makes suggestions on the type of land use(s). When the difference between EBI scores is 0, the current land use is optimal. In areas with positive EBI difference, a shift toward the optimal land use(s) as suggested by the model will improve service delivery.

In this example case, 45% of the study area's nonurbanized surface is under optimal conditions. Every pixel has, attached to its optimal EBI score, one or several optimal land-use scenarios to attain the maximum service bundle delivery. There are zones where several land-use scenarios can generate the maximum score. Wetlands are logically the most commonly suggested land use in the valleys along the rivers and in local depressions outside of the valleys (Figure 20-4a). On the other hand, intensive cropland is an optimal land-use alternative on well-drained ridges, with clustered blocks toward the southernmost parts where loamy and clayey soils prevail (Figure 20-4b). Production forests are proposed in a similar way by the model, but the central and northern parts where sandy soils dominate are more represented (Figure 20-4c) as these are less suitable for cropland. Grasslands, as the optimal combination of food production and carbon storage, are the suggested land use in the wetter and sandier parts of the study area (Figure 20-4d).

To illustrate the index's application, we go back to the example of the local community looking for an informed land-use strategy. Figure 20-5a shows the actual land use, with indication of three characteristic zones: one intensively drained cropland area (A), a cropland area without drainage ditches (B), and

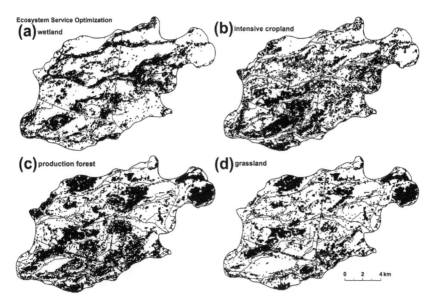

FIGURE 20-4 **Locations where the model suggested optimal land use to be wetland (a), intensive cropland (b), production forest (c), and grassland (d).**

a forest (C). Map b shows opportunities for improved supply of the EBI (EBI score for most optimal land use minus EBI score for actual land use). Areas B and C have little potential to increase their EBI score; that is, the actual land use is close to its maximum potential of ecosystem service delivery (for the services carbon sequestration, agricultural production, and wood production). Area A has large potential to increase its service delivery; more specifically, this can be achieved by converting the actual cropland into wetland (map c). This is explained by the suboptimal agricultural productivity resulting from the high soil water content and the low carbon sequestration under cropland. If the area were to be converted into wetland, agricultural production would largely be lost but carbon sequestration would be optimized, resulting in an overall increase in service delivery according to the model. Service delivery in areas B and C is close to its optimum, meaning that land use conforms with the biophysical potential of the landscape.

A more fundamental application of the EBI is the examination of trade-offs arising from land-use decisions. One of the most common trade-offs in freshwater ecosystems arises from the conversion of wetlands into agricultural land by drainage. The model allows calculation of the impact of reduced soil water content on agricultural production on the one hand and the effect on soil organic carbon sequestration potential on the other hand. When more services are incorporated, tipping points in overall service delivery can be determined, as well as overall sensitivity. These experimental model runs contribute to the fundamental understanding of systems' resilience and supporting capacity.

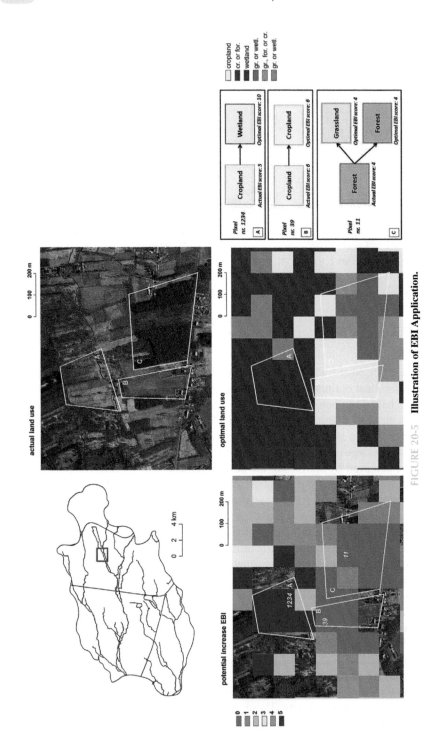

FIGURE 20-5   Illustration of EBI Application.

4. DISCUSSION

The EBI was developed in response to the critical need for transparent and user-friendly modeling and mapping techniques for multiple services, based on the best available data. The model was built as a prototype to allow testing of the basic functioning, and a long road is still ahead:

1. In this pilot case, only three services were included. Application in real-life contexts requires incorporation of all relevant provisioning, regulating, and cultural ecosystem services, and a legitimate design of the weighting function.

2. In the first instance, all the user-defined causal relationships in the Bayesian network were kept deterministic (the probability that the relationship is true is 100%). However, a major strength of Bayesian modeling is the incorporation of uncertainty by assigning probabilities to the relationships between variables and services. This allows making predictions when data scarcity and uncertainty are relevant, which is often the case in ecosystem service research.

3. Precisely the lack of empirical data and the impossibility of measuring services directly is what makes classic validation in ecosystem service modeling very challenging: When model results differ from a measured proxy, this difference can be caused by model imperfections as well as by the shortcomings of using one single proxy. Validation of Bayesian models therefore often relies on internal model performance and expert judgment. Assembling objective expert opinions to verify the accuracy and robustness of these models is often difficult. This prototype EBI model relies on scientific expert evaluations of the network structure and visual assessments of the mapped model output. Validation at the appropriate application level (e.g., local stakeholders and experts of a subcatchment) is essential.

4. Stakeholder involvement is not only recommended for model validation but should also be applied in the model development, application, and scenario evaluation process. In fact, offering discussion support and facilitating stakeholder engagement are major advantages of Bayesian modeling. Moreover, the visualization of model output scenarios through mapping offers opportunities in the context of participatory modeling, which Fish [8] denoted as essential in ecosystem service assessment.

5. Although the use of Bayesian networks in ecological modeling still has some important limitations, recent growing scientific interest will probably lead to model improvements in the future [9]. Some shortcomings of Bayesian networks in regard to ecological modeling, such as the absence of feedback loops, are, for example, currently dealt with by the use of time-dependent nodes or time sliced models. Further technical advances will similarly increase the feasibility of Bayesian models to take into account spatial dependencies and interactions between ecosystem services and conditions.

Many policy makers, decision makers, and stakeholder groups have looked for ways to generate awareness for our dependency on natural systems and to stimulate sustainable resource use. Qualitative maps and illustrative monetary quantifications have provided very convincing and powerful arguments. This might cause decision makers to apply oversimplified mapping tools on their local case and select maps and quantification results to serve their organization's stakes. Negative consequences might only surface in the long run.

Although EBI is a prototype in an early stage of development, realistic results have been obtained. Application of this and similar approaches in real-life pilot cases will put this approach to the test.

REFERENCES

1. Landuyt, D., Broekx, S., D'hondt, R., Engelen, G., Aertsens, J., and Goethals, P.L.M. 2013. A review of Bayesian belief networks in ecosystem service modelling. *Environmental Modelling and Software* 46, 1–11.
2. Haines-Young, R. 2011. Exploring ecosystem service issues across diverse knowledge domains using Bayesian Belief Networks. *Progress in Physical Geography* 35(5), 681–699.
3. ProvincieAntwerpen. 1998. De digitale Bodemkaart als basislaag voor de open ruimte—Een interpretatie van de Bodemkaart ten behoeve van land- en tuinbouw, bosbouw en milieu met behulp van GIS, edited by D. I. GIS-cel. Antwerpen: Provincie Antwerpen, 88.
4. De Vos, B. 2000. Achtergrondsinformatie bij het bodemgeschiktheidsprogramma voor boomsoorten BoBo (edited by I. v. B. e. W. IBW).
5. Meersmans, J., F. De Ridder, F. Canters, S. De Baets, and M. Van Molle. 2008. A multiple regression approach to assess the spatial distribution of Soil Organic Carbon (SOC) at the regional scale (Flanders, Belgium). *Geoderma* 143,1–13.
6. Post, W.W.E., P.J. Zinke, and A.G. Stangenberger. 1982. Soil carbon pools and world life zones. *Nature* 298, 156–159.
7. Adhikari, S., R.M. Bajracharaya, and B.K. Sitaula. 2009. A review of carbon dynamics and sequestration in wetlands. *Journal of Wetlands Ecology* 2, 42–46.
8. Fish, R.D. 2011. Environmental decision making and an ecosystems approach. *Progress in Physical Geography* 35(5), 671–680.
9. Aguilera, P.A., A. Fernandez, R. Fernandez, R. Rumi, and A. Salmeron. 2011. Bayesian networks in environmental modelling. *Environmental Modelling & Software* 26(12),1376–1388.

Chapter 21

ES Thinking and Some of Its Implications: A Critical Note from a Rural Development Perspective

Frédéric Huybrechs, Johan Bastiaensen and Gert Van Hecken

Institute of Development Policy and Management (IOB), University of Antwerp

In this chapter—which intends to contribute to the BEES project's outreach to the broader societal context—we present some aspects of how ecosystem services (ES)-thinking has been related to rural development and land-use dynamics in developing countries. The key policy areas touched upon by the ES perspective in the context of development have been those related to the redirection of land-use patterns, in order to reduce the pressure of increasing agricultural and cattle raising activities on some of the most important remaining natural areas and resource bases around the world [1, 2]. ES—and in particular the creation of payments for ES (PES)—have attracted a lot of attention in development research and policy circles as a promising conservation and restoration approach [3–7]. In what follows, we first present the ES concept's gradual evolution from an awareness-raising concept to a legitimation for market-based policy tools (and in particular PES) that intend to redirect land-use and rural development pathways. Then, looking

Ecosystem Services. http://dx.doi.org/10.1016/B978-0-12-419964-4.00021-4

at the application of PES in practice, we discuss the example of the Regional Integrated Silvopastoral Management Project (RISEMP) and its implementation in Matiguás and Río Blanco in Nicaragua. From this case study, we derive insights into the importance of a set of considerations about the land users' motivation, as well as context-related information according to land-use management, that go beyond those taken into account by the project or mainstream applications of PES. To conclude, we link these findings to a broader set of recent evidence that calls for a more integrated institutional approach to PES, where more attention is paid to the complexity of the socioecological system in which it aims to intervene.

1. INFLUENCE OF ES THINKING ON DEVELOPMENT AND LAND-USE POLICY

Throughout its history, ES thinking has gradually evolved from an awareness-raising exercise to a foundation for market-like approaches to nature conservation and restoration. When the concept emerged in the late 1970s, the coupling of biophysical ecosystem functions and the services they provide to human well-being and the economy was intended to report on biophysical limits to the wider public and policy makers (Patterson and Coelho [8], referring to Daly [9]). Nowadays the ES concept is more often related to the work of Costanza et al. [10] and the Millennium Ecosystem Assessment (MA) [11]. Costanza et al.'s contribution represents a starting point for the practice of economic valuation of ES by putting an aggregate value on the yearly services provided to the world economy. The MA, in turn, can be considered the real springboard for mainstreaming the concept of ES and their valuation, with a number of important international efforts and projects—such as TEEB[1] (The Economics of Ecosystems and Biodiversity) and IPBES[2] (Intergovernmental Panel on Biodiversity and Ecosystem Services)—developed since its publication. From the ES concept's incipience to its current use, there seems to have been an evolution toward monetization and commodification of ES [8], which Gomez-Baggethun et al. [12] link to a shift from incommensurable use values of nature's services to tradable exchange values.

In both practice and the literature on land-use changes and rural development, it seems that ES thinking is increasingly conflated with market-based instruments of environmental governance, such as Payment for Ecosystem Services (PES). The mainstream definition of this policy instrument is provided by Wunder [5 p.3], who defines PES as: "a voluntary transaction where a well-defined ES (or a land-use likely to secure that service) is being 'bought' by a (minimum one) ES buyer from a (minimum one) ES provider, if and only if the ES

---

1. With a strong emphasis on the valuation of the global economic costs and benefits of biodiversity and ES loss and conservation, TEEB examines and reports on matters related to ES and biodiversity to a wide range of end-users (see http://www.teebweb.org).
2. The IPBES aims to facilitate the science–policy interface at a global level (see http://www.ipbes.net).

provider secures ES provision (conditionality)." This definition emphasizes the main novelties of this approach compared to other practices of natural resource management. PES is defined as voluntary, which is in contrast to command and control approaches [5], and the conditionality criterion sets it apart from indirect incentive mechanisms [13], such as education and alternative economic activities promoted in Integrated Conservation and Development Programs [14].

The rationale underlying PES is based on the belief that environmental degradation is often caused by the failure of markets, which do not account for the environmental externalities[3] of private economic activities. PES proponents build on an interpretation of the Coase theorem, which asserts that under certain conditions individual and voluntary bargaining through the market will lead to the most efficient allocation of externalities [15]—which in this case are taken to be the ES. Payments for positive externalities generated by the adoption of environmentally sound land-use practices thus intend to make economic actors include the otherwise unvalued externalities (i.e., ES) in their decision making [4].

In the following section, we discuss an example of the application of the PES approach. It sheds light on a PES scheme in practice and makes us reflect on a number of elements related to land-use change. It reveals some potential constraints on the effectiveness of the mainstream PES approach as presented in the previous paragraph. After the case study, we will therefore again pick up on the evolution in (P)ES thinking in rural development circles by presenting emerging conceptual and practical adaptations to the policy instrument.

2. EXAMPLE OF THE PES APPROACH AND MOTIVATIONS RELATED TO LAND-USE MANAGEMENT

A leading pilot experiment of PES in a rural development context has been the RISEMP; funded by the World Bank and the Global Environment Facility (GEF) and implemented in three Latin American countries. In Nicaragua, this project took place in the region of Matiguás-Río Blanco[4] (see Figure 21-1).

Matiguás-Río Blanco is part of the "old agricultural frontier" in Nicaragua; the area colonized for subsistence food production and extensive cattle raising since the 1950s. The expansion of agriculture and cattle not only resulted in a transition from tropical forest to cultivated area, but also translated into an ongoing process of destruction of remaining forest patches and soil degradation

---

3. Externalities are effects ensuing from one economic actor's activities that influence other actors' costs and benefits without this being reflected in the costs or benefits of the initial actor.
4. Gert Van Hecken conducted intense fieldwork there between 2008 and 2010 for his PhD research—funded by the Flemish Interuniversity Council for Development Cooperation (VLIR-UOS)—from which the following example is compiled. For more information on the research results, see Van Hecken [16], Van Hecken and Bastiaensen [17], Van Hecken and Bastiaensen [18], and Van Hecken et al. [19].

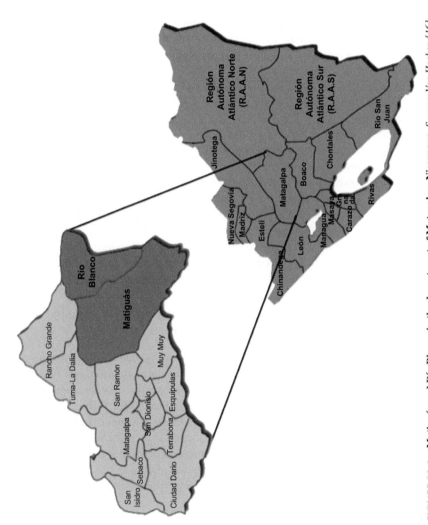

FIGURE 21-1   **Matiguás and Río Blanco in the department of Matagalpa, Nicaragua.** *Source: Van Hecken [16].*

within the colonized areas. This policy has caused increasing environmental problems and reduced supply of ES (water quantity and quality, carbon storage, biodiversity). Locals in Matiguás estimate a loss of more than 40% in forested area over the last 20 years. An unequal and extensive model of economic development, based mainly on large-scale cattle raising with a very low number of animals per hectare, has been—and still is—the most important driver in that evolution. The establishment of fresh milk collection centers, following the construction of an all-weather road to the capital and other cities, has recently generated a significant shift from beef to dairy cattle, and a surge in local economic growth. Contrary to expectations [20], these evolutions in general did not contribute to more intensified cattle activity since the increased cattle activities contributed to a further expansion of cultivated land of the richer farmers. They reduced forest patches on their farm and bought more land from poorer farmers, without any significant increase in overall output per unit of land [21], spurring migration of poorer farmers further down the new agricultural frontier. This process of land concentration for extensive cattle raising is one of the main drivers of deforestation in Nicaragua [2, 22].

It is in this context that, between 2002 and 2008, the RISEMP project was implemented in the area. This experimental pilot project sought to intervene in the trend of extensive cattle farming by offering conditional payments and technical assistance (TA) for the adoption of more intensive silvopastoral practices[5] and the adoption of a number of more environment-friendly practices (such as cocoa and shaded coffee as compared to annual crops). By doing so, the generation of local and global environmental benefits was expected; with a specific focus on the increase in CO_2 capture and biodiversity conservation induced by the intended practices [24]. The project designers also expected that the intensified cattle raising practices, as compared to the incumbent extensive cattle raising practices, would reduce the pressure on forests by increasing the cattle carrying capacity of available land [25].

Although potentially profitable in the long run, intensified silvopastoral practices are often deemed unattractive by farmers in the short term. The main barriers to implementation are assumed to be the significant investment of capital and labor required (for activities such as the implementation and maintenance of improved pastures, fodder banks, and tree cover); the time lag between new practices and higher productivity; and the lack of know-how [23]. Therefore, payments and TA intended to tip the balance toward adoption of the environment-friendlier practices [24, 26].

The project was designed as an experimental action-research project, as it sought to assess the effectiveness of different packages of incentives. The research methodology was based on a randomized experimental design, with

---

5. Defined by Dagang and Nair [23 p.149], as "systems integrat[ing] trees into livestock systems for multiple purposes including soil amelioration, shade, fodder, fruit, wood, and habitat for fauna."

various participant groups receiving different incentives (payments with or without TA) or no intervention at all (the control group). This setup would—at least in theory (see below)—separate the effect of PES and TA from other contextual variables in order to assess their relative effectiveness.

Payments to participants were based on ecological improvement in land use, valued by increases in an environmental service index (ESI) score of the participant's farm. This ESI score was linked to 28 different land-use types representing different levels of provision of biodiversity and carbon capture services (see Table 21-1). The score reflects the assumed relative ability of land uses to provide services of biodiversity conservation or carbon sequestration, with a maximum score of 1 being given to primary forest and a minimum of 0 to degraded pasture and annual crops (for detailed information, see Murgueitio et al. [27]). Farmers were paid an annual sum of US$75 (under a four-year scenario) or US$110 (under a two-year scenario) per incremental ESI point/ ha compared to their baseline ESI balance (monitored and calculated based on remote sensing imagery). The amounts of the payments were established on the basis of the estimated opportunity costs of the environmentally more attractive land uses [26].

The results of the project were considered to be positive and encouraging [26]. The relative area of degraded pastures decreased from 30.9% to 10.1% in five years. They were replaced by improved pastures with trees (up from 9% to 23.8% of the total area) and fodder banks (which more than tripled in area). During the project, the area occupied by annual crops halved, while the quantity of living fences almost quadrupled. Forests and scrub habitats remained stable, covering about 2% of the total area [18]. Pagiola et al. [26] attribute this success mainly to the payments made to the farmers, which would have tipped the balance in favor of the desired intensification through improved silvopastoral practices and away from the continued prevalence of degraded pasture and annual crops.

As part of a project evaluation, and with the purpose of taking into consideration broader social dynamics underlying decision-making processes regarding adopted land-use practices, we conducted additional qualitative research in the area. The results show a more nuanced conclusion concerning motivations underlying the adoption of the promoted practices. The first element that casts doubt on Pagiola et al.'s conclusion was that the control group, which received none of the payments or TA, engaged in quite similar land-use changes as compared to the treatment groups. Furthermore, interviews conducted with participants revealed that, although payments did indeed help to overcome the increase in labor and capital investments and the foregone income in the first years of implementation, their impact was only one among many other reasons for the adoption of targeted land uses. For instance, higher relative milk prices and the expectation of their further increase were important drivers too. Indeed, this affects the profitability of silvopastoral approaches—which have a positive effect on milk productivity [20]—and the benefits of intensification in terms of easier accessibility of the cows for milking.

TABLE 21-1 **Environmental Service Index (ESI) Used by the RISEMP**

Land use	Biodiversity index	Carbon index	ESI
1. Annual crops	0	0	0
2. Degraded pasture	0	0	0
3. Natural pasture without trees	0.1	0.1	0.2
4. Improved pasture without trees	0.1	0.4	0.5
5. Semipermanent crops (plantain, sun coffee)	0.3	0.2	0.5
6. Natural pasture with low tree density (< 30/ha)	0.3	0.3	0.6
7. Natural pasture with recently planted trees (> 200/ha)	0.3	0.3	0.6
8. New living fence (per km)	0.3	0.3	0.6
9. Improved pasture with recently planted trees (> 200/ha)	0.3	0.4	0.7
10. Monoculture fruit crops	0.3	0.4	0.7
11. Fodder bank	0.3	0.5	0.8
12. Improved pasture with low tree density (< 30/ha)	0.3	0.6	0.9
13. Fodder bank with woody species	0.4	0.5	0.9
14. Natural pasture with high tree density (> 30/ha)	0.5	0.5	1
15. Diversified fruit crops	0.6	0.5	1.1
16. Wind break (living fence) (per km)	0.6	0.5	1.1
17. Diversified fodder bank	0.6	0.6	1.2
18. Monoculture timber plantation	0.4	0.8	1.2
19. Shaded coffee	0.6	0.7	1.3
20. Improved pasture with high tree density (> 30/ha)	0.6	0.7	1.3
21. Bamboo forest	0.5	0.8	1.3
22. Diversified timber plantation	0.7	0.7	1.4
23. Scrub habitats (*tacotales*)	0.6	0.8	1.4
24. Riparian forest	0.8	0.7	1.5

Continued

TABLE 21-1 Environmental Service Index (ESI) Used by the RISEMP—cont'd

Land use	Biodiversity index	Carbon index	ESI
25. Intensive silvopastoral	0.6	1	1.6
26. Disturbed secondary forest	0.8	0.9	1.7
27. Secondary forest	0.9	1	1.9
28. Primary forest	1	1	2

Note: The ESI is the sum of the biodiversity and carbon indices. Points per hectare, unless otherwise specified.
Source: Pagiola et al. [26].

In a quantitative comparison between the treatment groups, the TA is not found to have a significant impact [26]. During the fieldwork, however, farmers repeatedly reported it as being a very relevant encouragement. A likely reason for its nonsignificance in the comparison between treatment groups is that TA appears to have gone beyond selected participants as it spread to others via existing social ties or unplanned participation in workshops. This invalidated a comparison of groups with and without TA. In other words, TA "leaked" knowledge and motivation to engage with silvopastoral practices into the entire local community. The changes in perception, knowledge, and practices that occurred in the area thereby generated a collective momentum for a more sustainable silvopastoral dairy development pathway [28]. Thus, although there is an increased adoption of improved silvopastoral approaches in Matiguás-Río Blanco, the underlying motivations and causal relationships turn out to be more complex than a simple, individually rational reaction to monetary incentives paid for the ES provided.

As we take a perspective that goes beyond the group of participants or the time span of the project, certain specificities of dealing with payments lead to additional issues. One of them is the potential danger of perverse incentives. To avoid having participants a priori destroying trees to obtain higher payments through more incremental ESI points, a payment was introduced for baseline points [26]. Yet, farmers of the same area, but excluded from participation in the project, might voice the same threat to claim a right to the same payments. Similarly, this can happen in neighboring areas that take some notice of the workings of the existing scheme. Van Hecken and Bastiaensen report worrying indications of this during fieldwork in an area close to where the RISEMP project had taken place [18]. A payment scheme's focus on monetary or economic valuation might thus lead to a self-fulfilling prophecy of individual economic rationality, neglect other types of motivation, and downplay the potential importance of moral principles and social markets in the success and failure of ES-governance [29–31].

Two other potential limitations of an approach that focuses on individualized incentives are a lack of a landscape approach and instances of leakage. A landscape approach recognizes the need to create a "corridor" of interconnected patches of forested land for biodiversity and also acknowledges that the spatial configuration of land uses is important; yet no collective approach or targeting was adopted in the RISEMP. Leakage, in turn, refers to the problem whereby the project might lead to pressures in other places, which offsets part of the achieved benefits of the project. In that context, a particular problem with the individual approach was that the larger farmers also captured most of the PES payments, as they had the capacity to introduce more significant land-use changes. This, in turn, further enhanced their capacity to acquire additional land from the poorer peasants, often making their overall production systems less intensive [21]. In this way, the dominant model of extensive cattle raising based on the concentration of land by larger farmers and the expulsion of poorer peasants was not at all challenged by the PES subsidies, directed to individual plots rather than the cattle model as a whole [32]. Additionally, the migration of poorer farmers to the agrarian frontier might correlate some of the improvements in the project area to further destruction of tropical forests elsewhere.

The reassessment of the RISEMP project shows that motivations behind the adoption of silvopastoral practices are not straightforward and cannot be confined to mere individual payment incentives. The broader social, economic, and institutional context is key to understanding motivations and incentives underlying land-use changes and to envisaging the bigger picture of land-use change dynamics. It also demonstrates the need to adopt a more conscious and integrated planning approach when seeking to obtain more desirable land uses.

3. ECOSYSTEM SERVICES AND SOCIOECOLOGICAL SYSTEMS

A growing body of empirical research similarly reports on the importance of recognizing that, regarding land use, there is more to decision making than the individual economic rationale. In its current configuration, it seems that PES might be ill suited for solving the issue of environmental degradation [6, 33, 34] and might not score well on accounts of poverty alleviation and fairness either [17, 35, 36]. Muradian et al. [34] note a gap between what the conventional theory of PES prescribes as a policy tool and the reality of land management and rural development encountered by practitioners. As such, they call for a more prominent incorporation of the characteristics of the local social and environmental context into the setup and interpretation of PES schemes. An important message in Muradian et al's reconceptualization of PES is that it should be seen more as part of a policy mix and less as a stand-alone alternative to public-communitarian governance of natural resources.

The concerns raised, and the ensuing plea for a more integrated approach to PES or natural resource management, point to some simplifications that have guided the mainstream (P)ES practice in (rural) development and that turn out to have some undesired effects [37, 38]. Ostrom and Cox [39] aptly state that society and nature, and their interaction in socioecological systems, are very complex, and that reducing this complexity in order to make it fit into a market-based model can only be done at the risk of overlooking the broader complex institutional framework from which it cannot be separated. As we saw in the first section, the ES concept has awareness-raising potential and may clearly underline linkages between the natural and human world. For the analysis and understanding of socioecological systems—which puts emphasis on exactly these interactions between humans, nature, and governance systems [39]—ES remains a valuable concept. It will, however, need to trace itself back to its origins as a tool for a better understanding, communication, and articulation of the problems at hand rather than a precursory element in a market-based economic exchange scheme.

REFERENCES

1. Angelsen, A., Kaimowitz, D., and Center for International Forestry Research. 2001. *Agricultural Technologies and Tropical Deforestation.* Wallingford, Oxon, UK; New York; CABI Pub. in association with Center for International Forestry Research.
2. López, M. 2012. Análisis de las causas de la deforestación y avance de la Frontera Agrícola en las zonas de Amortiguamiento y Zona Núcleo de la Reserva de Biósfera de BOSAWAS-RAAN, Nicaragua.
3. Asquith, N., and Wunder, S. 2008. Payments for watershed services: The Bellagio conversations. Fundación Natura Bolivia, Santa Cruz de la Sierra.
4. Engel, S., Pagiola, S., and Wunder, S. 2008. Designing payments for environmental services in theory and practice: An overview of the issues. *Ecological Economics* 65, 663–674.
5. Wunder, S. 2005. Payments for environmental services: Some nuts and bolts. *Occasional paper,* Vol. 42, CIFOR, Bogor.
6. Redford, K.H., and Adams, W.M. 2009. Payment for ecosystem services and the challenge of saving nature. *Conserv Biol* 23, 785–787.
7. Tacconi, L. 2012. Redefining payments for environmental services. *Ecological Economics* 73, 29-36.
8. Patterson, T.M., and Coelho, D.L. 2009. Ecosystem services: Foundations, opportunities, and challenges for the forest products sector. *For. Ecol. Manage.* 257, 1637–1646.
9. Daly, H.E. 1977. Steady-state economics. In *Thinking about the Environment: Readings on Politics, Property, and the Physical World* (Cahn, M.A., and O'Brien, R., eds.), pp. 250–255. New York: M.E. Sharpe.
10. Costanza, R., dArge, R., deGroot, R., Farber, S., Grasso, M., Hannon, B., Limburg, K., Naeem, S., ONeill, R. V., Paruelo, J., Raskin, R. G., Sutton, P., and vandenBelt, M. 1997. The value of the world's ecosystem services and natural capital. *Nature* 387, 253–260.
11. MA. 2005. *Ecosystems and Human Well-Being: Current State and Trends*, Washington, DC: Island Press.
12. Gomez-Baggethun, E., de Groot, R., Lomas, P.L., and Montes, C. 2010. The history of ecosystem services in economic theory and practice: From early notions to markets and payment schemes. *Ecological Economics* 69, 1209–1218.

13. Ferraro, P.J., and Simpson, R.D. 2002. The cost-effectiveness of conservation payments. *Land Econ* 78, 339–353.
14. Berkes, F. 2007. Community-based conservation in a globalized world. *P Natl Acad Sci USA* 104, 15188–15193.
15. Coase, R.H. 1960. The Problem of Social Cost. *J Law Econ* 3, 1–44.
16. Van Hecken, G. 2012. Payments for environmental services and governance of natural resources for rural communities: Beyond the market-based model: An institutional approach: case studies from Nicaragua. In *Instituut voor Ontwikkelingsbeleid -en beheer,* p. 219, PhD Dissertation. Universiteit Antwerpen, Antwerpen.
17. Van Hecken, G., and Bastiaensen, J. 2010. Payments for ecosystem services: justified or not? A political view. *Environ Sci Policy* 13, 785–792.
18. Van Hecken, G., and Bastiaensen, J. 2010. Payments for Ecosystem Services in Nicaragua: Do Market-based Approaches Work? *Dev Change* 41, 421–444.
19. Van Hecken, G., Bastiaensen, J., and Vasquez, W.F. 2012. The viability of local payments for watershed services: Empirical evidence from Matiguas, Nicaragua. *Ecological Economics* 74, 169–176.
20. Yamamoto, W., Dewi, I.A., and Ibrahim, M. 2007. Effects of silvopastoral areas on milk production at dual-purpose cattle farms at the semi-humid old agricultural frontier in central Nicaragua. *Agr Syst* 94, 368–375.
21. Polvorosa, J.C. 2013. Contract choice and Livelihood Strategies, the opportunities and constraints of poor rural households' livelihoods in the context of the booming dairy chain in Matiguás, Nicaragua. In *Institute of Development Policy and Management (IOB)*. Draft PhD thesis, University of Antwerp.
22. Maldidier, C. 1993. Tendencias actuales de la frontera agrícola en Nicaragua. In *Informe de consultoria para ASDI*, Instituto Nitlapán, Managua.
23. Dagang, A.B.K., and Nair, P.K.R. 2003. Silvopastoral research and adoption in Central America: Recent findings and recommendations for future directions. *Agroforest Syst* 59, 149–155.
24. Pagiola, S., Agostini, P., Gobbi, J., De Haan, C., Ibrahim, M., Murgueitio, E., Ramirez, E., Rosales, M., and Ruiz, J.P. 2004. Paying for Biodiversity Conservation Services in Agricultural Landscapes. In *Environmental Economics Series* 96 (Bank, W., ed.). Washington, DC: World Bank.
25. World Bank. 2002. Colombia, Costa Rica, and Nicaragua regional integrated silvopastoral approaches to ecosystem management project: Project appraisal document. Washington, DC: World Bank.
26. Pagiola, S., Ramirez, E., Gobbi, J., De Haan, C., Ibrahim, M., Murgueitio, E., and Ruiz, J. P. 2007. Paying for the environmental services of silvopastoral practices in Nicaragua. *Ecological Economics* 64, 374–385.
27. Murgueitio, E., Ibrahim, M., Ramirez, E., Zapata, A., Mejía, C.E., and Casasola, F. 2003. Usos de la tierra en fincas ganaderas: Guía para el pago de servicios ambientales en el Proyecto Enfoques Silvopastorales Integrados para el Manejo de Ecosistemas. (Nitlapán, CIPAV-CATIE, ed.), Cali, Colombia.
28. de Haan, L., and Zoomers, A. 2005. Exploring the frontier of livelihoods research. *Dev Change* 36, 27–47.
29. Vatn, A. 2010. An institutional analysis of payments for environmental services. *Ecological Economics* 69, 1245–1252.
30. Frey, B.S., and OberholzerGee, F. 1997. The cost of price incentives: An empirical analysis of motivation crowding-out. *Am Econ Rev* 87, 746–755.

31. Muradian, R., Arsel, M., Pellegrini, L., Adaman, F., Aguilar, B., Agarwal, B., Corbera, E., de Blas, D. E., Farley, J., Froger, G., Garcia-Frapolli, E., Gómez-Baggethun, E., Gowdy, J., Kosoy, N., Le Coq, J. F., Leroy, P., May, P., Méral, P., Mibielli, P., Norgaard, R., Ozkaynak, B., Pascual, U., Pengue, W., Perez, M., Pesche, D., Pirard, R., Ramos-Martin, J., Rival, L., Saenz, F., Van Hecken, G., Vatn, A., Vira, B., and Urama, K. 2013. Payments for ecosystem services and the fatal attraction of win-win solutions. *Conserv Lett* 6 (4), 274–279.

32. Harvey, C.A., Komar, O., Chazdon, R., Ferguson, B.G., Finegan, B., Griffith, D.M., Martinez-Ramos, M., Morales, H., Nigh, R., Soto-Pinto, L., Van Breugel, M., and Wishnie, M. 2008. Integrating agricultural landscapes with biodiversity conservation in the Mesoamerican hotspot. *Conserv Biol* 22, 8–15.

33. Corbera, E., Kosoy, N., and Tuna, M.M. 2007. Equity implications of marketing ecosystem services in protected areas and rural communities: Case studies from Meso-America. *Global Environment Change-Hum Policy Dimensions* 17, 365–380.

34. Muradian, R., Corbera, E., Pascual, U., Kosoy, N., and May, P.H. 2010. Reconciling theory and practice: An alternative conceptual framework for understanding payments for environmental services. *Ecological Economics* 69, 1202–1208.

35. Pascual, U., Muradian, R., Rodriguez, L. C., and Duraiappah, A. 2010. Exploring the links between equity and efficiency in payments for environmental services: A conceptual approach. *Ecological Economics* 69, 1237–1244.

36. McAfee, K. 1999. Selling nature to save it? Biodiversity and the rise of green developmentalism. *Environment and Planning D: Society and Space* 17, 133–154.

37. Norgaard, R.B. 2010. Ecosystem services: From eye-opening metaphor to complexity blinder. *Ecological Economics* 69, 1219–1227.

38. Kosoy, N., and Corbera, E. 2010. Payments for ecosystem services as commodity fetishism. *Ecological Economics* 69, 1228–1236.

39. Ostrom, E., and Cox, M. 2010. Moving beyond panaceas: A multi-tiered diagnostic approach for social-ecological analysis. *Environ Conserv* 37, 451–463.

Chapter 22

Enhancing Ecosystem Services in Belgian Agriculture through Agroecology: A Vision for a Farming with a Future

Alain Peeters[1], Nicolas Dendoncker[2] and Sander Jacobs[3,4]

[1]*RHEA Research Centre rue Warichet, Natural Resources, Human Environment and Agronomy (RHEA),* [2]*Department of Geography, University of Namur (Unamur). Namur Research Centre on Sustainable Development (NAGRIDD). Namur Centre for Complex Systems (naXys),* [3]*Research Institute for Nature and Forest (INBO),* [4]*University of Antwerp. Department of Biology, Ecosystem Management Research Group (ECOBE)*

Chapter Outline

1. INTRODUCTION

The concept of ecosystem services was initiated in the 1990s [1, 2, 3]. It is now developing very fast, especially after the publication of Millenium Ecosystem Assessment [4], Stern [5], and the TEEB reports. The application of the concept of ecosystem services to agriculture is still limited [6], but it is a promising concept that should be consolidated and deepened. Ecosystem service management is undoubtedly at the heart of the future of postindustrial societies.

Ecosystem Services. http://dx.doi.org/10.1016/B978-0-12-419964-4.00022-6

However, in postindustrial societies, most citizens and decision makers do not recognize the dependence of the economy on biodiversity and ecosystem processes. For nonspecialists, only some economic sectors like fishery, extensive farming, and agritourism seem to be concerned about the delivery of ecosystem services. The feeling that citizen wealth no longer depends on natural processes and biodiversity is not justified because ecosystems are still providing a wide range of essential services. These services are not acknowledged because they are not properly understood or considered as "normal" and "guaranteed," such as organic matter degradation in soil and water.

Since the middle of the 19th century, a large part of the services provided by ecosystems before the Industrial Revolution has been replaced by techniques relying on amassive use of fossil fuel. For instance, in agriculture, the artificial synthesis of nitrogen has replaced symbiotic nitrogen fixation by legumes, crop protection by pesticides has replaced the effect of pest and disease regulation by complex living communities, and mechanization has replaced manpower and draught animals.

The use of these artificial inputs and techniques has not only replaced some ecosystem services, but they have also increased production and induced negative impacts on the environment. They have provoked pollution and biodiversity losses that in turn have decreased the supply of ecosystem services essential to farming itself.

Now more than a century later, the massive use of fossil fuels has revolutionized human societies. In the last period of human history, scientific knowledge and technical performance progressed considerably and improved the living standards of a significant part of the world's human population. Historians qualify this period as the Great Acceleration [7]. The beginning of the 21st century is at a turning point of this evolution. For the first time in recent history, the consumption of fossil fuels almost exceeds the production capacity. According to several estimations, this so-called peak oil moment has already been reached or will be reached by 2020 [8, 9]. Nevertheless, in the next decades, the progressive exhaustion of oil stocks and the emergence of new economic powers and industrial development will induce important increases in energy prices. The extraction and massive utilization of shale gas could temporarily change price evolutions, but these processes are not desirable for environmental reasons (there is risk associated with hydraulic fracturing and GreenHouse Gas (GHG) emissions [10]. A strong inflexion of the dominant economic development model will thus be inevitable. Evolution to a more sustainable use of resources and ecosystem services is thus necessary. Since it will take time, it should start immediately.

In agriculture, this evolution could take several forms. A larger use of nitrogen fixing legumes and a better integration of these plants in farming systems will be desirable because the chemical synthesis of nitrogen is one of the most energy-demanding processes of the agricultural sector. That will require a higher diversity of crops (polyculture, including multicropping, intercropping, and companion planting) and longer crop rotations that will also be desirable afor breaking the life cycles of insect pests, diseases, and weeds, which could potentially reduce pesticide use. New forms of arable/livestock system integration will be necessary. Animal integration in agroecosystems will aid in achieving high crop yields if nutrient cycling is optimized [11]. In addition, systematic use of cover crops, adoption of reduced or no tillage techniques, integrated fertilization and crop protection, optimum management of organic matters, and development and diversification of an agroecological infrastructure including agro-forestry (ex.: fruit trees, trees for quality timber, traditional hedges, hedges of second-generation agro-fuels) will be the main strategies for restoring and enhancing ecosystem services [12]. Some transport costs will become prohibitive: Local product consumption will increasingly be a priority compared to overseas products. Product preservation over long periods by frost and heating glasshouses will have to face increasing costs. These evolutions should not lead to a return to the past. Research should define ways to progressively replace fossil fuels by restoring ecosystem services. A new technological revolution is needed [13]. The most up-to-date scientific knowledge is essential to achieve this transition. Research should also aim at developing tools and methods for better understanding ecosystem services, evaluating their importance, optimizing natural processes, developing socioeconomic systems for paying their production, and integrating ecosystem services in the intrinsic mechanisms of the society of tomorrow. Above all, research has to define the ways to develop new farming systems that will restore ecosystem services and make better use of them while controlling production costs and maintaining or improving farmer's income.

2. AGROECOLOGY, INTEGRATED FARMING, AND ECOSYSTEM SERVICES

Although the concept of ecosystem service is relatively recent, the study and the management of the components of the agricultural ecosystem and the services it provides, by the science of agroecology ,are not new. The term *agroecology* has different meanings: It can be considered as a set of techniques or practices, a scientific discipline, or a sociopolitical movement [14]. From its origin, it tries to apply ecological concepts and principles to the design and management of farming systems [15–18]. More recently, its objectives have been broadened. As a scientific discipline, it tends to integrate "the study of the ecology of the entire food systems, encompassing ecological, economic and social dimensions, or more simply the ecology of food systems" [19, p100].

As a practice or a technique, agroecology is close to concepts such as Integrated Farming systems [12, 20–23], Evergreen Revolution [24], Conservation Agriculture [25], Ecologically Intensive Agriculture (after 13), and Ecoagriculture [26, 27]. All these farming types, as well as organic farming [28, 29], permaculture [30, 31], and natural farming [32], differ from conventional agriculture by a search of harmony with nature. These systems try as much as possible to mimic and use natural processes such as biological control or biological nitrogen fixation. By doing so, they use a holistic approach instead of the reductionist approach of conventional farming. They look at interactions between problems (ex.: soil fertility, weeds, pests and diseases) and synergies between techniques (ex.: organic matter management, crop rotation, and diversity). These farming systems distinguish themselves from conventional farming, which often consider these processes separately. Synergies between techniques, together with the maintenance and restoration of biodiversity in agroecosystem (diversity of domesticated animals and crops, association between livestock and crops, functional diversity), promote soil fertility, productivity, and crop protection [33]. These holistic farming types thus offer a way to decrease dependence on agrochemical and energy inputs. They are mainly ecosystem-based in contrast with conventional agriculture, Green Revolution [34], Doubly Green Revolution [35] and sound farming (*agriculture raisonnée* in French) [36] that are mainly fossil-fuel-based (Table 22-1).

The term *agroecology* has the merit of being clear compared to many other terms, even for nonspecialists. It is quite similar to *ecoagriculture*, even if this latter term has a broader meaning. It would be desirable to merge the two approaches (see below). The technical content of agroecology should, however, be precisely defined since the term is used in different ways by different authors.

Integrated or agroecological farming systems are characterized by the following features [12, 33]:

(i) maintenance of a vegetative cover as an effective soil- and water-conserving measure. It is met through the use of no-tillage practices, mulch, and cover crops;

(ii) provision of a regular supply of fresh organic matter by the addition of manure, compost, and crop residues for the promotion of soil biotic activity;

(iii) enhancement of nutrient recycling mechanisms: use of livestock systems based on nitrogen-fixing legumes, optimum use of manure, and reduction of nutrient losses;

(iv) promotion of pest regulation by biological control agents. It is achieved by introducing and/or conserving natural enemies and antagonists and by developing an ecological network.

Integrated or agroecological farming systems are thus based, as much as possible, on the services provided by agroecosystems. These services are particularly the result of the activity of functional biodiversity [37]. Agro-forestry (in its wide acceptance) and minimum or zero tillage are often components of these farming systems.

One condition that may enhance ecosystem services is to recapitalize ecosystems by investing in soil organic matter (SOM) (humus) and ecological infrastructures. SOM is at the center of agroecological systems; it is the node wherein many relationships cross. Flows of nutrients and plant nutrition start with humus mineralization; plant dead tissues and animal effluents finish the cycle and are incorporated in SOM through humification.

TABLE 22-1 Classification of Farming Types According to a Gradient of Commercial Input use (High on the Left, Low on the Right) (Inputs Considered are Mainly Synthesis Nitrogen Fertilizer and Pesticide Transformed into Fossil-Fuel Consumption Necessary for their Synthesis and Fuel for Tractors)

Farming System Types

A	B	C	D	E
Large use of inputs (ex.: fertilizers, pesticides)	Cutting wastes and excesses in input use, decrease of pollution and costs	Minimum use of synthesis fertilizers and pesticides	No synthesis fertilizers and pesticides	No synthesis fertilizers and pesticides, no tillage
Intensive conventional farming, Green Revolution	Sound farming, Doubly Green Revolution[4]	Integrated farming,[5] **Agroecology,**[1] Evergreen Revolution, Conservation Agriculture,[3] Ecologically Intensive Agriculture, Ecoagriculture	Organic farming (including biodynamic), **Agroecology**[2]	Permaculture, Natural farming, **Agroecology?**[6]

<------------------------------------> <--->
Fossil-fuel-based   **Ecosystem-based**

According to authors, the definition of agro-ecology varies in relation to commercial input use.
[1]*Agroecology: minimum use of external inputs (after, for instance, Altieri, 1989 and 1995; http://www.agroecology.org/).*
[2]*Agroecology: no synthesis fertilizers and pesticides (after http://fr.wikipedia.org/wiki/Agroécologie).*
[3]*Conservation agriculture: Soil interventions such as mechanical soil disturbance are reduced to an absolute minimum or avoided, and external inputs such as agrochemicals and plant nutrients of mineral or organic origin are applied optimally and in ways and quantities that do not interfere with, or disrupt, the biological processes (FAO, 2006 [25]).*
[4]*Doubly Green Revolution: based on genetic engineering, synthetic nitrogen fertilizer, and better pesticide use (Conway, 1997 [35]).*
[5]*Integrated farming: system approach of land use for agricultural production which aims at reducing the use of external inputs (energy, chemical products) by using and developing as well as possible natural resources and encouraging natural regulation processes (Viaux, 1999 [23]).*
[6]*Possible definition of agroecology.*

In Belgium, most arable soils are poor in organic matter. When no fertilizers are applied, the production potential of a ryegrass crop is typically about 2 t dry matter (DM) /ha on arable soils of the Loamy Region and about 8 t DM/ha after the destruction of a permanent grassland plot in the Ardennes [38–40]. This difference illustrates that arable soils of the Loamy Region have lost their ability to support crop growth if not artificially fertilized. Today high crop yields in this region are possible only with high commercial fossil-fuel-based inputs. These soils have also lost other qualities like water infiltration and prevention of soil erosion. As conventional agriculture and agricultural research have often under-estimated the importance of SOM, it is not surprising that the SOM dynamic is one of the least understood aspects of the functioning of agroecosystems.

Enhancing SOM in specialized arable cropping farms implies, in most cases, developing new forms of livestock/crop combinations. Farmyard manure (FYM) and temporary grasslands in crop rotations are the two most efficient ways to build up humus. New livestock/crop combinations can be designed by exchanges (*i.a.* contracts) of FYM and green forage from temporary grasslands between farmers of the same region. Reduced tillage and no-tillage, the use of green manure, and the incorporation of crop residues are also important tech-niques for better SOM management. Investment in ecological networks should go beyond the agri-environmental measure approach by a concerted action at the landscape level of all land managers and owners.

Only a landscape approach can optimize the location of seminatural ele-ments (hedges, herbaceous strips), crops, and grasslands and maximize the synergies between them. This is particularly important for developing an infra-structure that can control erosion or develop dispersion corridors for fauna and flora. The optimization of geographical location of field margins and cropping strips is a prerequisite for an efficient biological control of pests such as aphids. Predators and parasitoids of crop pests are indeed often limited in their move-ments. They cannot move more than some dozen meters from an herbaceous strip into the cropped area. The space diversity of crop and grassland types should also be designed at the landscape level for enhancing mutual benefits. Such a spatial coordination between land managers can only be implemented through a (new) organization supported by public authorities. This organization of land-use planning (land consolidation) should operate in a democratic way; farmers should be able to join it on a voluntary basis and be the partners of the design of new landscapes, but farmer involvement would be strongly encour-aged by linking the access to agricultural subsidies with their participation in the process.

3. AGROECOLOGICAL RESEARCHES

Agroecology as a science has been actively developed in the United States in the 1980s and the 1990s, particularly in California and in orchard production [41]. It spread from there to Central and South America and was applied to tropical farm-ing systems through inspiration by or in collaboration with American research-ers. In Europe, similar research has been developed and named integrated farming [42]. More research was devoted to organic farming in Europe than in the United States. In contrast with American conditions, much research focused on arable systems in Europe. Among European researchers on integrated arable farming systems, the names of Adel El Titi (Lautenbach project, Germany), Vic Jordan (Long Ashton, UK), Pieter Vereijken (Nagele experimental farm in the Netherlands and leader of the Working Group on Integrated and Ecological Arable Farming Systems [I/EAFS] for EC- and associated countries'), and Philippe Viaux (Arvalis—Institut du végétal, France) must be cited [43, 44]. Their research in experimental farms demonstrated that, compared to conventional farming, integrated farming systems result in a significant reduction in pesticide and nitrogen fertilizer consumption, a lower nitrate content in the soil profile, a positive effect on the soil fauna, similar yields, and similar or even higher economic margin (see, for instance, [45]). Other research was developed with a holistic and participatory approach in pilot farms (commercial farms), including those of Pieter Vereijken and the members of the I/EAFS Working Group [46].

Many agroecology research programs are involved in action. They are sometimes called farming systems research or farmer participatory research. On-farm action research is an efficient method for helping farmers adapt to changing external factors (ex.: decrease of subsidy levels, price variations) and take into account new objectives and new challenges (ex.: pollution reduction, biodiversity restoration, enhancement of ecosystem services). In the European experience, researchers and farmers were associated, as partners, for the design, development, and dissemination of prototypes of farms. A research methodology has been formalized in five steps [44, 46] (see Annex). The methodology has been successfully applied in several European Union (EU) countries and in one tropical country [47]. It proved to be very effective but is still insufficiently known and used. The main methods were related to Multifunctional Crop Rotation, Ecological Nutrient Management, Integrated Crop Protection methods, Development of an Infrastructure for Nature and Recreation, and Farm Structure Optimization. They aimed at improving crop yield, product quality, and farmer's income while reducing pollution, improving landscape, and restoring biodiversity. They should now be adapted to other farm types such as livestock and mixed farms.

Outside the EU, in Switzerland, the agricultural policy supports integrated agricultural production and organic farming. As of 2002, Swiss farmers can obtain agricultural subsidies only if they comply with the rules of integrated production (Federal Office for Agriculture, FOAG). The experience of this country demonstrates that the transition toward agroecological systems can be achieved relatively fast.

In Belgium, much research has been developed on topics useful for implementing integrated farming systems or agroecology such as biological control, biological nitrogen fixation, integrated weed management, reduced and zero tillage, and sustainability indicator systems. The efforts of the Royal Belgian Institute for Beet Improvement (IRBAB/KBIVB) for a sound use of fertilization and weeding techniques of the sugar beet crop can be mentioned, for instance. Only a few studies, however, adopted a holistic approach to the farming systems. Nevertheless, two projects can be cited: the European research project (AIR, FP3) Advanced Ecological Farming Systems (EFS), based on best practices with organic farmers pilot groups: ECOFARM' [49] and the Walloon project PROjet-pilote pour la Protection des EAUx de la nappe des SABLEs bruxelliens – Prop'EauSable [50]. The SSTC (Belspo) project Framework for Assessing Sustainability Levels in Belgian Agricultural Systems (SAFE) was an attempt to develop an assessment system of integrated and ecological farming systems [51, 52]. The INTERREG project Boeuf des prairies gaumaises (Beef from the Gaume grasslands)' [53] can be considered an agroecology experience in the livestock sector. There is scope to increase research efforts in Belgium and to diversify research topics. Groups of pilot farmers could be formed by farm types (ex.: specialist dairying; specialist cattle-rearing and fattening; cattle-dairying, rearing, and fattening combined; field crops-grazing livestock combined). In each of these farm types, the best techniques for restoration and use of ecosystem services should be identified and promoted. Pilot farms should be used as examples for stimulating a large number of farmers to convert to agroecology. Local adaptation and dissemination of new technologies as well as farmer training are additional challenges to successfully transition to ecosystem-service-based farming systems. In addition, sociological research should identify the key factors impeding the agricultural transitions (including psychological, sociological, and cultural factors). The inertia of the system is indeed not only caused by the investments made but also by other more subtle factors belonging to the sociocultural realm.

Agroecology and integrated farming systems require larger investment in the farmer's knowledge and skills, and less investment in commercial inputs than conventional farming. In the first experiences, only scientists and farmers were associated in the prototyping of sustainable farms. Because farmers' training is essential for success, in more recent experiences, stakeholders other than farmers and scientists were associated with the process, including representatives of farmer's advisory services. That also facilitates the dissemination to larger groups of farmers and to a wider public of nonfarmers [47].

4. STRENGTHS AND WEAKNESSES OF AGROECOLOGY

The strength of integrated farming systems and agroecology is undoubtedly that they improve farm efficiency, have a positive impact on the environment and biodiversity, and are more profitable for farmers than conventional systems. An improved use and supply of ecosystem services is then possible and desirable in agriculture.

There are three categories of biodiversity in agriculture (Table 22-2): (i) Biodiversity planned by farmers, (ii) functional biodiversity that has a decisive influence on the functioning of the agroecosystem, and (iii) diversity of wild species and communities that use agricultural landscapes as

TABLE 22-2 Agricultural Biodiversity Categories (after Altieri, 1999; Biala et al., 2005; Peeters et al., 2004)

Agricultural Biodiversity Categories			Examples
Planned (Productive species)			Crop species and cultivars, sown forage species and cultivars, livestock species and breeds, planted trees
Associated	Functional	Beneficial species	Pollinators, parasites and parasitoids of crop pests, decomposers, spontaneous forage grassland species
		Detrimental species	Weeds, insect pests, microbial pathogens
	Heritage		Farmland birds

their habitat and can have a high conservation value (heritage biodiversity) [54, 55]. Planned biodiversity is deliberately introduced by farmers in the agroecosystem. The two other categories of biodiversity are associated with the system and appear spontaneously. Most agroecology authors recognize only two types of biodiversity in agroecosystems: planned and functional [38]. The importance of heritage biodiversity is often underestimated in agroecology, and enhancement strategies of this biodiversity are rarely taken into account (e.g., see [56]). Agroecology provides ecological bases for biodiversity conservation in agriculture [16] but, because it does not consider heritage biodiversity, it often fails in managing it adequately and restoring it. Wild biodiversity and ecosystem services both require explicit consideration in agroecosystems.

Agroecology, integrated farming, and organic agriculture have focused mainly on the development of "healthy" agroecosystems, on the reduction of the use of chemical synthesis products, and on the reduction of pollution. For achieving these goals, they use farm-scale actions. **Ecoagriculture** has the same approach but also coordinates efforts among farmers and other citizens to achieve biodiversity benefits at a landscape scale [27, 28]. Ecoagriculture is a fully integrated approach to agriculture, nature conservation, and rural livelihoods within a landscape context. It aims at maximizing ecological, economic, and social synergies, and minimizing the conflicts. Ecoagriculture landscapes are mosaics of areas in natural or seminatural habitats and areas under agricultural production. Ecoagriculture designs landscapes for [27]:

- providing protection of nesting areas of wild species from disturbance;
- protecting them from predators with diverse perennial covers;

- providing adequate access to clean water and to food from diverse sources throughout the year;
- developing ecological networks between dispersed population groups to ensure minimum viable populations;
- ensuring viable populations of predators and prey and sufficient populations of interdependent species (ex.: plants and their pollinators);
- promoting biologically active soils.

This approach has to be done in patches and networks of natural habitat and in production areas. To achieve these attributes in production areas, strategies should be developed for minimizing agricultural pollution, managing conventional cropping systems in ways that enhance habitat quality, and designing farming systems to mimic the structure and function of natural ecosystems. Ecoagricultural systems also aim at improving the livelihood of farmers. Consequently, it is necessary to maintain or increase agricultural output, reduce overall production costs, or enhance the market value of the products.

Most publications on agroecology and integrated farming systems focus on fruit, vegetable, and arable crop productions. A lot of work remains to be done on livestock- and grassland-based systems as well as on systems where arable crops are combined with grazing livestock.

On the basis of the information presented earlier in this chapter, it can be concluded that the main weaknesses of integrated farming systems and agroecology are the lack of systematic consideration of heritage biodiversity and a still low application in livestock production systems.

5. DISCUSSION AND RECOMMENDATIONS

5.1. Some Important Elements

The future of Belgian agricultural development should be analyzed in a broader context. The following aspects are often underestimated in the design and assessment of agricultural policies, but are arguments for the development of agroecology.

It is estimated that 30 to 50% of the total food produced is wasted in Western societies [57–59]. Each year, about 90 million tonnes or 179 kg per capita (in 2006) of food wastes are generated in the EU-27 [60]. These values are expected to rise. Households are responsible for the largest food waste (about 42% of the total). Reducing waste should thus require at least as much attention as increasing yields.

In Belgium, about three quarters (72% in 1999) of plant production is used for animal feeding [61]. In addition, massive imports of cereals and soybean (cake) are needed. Belgium, as is true of other EU-15 countries, is highly dependent to protein imports for animal feeding. In 2007/2008, for instance, the European Union (EU) imported the equivalent of 19 million ha of virtual land (land necessary for producing a given tonnage of commodity on the basis of regional yields) of soybean [62]. This area is equivalent to about the size of the Utilized

Agricultural Area (UAA) of Germany. These imported feeds are largely used for pig and poultry feeding. They are also used for dairy cow feeding and beef cattle fattening in supplementing maize and cereal-based rations. Higher protein self-sufficiency is desirable for strategic, economic, environmental, and human health reasons. That can be achieved by better integration of grain and forage legumes in Belgian farming systems. This integration is characteristic of agro-ecological systems.

Although red meat consumption has regularly decreased in the last 60 years in favor of pig and poultry meats, Belgian citizens still eat too much meat that has unfavorable fatty acid profiles. In the EU, the total per-capita protein consumption is about 70% higher than recommended, and the average intake of saturated fatty acids is about 40% higher than recommended. In Belgium, recommendation of meat consumption is 75 to 100 g/day.person [63], while meat consumption increased from about 150 g/day.inhabitant in 1955 to more than 260 g/day.inhabitant in 2005 [64]. Such statistics lead to human health problems, including obesity and coronary heart diseases [65]. A reduction in meat consumption is desirable, which means that part of the land could be devoted to crops other than annual crops for animal feeding, which creates opportunities for diversification and for a larger share of (permanent) grasslands in the agricultural area.

These three elements show that there is an important buffer in terms of food security in Belgium.

Subsidies to exports on the world market are almost totally suppressed in the EU, but still a huge support is given to farming through the single-payment scheme. This money is used to ensure a fair standard of living for the agricultural community in the EU, stable markets, and an affordable supply of food at a reasonable price for European consumers as defined in the Treaty of Rome. However, it seems unjustified that taxes paid by European citizens should be used to help farmers and members of the agri-food sector, who would not be competitive on the world market without these subsidies, and to put on this world market agricultural commodities (ex.: cereals, powder milk) that have many adverse effects on the economies of developing countries.

6. THE WAY AHEAD

Increasing production should no longer be a priority in Belgium and the EU. On the world market, the European Union can be competitive for exporting high-quality products (ex.: wine, champagne, alcohols (whisky), quality cheese). On the EU market, the objective with regard to improving farmers' income should be to reduce production costs, ensure higher self-sufficiency (ex.: protein, energy) [66]. "A more self-sufficient and thrifty farming" is the translation of the name of the book. Increase the rate of food processing in farms, shorten marketing chains, diversify products and activities, increase quality of products, and improve fairness in farmer's support [67]. For the society, it is also desirable to reduce the impact of agriculture on biodiversity and the environment.

As mentioned earlier, the current Belgian (and European) agricultural systems are based on massive use of fossil fuels and imports of animal feed from other continents. These systems have to be reformed urgently for the reasons described above. The link between agriculture and fossil fuel should be progressively cut. As fossil fuel and consequently fertilizer, cereal, and soybean prices will increase in the future, as they have already in a recent past, farmers should rely much more on ecosystem services than on oil and commercial inputs and should be much more resource-saving and self-sufficient instead of dependent on global markets. This strategy, combined with a short marketing chain and a quest for excellence in product quality,[1] can increase the Belgian farmer's income and reduce farming's negative impacts on the environment and human health.

We should therefore organize a rapid transition from conventional to integrated (agroecological and organic) farming systems. The target could be to implement a full transition toward an integrated farming system in the next ten years. That will require a firm political will, corresponding budgets, and a strong involvement of many types of stakeholders: scientists, teachers of technical schools, farmers' advisers, traders of the food sector, and of course farmers. The Swiss experience demonstrates that this fast transition can be achieved. Even with a strong political will, some farmers will not be able to convert completely in a short period of time. For those, the solution could consist in integrating more progressively agroecological principles with conventional agriculture.

This ambitious goal implies a total change of paradigm. It will require, as a priority:

- new agricultural policies (see below);
- a fundamental reform of the training of future farmers, technicians, and farmers' advisers in technical schools;
- a reform of the specialization training (master) of agricultural scientists in higher education institutions;
- a restructuring of agricultural research that will define new priorities. The dominance of reductionist research should be inverted at the benefit of holistic and participatory research. In the short term, groups of pilot farmers should be created in each region (Flanders and Wallonia) and associated different types of stakeholders (ex.: holistic researchers, reductionist researchers, farmers' advisers, technical and higher education schools, consumers, and nature conservationists).

---

1. Food quality can be defined by five components: nutritional, hygienic, technological and look, organoleptic (hedonist aspect), and ethical (fair trade, green labeling, animal welfare aspects).

- a technological revolution for designing and developing agroecological methods and systems (including and especially in livestock farms), adapting them locally, disseminating them, and supporting farmers in their transition period;
- the design of a biodiversity-friendly agriculture: the integration of heritage biodiversity enhancement in methods and practices;
- giving priority to agroecological products in school and administration cafeterias (canteens).

7. LINKS WITH EU POLICY INSTRUMENTS

A strong inflexion of the Common Agricultural Policy (CAP) is needed. The structure of the CAP budget in two pillars (first pillar = income support; second pillar = rural development) is no longer justified. Direct payments, even if better distributed among and within member states, should be abandoned. They can no longer be socially justified in a context of public expense reduction and economic difficulties. Farmers should not be paid just for farming. They should earn income from their farming activities that should be as much as possible economically viable. When this viability cannot be guaranteed, it can be increased by complementary activities based on the reality of a market (ex.: agritourism, food processing, short marketing chains) when they can have access to these complementary activities. Farmers should also be rewarded for the positive actions they undertake for a sustainable management of natural resources and for the delivery of ecosystem services. CAP should initiate and organize a market for the production of these public goods and services, and this market should be largely financed by public money ("public money for public goods"). This market should prioritize biodiversity and landscape conservation, carbon storage, and water quality. On the supply side, initiatives could come from farmers advised by experts (research organizations, specialized NGOs, R&D offices) for proposing (offering) public goods and services. On the demand side, the CAP would be the main source of payments, but other public authorities (national, regional, local), private companies, individuals, or group of individuals could evaluate these proposals and decide to pay (or buy) them. That implies a real delivery of public goods, a balance between the public goods produced, and the implementation of an effective control system of this delivery. Most CAP expenditures should be redirected to this objective. Improved agri-environmental measures should remain the reference and source of inspiration for this market. This system, based on private initiative, creativity, and efficiency, would boost the protection of the environment and stimulate rural life. It would create a vibrant countryside, create new jobs, and increase contacts between the agricultural sector and other citizen types. That will give a long-term legitimacy to the CAP budget. The CAP will in that way be able to demonstrate good value for money to taxpayers. On the other hand, farming activities that are detrimental to the environment and human health should be strictly controlled and forbidden.

In addition, about 20% of the total CAP budget should be reserved mainly for stabilizing income in case of high price volatility (safety net).

Another option should be to transfer most budgets (about 80%) from the first to the second pillar of the CAP. A large part of the budget of the second pillar would then be available for paying farmers for producing ecosystem services.

The CAP should support:

- as a priority, those farming systems that provide the most environmental public goods and services. High Nature Value (HNV) farming systems are one of them. Their persistence is threatened by a low profitability. HNV farmland is often managed by small farmers who need more support than others;
- seminatural grasslands and other ecological networks;
- agroecological, integrated, and organic farming systems.

Agroecological and integrated farming practices should be considered as the minimum level of good agricultural practices (cross-compliance principle). A label or trade marks for agroecology or integrated farming systems should be created for distinguishing between agroecological EU products and non-EU products that are not produced by these farming systems. Specification and control systems should thus be defined and organized.

Existing policy tools should be improved if necessary; among these tools are the agri-environmental scheme, the High Nature Value farmland scheme, the EU product quality policy, the Nitrate Directive, the Water Framework Directive, and the NATURA 2000 network. The Regulation (EC) No 1107/2009 on plant protection products is certainly a major instrument that could eliminate the most dangerous pesticides in the short or the mid-term [68].

The current agri-environmental scheme is a system for paying farmers for ecosystem services. It has had a positive effect on the environment by slowing down degradation, by maintaining a situation, or by restoring biodiversity and landscapes. Its effect is insufficient, however, as recognized in the EU Biodiversity Strategy document. There are several reasons for this, including inadeqate budget. The budget for Agri-Environmental Measures (AEM) should therefore be increased. Improving AEM's efficiency is also necessary. Result-oriented AEM instead of mean obligation measures are likely to be more efficient when they are applicable [69–72]. AEM should also be better targeted. "From an environmental perspective, the more tailored the measures are to specific environmental needs and the more they are targeted at the locations in which action is needed, the more effective and efficient they are likely to be in achieving their objectives" [73, p. 6].

The High Nature Value (HNV) farmland concept offers real added value for nature conservation outside the Natura 2000 areas, where agri-environmental payments would anyway be available for farmers affected by the compulsory implementation of nature management practices. Additional programs could be developed for such HNV farmland areas. It is important that such measures are realistic. In Wallonia, for example, the payment per ha for extensive grazing seems to be too low compared to the actual income loss incurred

[74]. Specific measures are needed for encouraging extensive practices, for developing innovative extensive systems, and for managing seminatural farmland features. Little is known about the socioeconomics of these farms and what tools, coaching promotion, or financial incentives are necessary to render HNV farming a genuine part of profitable agriculture in Belgium [75].

The EU product quality policy [76] is dealing with the protection and promotion of products of local origin[2] and organic farming. Local origin products can certainly help farmers get higher prices and increase the market for their products. They can have an indirect beneficial effect on the environment if and only if production guidelines include criteria related to biodiversity conservation (local breeds and cultivars, species-rich grassland, ecological network) and to the principles of agroecology.

8. CONCLUSION

Current degraded landscapes are the result of past activities. Many elements were destroyed during the intensification phase of agriculture because their functional role was no longer recognized, or because they interfered with the mobility of machinery, or just because they took space that could be used for production. Some elements, however, survived. Priority should be given to the conservation of their high cultural and biological value. However, if ecosystem services are to be optimized, a new design and transformation of landscapes is necessary. This is a very important aspect of the transition phase. A participatory organization of land-use planning (land consolidation) is necessary for implementation of these new landscapes and for recapitalizing agroecosystems with ecological structures.

The scientific sphere should also play its role. The creation of sustainable agroecosystems necessitates a wider mobilization of knowledge: not only from agronomy but also from ecology, geography, and social sciences. Agroecological research should be stimulated. Universities and research centere should bridge their skills, as the aim is to solve a societal problem requesting a transdisciplinary approach.

Both an increase in the grassland area and the implementation of an efficient system for biodiversity conservation in agriculture are also highly desirable. More grasslands in the agricultural area will provide better landscapes and more ecosystem services. Other challenges such as protein self-sufficiency for animal feeding, food waste reduction, and improvement of the fatty acid composition of animal products and its consequences for human health should be taken into account in CAP reform proposals. While the CAP's objective is not to contribute to solving all these problems, it is influenced by them. It can contribute to solving a part of these problems, and its reform has to be considered in this

---

2. PDO—Protected Designation of Origin, PGI—Protected Geographical Indication and TSG - Traditional Speciality Guaranteed (EC No 1898/2006)

context. The CAP is no longer an instrument for food security and income support of farmers, rather, it has progressively become a tool for answering wider challenges. Food waste reduction should notably receive more attention than yield and production increases. That demonstrates that the present use of fertilizers and pesticides could be reduced. A holistic approach has to be adopted at all decision levels, ranging from farm to global level.

ACKNOWLEDGMENT

This research was funded by the Belgian Science Policy Office (BELSPO) within a cluster project called Belgian Ecosystem Services (BEES). The author thanks the reviewers for their useful comments: Mrs. Elisabeth Simon and Prof. Guido Van Huylenbroeck (Department of Agricultural Economics, University of Ghent).

REFERENCES

1. Costanza, R., d'Arge, R., de Groot, R., Farber, S., Grasso, M., Hannon, B., Limburg, K., Naeem, S., O'Neill, R.V., Paruelo, J., Raskin, R.G., Sutton, P., and van den Belt, M. 1997. The value of the world's ecosystem services and natural capital. *Nature* 387, 253–260.
2. Daily, G. (ed.). 1997. *Nature's Services: Societal Dependence on Natural Ecosystems*. Washington, DC: Island Press.
3. Pearce, D., and Moran, D. 1994. *The Economic Value of Biodiversity*. London: IUCN and Earthscan Publications.
4. MEA (Millennium Ecosystem Assessment). 2005. *Ecosystems and Human Well-being: Synthesis*. Washington, DC: Island Press, 137 pp.
5. Stern, N.H. 2007. *The Economics of Climate Change: The Stern Review*. Cambridge: Cambridge University Press, 692 pp.
6. DEFRA. 2008. Accounting for the positive and negative environmental impacts of UK agriculture. DEFRA, 2 pp.
7. Roberts, J.M. 1976. *History of the World*. New York: Knopf, 1054 pp.
8. Koppelaar R.H.E.M. 2006–2009. World Production and Peaking Outlook. Peakoil Nederland. http://peakoil.nl/wp-content/uploads/2006/09/asponl_2005_report.pdf. Retrieved on July 27, 2008.
9. Zittel, W., and Schindler, J. 2007–2010. Crude Oil: The Supply Outlook. Energy Watch Group. Downloadable from: http://www.energywatchgroup.org/fileadmin/global/pdf/EWG_Oilreport_10-2007.pdf.
10. Broomfield, M. 2012. Support to the identification of potential risks for the environment and human health arising from hydrocarbons operations involving hydraulic fracturing in Europe. European Commission, DG Environment and AEA Technology, 276 pp.
11. Pearson, C.J., and Ison, R.L. 1987. *Agronomy of Grassland Systems*. Cambridge: Cambridge University Press.
12. Maljean, J.F., and Peeters, A. 2003. Integrated farming and biodiversity: Impacts and political measures. In Towards integrating biological and landscape diversity for sustainable agriculture in Europe, Proceedings of the High-Level Pan-European Conference on Agriculture and Biodiversity, Paris, France. Council of Europe Publishing, *Environmental Encounters* 53, 119–132.
13. Griffon, M. 2011. Pour des agricultures écologiquement intensives, des territoires à haute valeur environnementale et de nouvelles politiques agricoles. *Editions de l'Aube*, 136 pp.

14. Wezel, A., Bellon, S., Doré, T., Francis, C., Vallod, D., and David, C. 2009. Agro-ecology as a science, a movement or a practice. *Agronomy for Sustainable Development* 29, 503–515.

15. Altieri, M.A. 1989. Agroecology: A new research and development paradigm for world agriculture. *Agriculture, Ecosystems and Environment* 27, 37–46.

16. Altieri, M.A. 1995. *Agroecology: The Science of Sustainable Agriculture.* Boulder,CO: Westview Press, 433 pp.

17. Carrol, C.R., Vandermeer, J.H., and Rosset, P.M. 1990. *Agroecology.* New York: McGraw-Hill Publishing.

18. Gliessman, S.R. 1998. *Agroecology: Ecological Processes in Sustainable Agriculture.* Ann Arbor, MI: Ann Arbor Press.

19. Francis, C., Lieblein, G., Gliessman, S., Breland, T.A., Creamer, N., Harwood Salomonsson, L., Helenius, J., Rickerl, D., Salvador, R., Wiedenhoeft, M., Simmons, S., Allen, P., Altieri, M., Flora, C., and Poincelot, R. 2003. Agroecology: The ecology of food systems. *Journal of Sustainable Agriculture* 22(3), 99–118.

20. El Titi, A. 1992. Integrated farming: An ecological farming approach in European agriculture. *Outlook on Agriculture* 21(1), 33–39.

21. Vereijken, P. 1989. From integrated control to integrated farming, an experimental approach. *Agriculture, Ecosystems and Environment* 26(1), 37–43.

22. Vereijken, P., and Viaux, P. 1990. Vers une agriculture intégrée. *La Recherche* 227, 22–25.

23. Viaux, P. 1999. Une 3ème voie en Grande Culture. Environnement—Qualité—Rentabilité. *Agridécisions*, 211 pp.

24. Swaminathan, M.S. 1996. *Sustainable Agriculture, Towards an Evergreen Revolution.* Delhi: Konark Publishers, 219 pp.

25. FAO 2006. (conservation agriculture website). Downloadable from: http://www.fao.org/ag/ca/1a.html

26. McNeely, J.A., and Scherr, S.J. 2003. *Ecoagriculture: Strategies for Feeding the World and Conserving Wild Biodiversity.* Washington, DC: Island Press.

27. Scherr, S.J., and McNeely, J.A. 2012. Biodiversity conservation and agricultural sustainability: Towards a new paradigm of "ecoagriculture" landscapes. *Philosophical Transactions of the Royal Society* B: Biological Sciences 2008 363, 477–494.

28. Lampkin, N. 1994. Organic farming: Sustainable agriculture in practice. In Lampkin, N., and Padel, S. (eds.), *The Economics of Organic Farming. An International Perspective.* Oxford: CABI.

29. IFOAM. 1998. Basic Standards for Organic Production and Processing. IFOAM Tholey-Theley, Germany.

30. Mollison, B.C., and Holmgren, D. 1978. *Permaculture One, A Perennial Agriculture for Human Settlements.* Melbourne: Transworld Publishers.

31. Mollison, B.C. 1990. *Permaculture: A Practical Guide for a Sustainable Future.* Washington, DC: Island Press, 568 pp.

32. Fukuoka, M. 1978. *One-Straw Revolution: An Introduction to Natural Farming.* Emmaus: Rodale Press, 212 pp.

33. Altieri, M.A., and Rosset, P. 1995. Agroecology and the conversion of large-scale conventional systems to sustainable management. *International Journal of Environmental Studies* 50, 165–185.

34. Gaud, W.S. 1968. The Green Revolution: Accomplishments and Apprehensions. Conference at Shoreham Hotel, Washington, DC, March 8, 1968. Downloadable from: http://www.agbioworld.org/biotech-info/topics/borlaug/borlaug-green.html

35. Conway, G. 1997. *The Doubly Green Revolution: Food for All in the Twenty-First Century.* Comstock Pub. Associates, 334 pp. [City?]

36. Paillotin, G. 2000. L'agriculture raisonnée. Rapport au ministre de l'Agriculture et de la Pêche: 41 p.

37. Altieri, M.A. 1999. The ecological role of biodiversity in agroecosystems. *Agriculture, Ecosystems and Environment* 74(1–3), 19–31.

38. Peeters, A. 2004. *Wild and Sown Grasses. Profiles of a Temperate Species Selection: Ecology, Biodiversity and Use.* Rome: FAO and Blackwell Publishing, 311 pp.

39. Peeters, A., and Janssens, F. 1996. Concilier conservation de la biodiversité et production agricole performante en prairie: est-ce possible? *Annales de Gembloux* 101, 127 Melbourne. 147.

40. Toussaint, B., and Lambert, J. 1984. Contribution à l'étude d'une stratégie de la fertilisation azotée en prairie temporaire. *Revue de l'agriculture* 1(37), 29 Melbourne.41.

41. Wezel, A., and Soldat, V. (2009). A quantitative and qualitative historical analysis of the scientific discipline of agro-ecology. *Journal of Agricultural Sustainability* 7(1), 3 Melbourne.-18.

42. EISA (European Initiative for Sustainable Development in Agriculture). (No date). A Common Codex for Integrated Farming. Downloadable from: http://ec.europa.eu/environment/ppps/pdf/ilubrochure.pdf

43. Holland, J.M., Frampton, G.K., Çilgi, T., and Wratten, S.D. 1994. Arable acronyms analysed— a review of integrated arable farming systems research in Western Europe. *Annals of Applied Biology* 125, 399–438.

44. Vereijken, P. 1997. A methodical way of prototyping integrated and ecological arable farming systems (I/EAFS) in interaction with pilot farms. *Developments in Crop Science 25*, 293–308.

45. El Titi, A., and Landes, H. 1990. Integrated farming system of Lautenbach: A practical contribution toward sustainable agriculture in Europe. In Edwards, C.A., Lal, R., Madden, P., Miller, R.H., and House, G., *Sustainable Agricultural Systems.* Ankeny, Iowa: Oil and Water Conservation Society, pp. 265–286.

46. Vereijken, P. 1999. Manual for prototyping integrated and ecological arable farming systems (I/EAFS) in interaction with pilot farms. *AB-DLO*, 53 pp. + annexes.

47. Sterk, B., van Ittersum, M.K., Leeuwis, C., and Wijnands, F.G. 2007. Prototyping and farm system modelling—Partners on the road towards more sustainable farm systems? *European Journal of Agronomy* 26(4), 401–409.

48. Federal Office for Agriculture, FOAG. http://www.blw.admin.ch/index.html?lang=en

49. Peeters, A., and Van Bol, V. 2000. ECOFARM: A research/development method for the implementation of a sustainable agriculture. FAO, REU Technical Series 57, 41–56.

50. Lambert, R., Van Bol, V., Maljean, J.F., and Peeters, A. (2002). Prop'eau-sable'. Recherche-action en vue de la préparation et de la mise en œuvre du plan d'action de la zone des sables bruxelliens en application de la directive européenne CEE/91/676 (nitrates). Final activity report March 1997—March 2002 (Walloon Region, DGRNE contract). Laboratory of Grassland Ecology (UCL), Belgium, 107 p.

51. Sauvenier, X., Valckx, J., Van Cauwenbergh, N., Wauters, E., Bachev, H., Biala, K., Bielders, C., Brouckaert, V., Garcia-Cidad, V., Goyens, S., Hermy, M., Mathijs, E., Muys, B., Vanclooster, M., and Peeters, A. 2005. Framework for assessing sustainability levels in Belgian agricultural systems (SAFE). Research contract no. CP/10/281. Final report, 119 pp.

52. Van Cauwenbergh, N., Biala, K., Bielders, C., Brouckaert, V., Franchois, L., Garcia, Cidad, V., Hermy, M., Mathijs, E., Muys, B., Reijnders, J., Sauvenier, X., Valckx, J., Vanclooster, M., Van der Veken, B., Wauters, E., and Peeters, A. 2007. SAFE—A hierarchical framework for assessing the sustainability of agricultural systems. *Agriculture, Ecosystems and Environment* 120(2–4), 229–242.

53. Stassart, P., and Stilmant, D. 2012. Lorsqu'une filière s'identifie à son territoire: que nous apprend l'expérience du "Boeuf des prairies gaumaises"? CRA-W & GxABT–Carrefour Productions animales, 15 pp.

54. Biala, K., Peeters, A., Muys, B., Hermy, M., Brouckaert, V., Garcia, V., Van Der Veken, B., and Valckx, J. (2005). Biodiversity indicators as a tool to assess sustainability levels of agroecosystems with a special consideration of grasslands areas. *Options méditerranéennes* 67, 439–443.

55. Peeters, A., Maljean, J.F., Biala, K., and Brouckaert, V. 2004. Les indicateurs de biodiversité pour les prairies: un outil d'évaluation de la durabilité des systèmes d'élevage. *Fourrages* 178, 217–232.

56. Altieri, M.A. 1994. *Biodiversity and Pest Management in Agroecosystems.* New York: Haworth Press, 185 pp.

57. Parfitt, J., Barthel, M.. and Macnaughton, S. 2010. Food waste within food supply chains: quantification and potential for change to 2050. *Phil. Trans. R. Soc.* B 365, 3065–3081.

58. WRAP. 2009. *Household Food and Drink Waste in the UK.* Banbury, UK.

59. WRAP. 2010. *A Review of Waste Arising in the Supply of Food and Drink to UK Households.* Banbury, UK.

60. European Commission. 2010. Preparatory study on food waste across EU 27. Final report. DG Environment Directorate C: 210 pp.

61. Genot, L. (ed.). 2005. Les ressources génétiques des animaux d'élevage en Belgique. Contribution de la Belgique au Premier Rapport sur l'État des Ressources Zoogénétiques dans le Monde. Rapport national à la FAO. 58 pp. Downloadable from: http://agriculture.wallonie.be/apps/spip_wolwin/IMG/pdf/RapportNationalFAO.pdf

62. Von Witzke, H., and Noleppa, S. 2010. EU Agricultural Production and Trade: Can More Efficiency Prevent Increasing "Land-Grabbing" Outside of Europe? Agripol, OPERA—Research Centre, 36 pp.

63. ISSP. 2006. Enquête de santé par interview Belgique, 2004. IPH/EPI Reports N°2006-036, N° de dépôt: D/2006/2505/5. Brussels: Institut scientifique de santé publique. Downloadable from: http://www.iph.fgov.be/epidemio/epifr/crospfr/hisfr/his04fr/his32fr.pdf.

64. Task force développement durable. 2009 Indicateurs, objectifs et visions de développement durable. Rapport fédéral sur le développement durable 2009. Bureau Fédéral du Plan, 224 pp.

65. World Health Organization (WHO). 2012. Obesity and overweight. Donwloadable from: http://www.who.int/mediacentre/factsheets/fs311/en/index.html

66. Poly, J. 1980. Pour une agriculture plus économe et plus autonome. Paris INRA: 65 pp.

67. Peeters, A. 2012. Past and future of European grasslands. The challenge of the CAP towards 2020. *Grassland Science in Europe* 17, 17–32.

68. EUROPA. 2009. Sustainable Use of Pesticides. Downloadable from: http://ec.europa.eu/environment/ppps/home.htm. Retrieved on 2009-06-04.

69. Matzdorf, B., T. Kaiser, and Rohner, M.S. 2008. Developing biodiversity indicator to design efficient agri-environmental schemes for extensively used grassland. *Ecological Indicators* 8, 256–269.

70. Oppermann, R. 2003. Nature balance scheme for farms—evaluation of the ecological situation. *Agriculture, Ecosystems and Environment* 98, 463–475.

71. de Sainte-Marie, C. 2009. Favoriser la biodiversité par des mesures agri-environnementales à obligation de résultat. Les prairies fleuries du Massif des Bauges (Savoie). INRA ed. Downloadable from: http://www.inra.fr/layout/set/print/presse/mesures_agri_environnementales_avec_engagement_de_resultat_ecologique

72. Wittig, B., Kemmermann. A.R. gen, and Zacharias, D. 2006. An indicator species approach for result-orientated subsidies of ecological services in grasslands: a study in Northwestern Germany. *Biological Conservation* 133, 186–197.

73. Hart, K., and Baldock, D. 2011. Greening the CAP: delivering environmental outcomes through pillar one. Institute for European Environmental Policy (IEEP): 26 pp.

74. Turlot, A., Remence, V., Picron, P., Barthiaux-Thill, N., Stilmant, D., and Rondia, P. 2010. La gestion des milieux de haute valeur écologique: une diversification pour les exploitations agricoles (The management of high ecological value habitats: a diversification for agricultural holdings) (Acronyme: ECOGEST). Rapport final. Centre wallon de Recherches agronomiques, 70 pp. + annexes.

75. Danckaert, S., Derijck, K., Mulders, C., and Peeters, A. 2012. Belgium in *High Nature Value Farming in Europe* (Oppermann, R., Beaufoy, G., and Jones, G., eds.). Ubstadt: Verlag Region- alkultur.

76. European Commission. 2007. European policy for quality agricultural products. Fact sheet. *European Communities*, 11 pp.

ANNEX: METHOD FOR THE DESIGN, DEVELOPMENT, AND DISSEMINATION OF PROTOTYPES OF FARMS

Step 1—Hierarchy of objectives. Developing a hierarchy of objectives of the system in which the shortcomings of current farming systems are corrected (ex.: Clean environment regarding nutrients).

Step 2—Parameters and methods. Major objectives are translated into mul- tifunctional indicators (or parameters). Subsequently, methods or system tech- nologies (ex.: Ecological Nutrient Management) are developed. Their efficiency to improve the system is checked in groups of pilot farmers (10–15 farms per group) and quantified by the multi-objective indicators (ex.: Nitrogen leaching). These indicators are able to measure, in each pilot farm, to what extent methods are efficient for reaching the objectives.

Step 3—Design of theoretical prototype and methods. Conflicts can appear between objectives and between methods. These conflicts are solved by iden- tifying trade-offs and by further designing methods. Methods and farmers' practices are thus improved in an iterative way by regular meetings between scientists and individual farmers in "kitchen table" meetings and field visits every second week in the crop-growing season and during four to six annual group meetings (all farmers and scientists together) in the winter period.

Step 4—Layout of prototype to test and improve. The result is a consistent package that can be described by a text and by a chart, the "prototype." It is tested and improved in relation to the objectives.

Step 5—Dissemination. When farms are sufficiently improved to be close to the objectives, the prototype can be disseminated to other farms at a regional or country level.

Part V

Ecosystem Service
Reflections from Practice

Chapter 23

Ecosystem Service Practices

Hans Keune[1,2,3,4], Nicolas Dendoncker[5] and Sander Jacobs[6,7]

[1]Belgian Biodiversity Platform, [2]Research Institute for Nature and Forest (INBO), [3]Faculty of Applied Economics – University of Antwerp, [4]naXys, Namur Center for Complex Systems – University of Namur, [5]Department of Geography, University of Namur (Unamur). Namur Research Centre on Sustainable Development (NAGRIDD). Namur Centre for Complex Systems (naXys), [6]Research Institute for Nature and Forest (INBO), [7]University of Antwerp. Department of Biology, Ecosystem Management Research Group (ECOBE)

1. INTRODUCTION

As was illustrated in the first parts of this book, a diversity of scientific actors have clearly picked up the concept of ecosystem services. Over the last years, the concept of ecosystem services has increasingly been finding its way to practice. This development reflects international developments in the field of ecosystem services: Now that the first international concepts of ecosystem services have been developed (e.g., Millennium Ecosystem Assessment) and awareness regarding its value for society has been raised (The Economics of Ecosystems and Biodiversity), local practical application seems to be the most prominent challenge. As the first parts of this book illustrate, many scientific and methodological challenges still lie ahead, part of which indeed focus on practical application. Reflections from practice, from other than scientific actors, are therefore crucial and form an important part of this book.

Ecosystem Services. http://dx.doi.org/10.1016/B978-0-12-419964-4.00023-8

We have asked a diversity of real-world actors,[1] that is, actors outside aca-demia, for their reflections on the usefulness of the ecosystem services con-cept for their practice. The fact that we are able to present quite a diversity of contributions from mainly local and some international nonacademic actors illustrates the current interest. Most of the local contributors are members of the BElgium Ecosystem Services community. The fact that most of them are policy actors seems to indicate that broader societal uptake of the ecosystem services concept still is rather limited. And the fact that most of these policy actors are working in (at least partly) nature-related policy domains (the domain of agricul-ture being second best) also seems to indicate that the larger part of policy still is unaware of or not explicitly interested in the concept. Moreover, most contribu-tions are from the Flemish part of Belgium, indicating the concept to be mainly picked up there. This diversity of contributors resulted from our professional contacts in Belgium: The result mirrors the situation in Belgium quite well.

We asked the contributors to reflect on the usefulness of the ecosystem ser-vices concept for their organization. We asked them to what extent and how it is being used in the practice of their organization and what are the advantages and disadvantages of its use. We also asked about the main challenges they see for the concept and about the added value of a local Ecosystem Services Com-munity of Practice to their work. We present here some key messages that came from the diversity of reflections.

2. USEFULNESS OF THE ECOSYSTEM SERVICES CONCEPT

Contributors generally consider the ecosystem services concept useful for tack-ling major societal challenges. Among the challenges they mention are societal problems such as a systemic environmental sustainability crisis and the fail-ure of governance to address this crisis adequately. Mentioned too are system characteristics such as complexity, the link between local and global dynamics, and the interconnectedness of system dynamics and governance: Some of the sustainability challenges are imposed on societies just because of the way we govern our world. This mutual society–nature dependency is quite explicit in several contributions from practice. Society's dependence on nature, for exam-ple, shows in evident in how our GDP depends on natural goods and services. Vice versa, ecosystems and biodiversity heavily depend on human actions, for example, on the extent to which financial institutions consider the effects on nature when they make investment decisions. This mutual dependence is not always recognized or explicit. In fact, some contributors point out the useful-ness of the ecosystem services concept for better informing policy makers when conflicts develop between proponents of nature conservation and economic development. If financial institutions would integrate the ecosystem services

---

1. The contributions do not necessarily represent the views of the institutions the contributors work for; the contributions are made on a personal title.

perspective in their investment decisions, this action "could be a catalytic force in transforming the impact on the ecosystem of the business world" (Chapter 25) and in the longer run would lead to more sustainable economies. This mutual society–nature dependency exists not only in the economic sense, but also with regard to public health: The benefits of nature for human health can be made more explicit through use of the concept of ecosystem services, which in turn can raise awareness and support for biodiversity conservation.

Apart from the concept of ecosystem services as providing useful information about these dependencies and managing the social relational aspects that partly steer them, the integrative capacity of the concept is highlighted in some contributions. First, we have the idea of integration in terms of actors: representatives from different scientific and expert fields, together with stakeholders and (other) representatives from practice. The concept of ecosystem services "provides a valuable framework to bring actors with different interests together" (Chapter 35). Second is the notion of bridging sectors and domains: "integrate nature conservation targets in other policy domains" and "bridging the gap between different sectors and enabling win-win situations" (Chapter 33). Third, we have integration for multifunctional land management: "When evaluating a bundle of ecosystem services, it might be possible to develop new, more multifunctional measures that have a more positive effect on the environment" (Chapter 36).

3. HOW IS THE ECOSYSTEM SERVICES CONCEPT USED IN PRACTICE?

The contributions do not clearly define or describe practice, nor is it easy to distinguish it from what one might *not* call practice but something else. One approach would be to describe the variety of practices represented by the contributors in this section. This would of course already limit our focus on practice by excluding scientific practice (the other parts of this book) from what we want to address here—that is, the main responsibility of contributors is not academic scientific research. We will not go into philosophical or epistemological depth here on science–society relations and demarcations. Let us for now demarcate practice as the use of the ecosystem services concept by nonacademic actors and describe what they present and represent as their practice(s).

First, the working domains in which contributors are principally active. Domains covered by policy contributors are quite diverse: EU environmental policy, national health–food–environmental policy, regional sustainability policy, regional environment–nature–agricultural policy, regional environmental policy, regional environment and nature policy, regional agriculture and fisheries policy, regional nature and forest conservation policy, regional land management, provincial nature education, and provincial environmental policy. Domains covered by nonpolicy contributors include biodiversity offset and habitat banking, corporate social responsibility rating, soil and groundwater knowledge, wildlife management, agriculture, nature conservation, and forest management. Diversity is present within these domains, but seem to be largely restricted to what we might call "the usual suspects": environmental and nature policy and management. The main exceptions are health and agricultural policy, land management, and the financial sector (banking and investment). These are examples of mainstreaming the concept beyond "traditional" nature- or environment-aware sectors. Again, this selection resulted from our professional contacts in the ecosystem services community.

Second, the type of ecosystem services knowledge application that contributors mention as being suitable to their practice, either as being effectively applied in practice or as part of their ecosystem-services-method wishlist. Named are mapping and assessing ecosystems and their services in general and economic valuation specifically. Quite a diversity of economic/monetary assessment/valuation methods/instruments in which the concept of ecosystem services may play a role are concretely mentioned: (1) Financial market instruments: market mechanism, investment indicators. (2) Policy instruments: agricultural-environmental payments and determination of environmental pollution fines. (3) Policy-relevant assessments methods: societal cost-benefit analysis and environmental impact assessment. (4) Nature education and awareness raising. Furthermore development of indicators is mentioned in general, be it not limited to economic assessment or evalution only: "to evaluate the potential impacts of these challenges/risks on biodiversity and ecosystems and their functions and services, as well as the potential links between impacts on biodiversity, ecosystem, ecosystem services, and public health" (Chapter28).

Several other methodological approaches are cited. Integrated approaches are described for connecting otherwise often disconnected fields, issues, or actors: scientific disciplines, social sectors, science and practice, science and social debate (analytical deliberative process, transition management), or a diversity of ecosystem services. Collaborative approaches are mentioned partly for the same reasons: "grouping competences and stakeholders towards better ecosystem management" and "quantitative/monetary valuation is currently not practical at our working scale. At our level, the ecosystem services approach can be practically implemented through building good relationships between services providers (farmers, foresters, land managers) and beneficiaries (hunters, naturalists, citizens)" (Chapter30). Finally, awareness-raising approaches and capacity-building are described as specific methods. In Part 6 of this synthesis you will find more details on capacity building.

Some practice-relevant methodological principles and key issues are presented first, with regard to practice relevance itself, by increasing practical experiments and involvement of end-users, and by taking into account both the context specificity of practical contexts and the opportunity for drawing generic lessons from a diversity of practical cases. Connection to similar approaches is also mentioned as a strategy; this may enhance opportunities for mainstreaming the concept of ecosystem services. As an interpretive methodological stance, use of the precautionary principle is proposed for specific cases. Either in the case of the availability of sufficient but perhaps not fully conclusive evidence: ". . . the implications of biodiversity loss (i.e., loss in genes, species, and ecosystems) for the functioning of ecosystems and delivery of ecosystem services, environmental thresholds and ecological tipping points beyond which adverse trends start to become irreversible. There is sufficient evidence in all these areas to warrant precautionary action" (Chapter 24). Or in the case of persistent scientific uncertainties: ". . . recognize the limits of science to deliver knowledge that is advanced and certain enough for clear-cut and unambiguous decision support. Scientific uncertainties will remain and deem us to consider the precaution principle in some cases" (Chapter 28).

Third, specific types of activities, projects mentioned by contributors as effectively being part of their ecosystem services practice(s). Not all contributors believe that the ecosystem services concept is currently being applied in their practice, or if it is being applied, only in a limited way. Part of the reason may be a matter of wording: Several contributors recognized similar concepts and approaches in their practice, even long before the concept took off internationally as a new and promising concept. With other wording, equally integrated, systemic nature conservation or awareness-raising approaches have been used for many years. Clearly, the ecosystem services concept offers momentum or is functional in the current momentum for raising interest to preserve ecosystems and biodiversity. Other contributors make a distinction between, on one hand, use of the concept in the vocabulary of their institutions, as in policy plans, and on the other, real practice, as in concrete actions on the ground. Apparently, the concept of ecosystem services is starting to find its way in some documents and discourses. Scientifically, it is far from uncommon to label these also as practices, like discursive practice. The same can be concluded regarding science as a practice; in fact, some contributors mention research or their involvement in research as practice examples. Clearly, on the ground application/action in practice is still in its infancy. Nevertheless, some interesting examples are presented: we will not list specific examples here, as this would overlap substantially with the above mentioned methods. We invite you to read more in the individual contributions from practice, to be found right after this synthesis.

4. RISKS OF THE USE OF THE ECOSYSTEM SERVICES CONCEPT

According to contributors from practice, use of the ecosystem services concept also poses risks. Clearly, some contributors show concern about the monetary valuation of ecosystem services. One often-mentioned risk centered on the so-called merchandizing nature or price-tagging nature. This may cause underestimation or neglect of ecosystem services or components whose importance is not straightforward or explicit in monetary terms. This problem may lead to the "risk that the valuations of ecosystem services assessment would be limited to those services that would have a material risk for the investor" (Chapter 25). Also, "Price-tagging nature elements can create a perception of a 'license to pollute' (or to destroy)" (Chapter 33). One may also run the risk of discarding "zero utility nature" (Chapter 31), stressing the purely anthropocentric and utilitarian turn taken by ecosystem services valuation. By incorporating concern for ecosystem services in investment culture, one runs "the risk that the concept of ecosystem services would only be reduced to a cosmetic marketing tool" (Chapter 25).

Moreover, there are methodological concerns: Monetary valuation is difficult to do in many respects and also risks denial of the importance of specific ecosystem elements not because of sheer lack of importance, but merely as one cannot grasp the importance in monetary terms. For example, "mere economic valuation of biodiversity services also is not always easy" or "quantification, especially economic quantification, of ecosystem services has the danger of trade-offs with biodiversity" (Chapter 28). The risk of monetary valuation is to some extent considered to be part of a general concern with focusing only on quantification: "some functions and services of biodiversity, like those linked to culture and mental health, are hard to quantify, but still they are important to human well-being" (Chapter 28).

Apart from methodological concerns, there may also be operational concerns: "It seems worthwhile to invest in systems that divert a part of the benefits for society to concerned landowners and other stakeholders as a way to compensate for efforts and income forgone. But still, there are a lot of questions: Who will pay? Who will be paid? What will be paid for? How much will be paid? . . . These questions regard efficiency and effectiveness, but also touch on ethics and transparency. Perverse effects are not unlikely and need to be averted" (Chapter 33).

Despite the risks of monetary valuation, contributors do not always dismiss its usefulness; it depends on how it is applied and in which contexts. It may also depend on how it is used relative to other valuation or assessment methods: "its use is not recommended in all cases; it must always be carried out in a controlled and transparent way and ideally be part of a broader assessment like multi-criteria analysis" (Chapter 35). Or should the ecosystem services concept have its place next to other concepts?: "ES is only "one" tool in a "toolbox" that must be widened" (Chapter 31).

Some contributors raise some questions regarding the status quo when applying the concept of ecosystem services. These questions focus on agriculture. There is a fear about "an increasing reference level" (Chapter 36) that would put further pressure on agriculture's economic performance: "In practice we note it is not always easy to translate these initiatives into economic value for the farmers because society is not always prepared to pay for the services, although they generate an economic loss for the farmer" (Chapter 37). Concern is also expressed about the high urbanization rate in Flanders compared to the European average: "34% of Flanders is urbanized (residential areas, industry, transport infrastructure) while the European average is 11%. This makes the pressure on land very high, leading to very high prices for Flemish agricultural land. This is one of the most important reasons that our Flemish agriculture is intensive, and it is more difficult than in other regions to find win-win situations that combine economic efficiency and ecological value" (Chapter 37). Finally, there are legal concerns: "If society/the government wants to stimulate farmers to take environmental measures, they should give farmers the necessary legal certainty that the farming activity is not threatened" (Chapter 37).

5. CHALLENGES REGARDING THE USE OF THE ECOSYSTEM SERVICES CONCEPT

Governance faces several key challenges. First is the challenge to overcome governance's failure to protect ecosystems and biodiversity and to apply an ecosystem services approach in practice. Other challenges involve ethical and legal issues related to ecosystem services approaches; the global perspective of local actions—the global impact on ecosystems and biodiversity; and the need to deal with limited and/or ambiguous scientific knowledge: considering the precautionary principle. Societal acceptability is mentioned as a crucial awareness-raising challenge. Part of this challenge involves translation of complex conceptual terminology into understandable communication material; part might include the role of stakeholders from the start in participatory processes in order to create more support; and yet another part of this challenge might have to do with bridging the functional ecosystem services approach with people's personal relation with nature.

Key challenges regarding scientific and methodological aspects are also highlighted. Scientific understanding of the limits of ecosystem resilience in the face of global and local changes/pressures, and the link to biodiversity represents one specific scientific challenge, and comparing sustainable ecosystems functions and their services and unsustainable technologies is another. Methodological challenges include (1) balancing scientific quality–pragmatic application in practice; (2) the science–policy-practice interface and a transdisciplinary approach; (3) the need for an inclusive approach: managing ecosystem services in relation to wider stakeholder values, needs, and priorities; (4) the challenge of taking into account qualitative information, as some ecosystem services are difficult to quantify; (5) the effects of (economic) quantification of ecosystem services on biodiversity; (6) the challenge of constructing baselines for comparative analysis when realizing that the status of biodiversity and ecosystem functions and services are not static, even in absence of human pressures; and (7) comparison of an ecosystem services-oriented approach to other integrated approaches.

Finally, several development challenges regarding tools and methods fit for purpose, context, scale, and end-user are mentioned. In this respect, methods are not necessarily useful for or applicable to all issues, scales, purposes, or practices. Also discussed is the need for indicators and assessment tools for evaluation of potential impacts on biodiversity and ecosystems and their functions and services.

6. THE IMPORTANCE OF A LOCAL ECOSYSTEM SERVICE COMMUNITY OF PRACTICE

As the ecosystem services field of expertise is developing rapidly, contributors consider a local *Ecosystem Service Community of Practice* such as the Belgium Ecosystem Services (BEES) community important for local capacity of ecosystem services expertise. Contributors present several specific functions for such a Community of Practice. First, it can help keep track of international and local (Belgian) developments. It can be supportive in regularly producing state-of-the art overviews, fit for local Community of Practice members. The need to enhance the link between science and practice, society, and policy is often underlined by the local policy representatives and stakeholder groups who contributed to this book. Therefore, a transdisciplinary approach (i.e., a close collaboration between academic scientific and nonacademic actors) is deemed necessary. The challenge lies not only in exchanging data between science and policy, but also in the collaborative development of practice-relevant tools and methods. A local *Ecosystem Service Community of Practice* such as the Belgium Ecosystem Services (BEES) community could function as a neutral and interdisciplinary umbrella and reference network, supporting translation from science to practice and vice versa. Second, the aim of further mainstreaming the ecosystem services approach is considered an important task for such a community. Linked to this aim are awareness raising and communication activities: supporting translation of complex conceptual and scientific issues to a broader audience. Third, the need to exchange implementation experiences between practitioners from diverse practical contexts is often mentioned as a task for such a Community of Practice. For all these interface activities, a local *Ecosystem Service Community of Practice* such as the Belgium Ecosystem Services (BEES) community can offer opportunities for networking, encompassing scientists, politicians, public officers, business, NGOs, and citizens. Finally, a local *Ecosystem Service Community of Practice* can serve as an interface between local (Belgium) and international activities and developments. This can work in two directions: informing the local (Belgian) Community of Practice about international dynamics, and vice versa. Informally, such a community is considered to be supportive to the development of a common understanding and joint (Belgian) viewpoints on international ecosystem services.

7. CONCLUSIONS

Before we invite you to read the series of the most interesting contributions from practice, we will briefly sketch some conclusions. This relation between the concept of ecosystem services and practice is a moving target: a dynamic field of operation, experimentation, discourse, and concern, presenting many challenges and ambitions. One dynamic frontier that may show crucial developments in coming years is that between discursive practice and on-the-ground application. While in some domains it is beginning to be welcomed on a discursive level, it still is awaiting more elaborate on-the-ground application. This reality check will also become a challenge for evaluation of practice. Another dynamic frontier is that of mainstreaming: expanding the domains of application from frontrunner actors and domains, such as nature-related practice fields, to other societal domains, such as agriculture, public health, the financial sector, and the general public at large. On those new grounds, awareness raising and nesting itself in discourse practice are only starting or have yet to be achieved. Finally, a methodological frontier displaying still many adventurous options and challenges. One of the most prominent features of this debate seems to be a tension between calculation and social debate. On the one hand, there is a strong tendency to look for proof of concept in the vocabulary of what politically seems to rule our world: economic discourse, money talk, and life in a quantified format. On the other hand, we note a strong movement away from this culture of math and money: This culture is seen as part of the disease of unsustainability that should be cured and that to some extent is being replaced by alternative methods and practices—those of social debate and social relations between humankind and nature. The tension may also lead to a coalition of diverse forces that join together for similar aims.

Let us now hear from our contributors who have already paved the way for the concept of ecosystem services.

Chapter 24

Reflections from Policy Practice

Anne Teller

DG Environment, European Commission

Anne Teller - European Commission, DG Environment. Anne is Biodiversity Policy Officer in the Directorate-General Environment of the European Commission. She has more than twenty years experience in European environmental policy. Her specific domain of interest is the strengthening of the evidence-based approach to environment policy making. Anne is currently coordinating a new initiative on Mapping and Assessment of the state of Ecosystems and of their Services (MAES) in Europe.

In the coming years, policy makers, businesses, and citizens will face complex societal challenges. A sound knowledge basis will be required for facilitating their understanding and decision making, which should fully reflect true social, economic, and environmental costs and benefits.

Over the past decades, there have been improvements in the way environmental information and statistics are collected and used, at both EU and member state level, as well as globally. Considerable progress has been made in strengthening the evidence for EU environmental policy based on environmental monitoring, data, indicators, and assessments linked to implementation of EU legislation, as well as scientific research and citizen science initiatives.

However, the pace of current developments and uncertainties surrounding likely future trends requires further steps to maintain and strengthen this evidence base to ensure that policy in the EU continues to draw on a sound understanding of the state of the environment, possible response options, and their consequences.

Ecosystem Services. http://dx.doi.org/10.1016/B978-0-12-419964-4.00024-X

Today, the vast bulk of research programs in Europe, while providing excellent delivery on specialized issues, are often run with little true interaction, especially between social, economic, and environmental sciences, leading to sometimes diverging views from science on a particular question. Significant gaps in knowledge still remain, and it is therefore essential to invest in further research to fill these gaps.

Horizon 2020, the EU research agenda for the coming years,[1] will need to remedy this situation, by pooling national, European, and international research efforts and resources, consolidating the views from science and closely interfacing with policy to tackle common European challenges in a more structured manner.

The proposed general Union Environment Action Program to 2020, entitled Living Well within the Limits of Our Planet,[2] is calling for advanced interdisciplinary research to fill gaps in knowledge and develop adequate modeling and foresight tools to better understand complex issues related to environmental change. Among these issues are climate change and disaster impacts, the implications of biodiversity loss (i.e., loss in genes, species, and ecosystems) for the functioning of ecosystems and delivery of ecosystem services, environmental thresholds, and ecological tipping points, beyond which adverse trends start to become irreversible. There is sufficient evidence in all these areas to warrant precautionary action; however, further research into multiple stressors, planetary boundaries, systemic risks, and extreme events, and society's ability to cope with risks and irreversibility is needed in order to develop appropriate responses. This should include investments in closing data and knowledge gaps, mapping and assessing ecosystem services, understanding the role of biodiversity in underpinning them, and learning how they adapt to climate change.

In 2012, a joint initiative between member states and the European Commission[3] was launched to map and assess the state of ecosystems and their services in their national territory by 2014, assess the economic value of such services, and promote the integration of these values into accounting and reporting systems at EU and national level by 2020. The objective of such assessments is to provide a critical evaluation of the best available information for guiding decisions on complex public issues. A precondition thereof is to ensure commitment to long-term monitoring, as long-term data series are indispensable to both global environmental change research and assessment of environmental policies.

The transition to an inclusive green economy requires proper consideration of the interplay between socioeconomic and environmental factors. Improving our understanding of what constitutes sustainable consumption and production,

---

1. See http://ec.europa.eu/research/horizon2020/index_en.cfm?pg=home&video=none
2. See http://ec.europa.eu/environment/newprg/index.htm
3. See the EU Biodiversity Strategy to 2020 and in particular its Action 5 under Target 2 http://ec.europa.eu/environment/nature/biodiversity/comm2006/2020.htm

how costs of action or inaction can be considered more accurately, how changes in individual and societal behavior contribute to environmental outcomes, and how Europe's environment is affected by global megatrends can help better target policy initiatives toward improving resource efficiency and relieving pressures on the environment.

Research activities will continue to have a crucial role in achieving the significant transformation required for Europe to achieve sustainable development based on balanced economic growth and price stability, a highly competitive social market economy aiming at full employment and social progress, and a high level of protection and improvement of the quality of the environment as spelled out under Articles 2 and 3 of the Lisbon Treaty. Technological, but also social, behavioral, organizational, and institutional innovations will be needed to generate the breakthroughs that will solve a range of environmental challenges, bringing solutions that respect the planet's ecological limits. We need to move toward a low-carbon and resource efficient green economy, and to maintain a secure and healthy natural environment. Tipping points need to be identified, and a deeper understanding of long-term trends that impact on the state of the environment is required. In every policy area, knowledge gaps need to be filled if policy success is to be assured. With the finite resources available, it is crucial for the gathering of that knowledge—and its dissemination to those who need it to improve policy and its implementation—to be logically and systematically organized.

It is now time to join forces at all levels to support and implement the necessary science that will strengthen the evidence base and ensure that policy in the EU continues to draw on a sound understanding of the state of the environment, possible response options, and their consequences. The concept is on a twin track: one track concerns better use of past and future environmental knowledge by developing better environmental information management systems and science-policy interfaces, so that those who need quality-assured credible information get it; and the other track involves better-targeted pursuit of new knowledge, through better procurement and better strategic orientation of future research given current policy priorities—another function of enhanced science-policy interfaces—and an accent on improved foresight of emerging issues and corresponding knowledge needs In this way, environment policy will fully deliver its part of the Europe 2020 Strategy.

Horizon 2020 will provide the opportunity to focus research efforts and to deploy Europe's potential by bringing together resources and knowledge across different fields and disciplines within the EU and internationally.

(how) Can Financial Institutions Contribute to Sustainable Use of Ecosystem Services?

Frederic Ghys

Sustainability analyst

Since April 2013 **Frederic** works as policy officer organic agriculture for BioForum vzw. Between 2008-2013 he has worked as senior sustainability analyst in Vigeo. His tasks included research on corporate social performance of companies based on a risk-approach. Furthermore he developed tailor-made ESG (Environmental Social Governance) investment strategies for institutional investors. He coordinated research on the impact of companies on community involement. Finally he managed the development of a research database on disputable activities.

In 2011, the United Nations Environment Programme (UNEP) issued a report titled "Why Environmental Externalities Matter to Institutional Investors?" It estimated the cost of environmental damage caused by the world's 3000 largest publicly listed companies in 2008 at a whopping 2.15 trillion USD. Large institutional investors are in fact "Universal Owners". Financial institutions such as pension funds and insurance funds often have highly diversified and long-term portfolios that are representative of global capital markets. Negatively, investors are the universal owners of an economy that is causing severe losses of ecosystems and biodiversity. Positively, financial institutions integrating ecosystem services in their investment decisions could be a catalytic force in transforming

Ecosystem Services. http://dx.doi.org/10.1016/B978-0-12-419964-4.00025-1

the impact on the ecosystem of the business world. The industries are largely dependent on a well-functioning and balanced environment. A decline in ecosystem services could pose a serious risk for the operations of corporations. Consequently, there is a direct financial incentive to value these services in the investment decisions. Currently, a significant number of institutional investors are already committed to principles of responsible investment. Increasingly, investors are starting to incorporate social and environmental factors in investment models. However, the development of the methodology to integrate environmental indicators in the investment process is still in its infancy. The economics and evaluation of ecosystem services could help investors to assess the costs of biodiversity loss and ecosystem degradation. However, there are some inherent risks linked to using the ecosystem services concept in financial valuation systems.

The United Nations supports a network of investors that uphold Principles for Responsible Investments (PRI). The initiative aims to enhance understanding and implementation of the impact of sustainability for investors. As of March 2013, 1175 financial institutions, representing more than 31 trillion USD, had already signed the principles for responsible investment. Signing the PRI implies committing to six principles that focus on the incorporation of Environmental Social Governance (ESG) issues into investment analysis and decision-making processes. The investors believe that ESG issues can affect the performance of investment portfolios. The PRI signatories explicitly acknowledge that investors act as "stewards of the world wealth," affirming that they have the duty to best the long-term interests of its beneficiaries and the society at large. At this stage the financial institutions have not yet taken up this responsibility. The reality remains that financial institutions are currently investing in corporations that are causing high and increasing environmental externalities. The environmental costs are caused by greenhouse gas emissions, overuse of water, pollution, and unsustainable natural resource use. These externalities can affect shareholder value because they lead to a more uncertain, rapidly changing economic environment and greater systemic risks. There is a solution; investors are not only impacting the environment, but they can also be a leverage of change. Institutional investors can exercise ownership rights and encourage the valuation of the ecosystems needed to maintain the economy and investment returns over the long term. The methodology to integrate the environmental impact is still in its infancy. Valuation of the economics of the ecosystem services could help investors to gain clarity on the hidden impact of their investments on ecosystems. At the same time, the link between the operations and impact on the environment could highlight the specific risks investments have on ecosystems.

As indicated above, the business and financial world is increasingly recognizing the impact of ecosystem services on operational performance and financial returns. This implies that the integration of ecosystem services in investment decisions presents some pitfalls. As a first example, there is the risk that the valuations of ecosystem services assessment would be limited to those services that would have a material risk for the investor. For instance, the investor would value the level of water withdrawal of a mining company operating in a water-scarce region. However, the investor would at the same time ignore the value of the water consumption of a mining site operating in a region with more abundance of water. The latter mining operation itself would not be directly endangered by the lack of water. But the water-intensive mining operation could put undue stress on the local ecosystem or the livelihood of surrounding farmers. Since the financial return would focus on the mining operation, the loss for the local community would not be integrated in the assessment methodology. The financial institutions, through for instance the PRI initiative, could take up the responsibility to ensure that the scope of beneficiaries of ecosystems would not be limited to the narrow short-term economic interest of the operations they are investing in.

There is also the danger of "dilution" when applying a global ecosystem service assessment tool. When using a multifactor value attribution model, it is likely that the impact on a critical ecosystem service could be ignored. For instance, a hypothetical model would calculate a global score based on the impact of a set of, for instance, 40 types of ecosystem services. The danger would be that the negative score of a few ecosystem services would be diluted in the overall score.

The reference to sustainability, green investments, or investments valuing ecosystem services could prove a reputation advantage for the financial institutions. Thus, there is the risk that the concept of ecosystem services would only be reduced to a cosmetic marketing tool. Increasing the general public's knowledge of the different types of ecosystem services should be requisite to ensuring a critical approach to the use of ecosystem services by their financial institutions.

When financial institutions integrate ecosystem services in their investment decisions, they can transform the market. They can raise the capital cost for companies that negatively impact ecosystems and alternatively actively select companies that enhance and value ecosystem services. Thus the financial institutions can positively influence the way business is conducted in order to reduce externalities and minimize their overall exposure to these costs. However, research is needed to ensure that a model of ecosystem valuation is at the same time meaningful for investors and provides an accurate overview of the benefits and costs of the financed activity on the ecosystems and biodiversity.

Chapter 26

Making Natural Capital and Ecosystem Services Operational in Europe through Biodiversity Offsetting and Habitat Banking

Guy Duke
Environment Bank Ltd

Guy Duke is Director Europe and Research, Environment Bank Ltd in which capacity he leads on business expansion in Europe, EU affairs and company engagement in EU research and innovation, including the current FP7 project OpenNESS. Guy also consults independently on ecosystem markets, for example leading research for the UK Ecosystem Markets Task Force. He holds a UK public appointment as an Independent Member of the UK Joint Nature Conservation Committee and is a Senior Visiting Research Associate at Oxford University's Environmental Change Institute.

One of the key challenges of the concepts of natural capital (stocks) and ecosystem services (flows) is to make these concepts operational in practice. Because biodiversity is a key element of natural capital, many of the conventional instruments by which we seek in practice to conserve it (e.g., protected areas, species conservation measures, invasive species controls) also serve, even if they were not explicitly designed to do so, to safeguard natural capital and ecosystem services. However, the increasing prevalence of the natural capital and ecosystem services paradigm is leading to the emergence of instruments that more explicitly seek to maintain or enhance natural capital and sustain or enhance ecosystem services.

Ecosystem Services. http://dx.doi.org/10.1016/B978-0-12-419964-4.00026-3

Biodiversity offsets are "measurable conservation outcomes resulting from actions designed to compensate for significant residual adverse biodiversity impacts arising from project development and persisting after appropriate prevention and mitigation measures have been implemented. The goal of biodiversity offsets is to achieve no net loss, or preferably a net gain, of biodiversity on the ground with respect to species composition, habitat structure, and ecosystem services, including livelihood aspects."[3]

Habitat banking is "a market where the credits from actions with beneficial biodiversity outcomes can be purchased to offset the debit from environmental damage; credits can be produced in advance of, and without ex-ante links to, the debits they compensate for, and stored over time."[4]

[3] Ecosystem Markets Task Force. 2013. Realising Nature's Value: Ecosystem Markets Task Force Final Report. Defra, UK. http://www.defra.gov.uk/ecosystem-markets/work/publications-reports/
[4] BBOP. 2009. *BBOP Biodiversity Offset Design Handbook*. Washington, DC: BBOP.

Use of the words "capital" and "services" tends to highlight the utilitarian values of nature and to foster greater recognition of these values, including in monetary terms. This increasing recognition of the value of natural capital stocks and of the ecosystem services that flow from these stocks has fostered increased interest from policy makers in financial instruments, including market mechanisms, to maintain or enhance natural capital and/or to sustain or enhance ecosystem services. This interest has been further increased by the fact that the economic crisis has considerably diminished the public purse available to safeguard nature. Governments and others (not least environmental NGOs) are increasingly looking to the private sector to deliver the necessary finance. Market mechanisms offer a means to achieve this objective.

The thinking on natural capital and ecosystem services thus underpins the emergence of businesses predicated on such financial instruments and market mechanisms. The Environment Bank Ltd. (EBL), a UK company that is increasingly expanding its interests in Europe, is one such business.[1] EBL is the leading trader in the UK in environmental assets (natural capital stocks), enabling and brokering deals between buyers (developers, corporate, investors) and sellers (landowners, farmers, conservation bodies, land management companies), thereby facilitating new markets to substantially increase investment in the natural environment.

EBL's current focus is on biodiversity offsetting through habitat banking (see Box 1 Definitions). This is timely, both in the UK and more widely in Europe. The high-profile business-led Ecosystem Markets Task Force, one of the top business opportunities for valuing nature, has recently identified

1. See for example www.teebweb.org

biodiversity offsets in the UK.[2] At the EU level, the European Commission is currently developing policy for a "no net loss initiative" scheduled for 2015, which is expected to promote biodiversity offsetting, including habitat banking, as one of its central measures.

Biodiversity offsetting accomplished through habitat banking is a potential business opportunity which, for England alone, has been estimated to be worth up to £470+ m/yr, with potential to deliver some 340,000 ha of improved habitat over 20 years. The potential EU market is €7.5 billion/yr, with potential to deliver > 5 m ha of improved habitat over 20 years.[5] Biodiversity offsetting is already widely practiced in Germany and is used to some extent in some other EU member states (e.g., the UK, France, and the Netherlands), but so far its use has been very limited in most of the EU.[6] The worldwide biodiversity offsetting market in 2010 was already worth at least $ 2.4–4 billion.[7]

EBL regards biodiversity as one of several natural asset classes that may in certain circumstances be traded; others include bio-carbon (a natural capital stock) in woodlands and peatlands, and water quality (an ecosystem service flow). EBL's developing business model and systems allow integration of these various asset classes. They provide for the "stacking" of payments in order to deliver several such assets in one location. In this way, funding for biodiversity gain, carbon sequestration, and water quality can be additive and thus more financially viable.

EBL has to date worked effectively in the UK to inform and influence the development of policy, which is increasingly supportive of biodiversity offsetting, and is currently informing and influencing the development of EU policy in the context of Commission plans to propose a No Net Loss (of biodiversity) initiative by 2015.[8] The company however, has focused mainly on developing and testing, often in partnership with government, NGOs, and others, systems for the delivery of biodiversity offsets. These systems include the following:

- metrics to measure both the extent of the impacts of development on biodiversity and the extent of biodiversity "uplift" delivered by habitat restoration or creation in offset "receptor sites";

2. www.environmentbank.com

5. eftec, IEEP et al. 2010. *The use of market-based instruments for biodiversity protection—The case of habitat banking. Technical Report.* http://ec.europa.eu/environment/enveco/index.htm

6. Duke, G., Conway, M., Dickie, I., Juniper, T., Quick, T., Rayment, M., and Smith, S. 2013. *EMTF Second Phase Research: Opportunities for UK Business that Protect and/or Value Nature.* Final Report. ICF GHK, London. http://www.defra.gov.uk/ecosystem-markets/work/evidence/

7. ICF GHK. (2013) *Exploring potential demand for and supply of habitat banking in the EU and appropriate design elements for a habitat banking scheme.* Final report submitted to DG Environment. http://ec.europa.eu/environment/enveco/taxation/pdf/Habitat_banking_Report.pdf

8. Madsen, B., Carroll, N., Kandy, D., and Bennett, G. 2011. *2011 Update: State of Biodiversity Markets.* Forest Trends, Washington, DC. http://www.ecosystemmarketplace.com/reports/2011_update_sbdm

- trading systems, including an online trading platform, the Environmental Markets Exchange (EME),[9] through which buyers (developers needing to offset impacts) can purchase credits from sellers (farmers, NGOs, and other landowners offering receptor sites) to discharge their liabilities; and
- delivery systems for the delivery of offsets, including legal, financial, and administrative provisions for habitat restoration/creation and long-term management, monitoring, and compliance.

EBL's approach has potential application in all EU countries; Belgium is no exception. The area of built-up land in Belgium rose over the 10-year period 2000 to 2009 by 390 km^2 (from 5640 km^2 to 6050 km^2).[10] Assuming a similar rate of development of c.4000 ha/yr going forward, applying a (conservative) 2:1 ratio of offset to a development area (generating an offset requirement of 8000 ha/yr) and assuming an average offset cost of €30,000–65,000/ha (based on UK estimates), the potential Belgian market for biodiversity offsets can conservatively be valued at €240–520 million/yr. Over 20 years this would deliver €4.8–10 billion and 1600 km^2 of improved habitat. These are clearly large figures, both in relation to total annual Belgian public sector spending on the environment (€1.615 bn in 2007[11]—NB: this relates to *all* environmental spend, not just nature) and in relation to current areas of natural habitat (e.g., the terrestrial Natura 2000 area in Belgium is 3870 km^2).[12]

The restoration/creation of habitats on this scale need not prove a threat to agricultural or forestry production. Total areas for these land uses in Belgium (2009) are 15,350 and 6970 km^2, respectively.[13] An increase of (semi-) natural habitats of 1600 km^2 over 20 years would involve around 7% of the total agricultural and forestry area, but much of the new habitat could be compatible with ongoing production (e.g., woodfuel from native woodlands, extensive grazing). It is also important to note that landowners and farmers would offer land as "receptor sites" on a voluntary basis and could achieve a price that would make offset provision an economically viable and attractive alternative to other forms of land use.

The costs of biodiversity offsetting (the price of the credits that the developer is required to purchase to discharge his liabilities for damages to biodiversity) do

9. European Commission (2011) Communication from the Commission to the European Parliament, The Council, The Economic and Social Committee and the Committee of the Regions. Our life insurance, our natural capital: an EU biodiversity strategy to 2020. COM(2011)244 final. http://ec.europa.eu/environment/nature/biodiversity/comm2006/pdf/2020/1_EN_ACT_part1_v7[1].pdf

10. http://www.environmentbank.com/environmentalmarketsexchange.html

11. Utilisation du sol par commune et région, la Belqique (1834–2011). http://statbel.fgov.be/fr/modules/publications/statistiques/environnement/fichiers_telechargeables/utilisation_du_sol.jsp

12. Eurostat—Environmental protection expenditure in Europe—detailed data. http://appsso.eurostat.ec.europa.eu/nui/show.do?dataset=env_ac_exp1&lang=en

13. Natura 2000 Barometer, Natura 2000 sites (January 2011). http://ec.europa.eu/environment/nature/natura2000/barometer/

not necessarily increase the costs of development. There is therefore no reason why they should act as a brake on development. Indeed, the costs largely accrue not to the developer, but to the landowner selling land for development. The same landowner will typically have benefited, as a result of a public decision to allow development on his land, from "windfall" uplift in land value. In Belgium, agricultural land has an average value of €27,190/ha (2007 value,[14]) whereas the value of land permitted for development is typically 30–40 times greater (average c. €700,000/ha in 2005, c. €1m/ha in 2010).[15] If a 1 ha development typically requires 2 ha of offset at an average cost of €30,000–65,000/ha, the total offset cost of €60,000–130,000 represents < 20% of the land uplift value; this is unlikely to discourage landowners from selling land for development. Moreover, as the development involves loss of a public good (biodiversity, natural capital, and ecosystem services), it is not unreasonable for the government to require that a part of this uplift value be deployed (via offsetting) to restore it elsewhere.

While landowners selling land for development will bear the costs of offsetting (though they still receive a very healthy windfall), farmers and landowners offering offset sites will benefit from a new and economically viable income source. A Belgian offsetting market of €240–520 m/yr would stimulate the creation and growth of a range of new businesses required to facilitate, deliver, and monitor offsets, and would generate thousands of new jobs, largely in rural areas.

Offsetting is in line with the compensation principle, one of the ten guiding principles enshrined in Belgium's biodiversity strategy,[16] and habitat banking offers a market mechanism to achieve offsetting.[17] The strategy also calls for limiting land conversion impacts on biodiversity, and promotes the wider use of market-based instruments. A well-designed offset system would retain conventional safeguards for important biodiversity (such as European, national, and regional protected area and species designations) and ensure that offsetting does not create a "license to trash." Moreover, a well-designed offset system delivers net gain for biodiversity. The vast majority of housing, industrial, and infrastructure development occurs on agricultural land that, due to current management practices, is of limited biodiversity value. Offsetting can permit trading up in such cases, for example delivering a 1 ha offset of high biodiversity value to compensate for development on 5 ha of low biodiversity arable land.

---

14. Utilisation du sol par commune et région, la Belqique (1834–2011). http://statbel.fgov.be/fr/modules/publications/statistiques/environnement/fichiers_telechargeables/utilisation_du_sol.jsp

15. Eurostat: Land prices and rents - annual data. http://appsso.eurostat.ec.europa.eu/nui/show.do?dataset=apri_ap_aland&lang=en

16. **Prix moyen des ventes des terrains**. http://statbel.fgov.be/fr/statistiques/chiffres/economie/construction_industrie/immo/prix_moyen_terrains/

17. Moreau, R. (ed.). 2006. *Belgium's National Biodiversity Strategy 2006–2016*. Environment Directorate-General, Federal Public Service of Health, Food Chain Safety and Environment. http://www.biodiv.be/implementation/docs/stratactplan/national_strategie_biodiversity_en.pdf

There are political, social, and scientific challenges in introducing biodiversity offsetting and habitat banking. There is a need to build understanding of the considerable potential (both economic and ecological) of offsetting among key stakeholders, including politicians, planners, developers, landowners, and environmental NGOs. There may also be a need to build social acceptance of offsetting; some critics worry that offsetting "commoditizes" nature and that it is wrong to put a financial value on wildlife. But in fact, offsetting is cost- and not value-driven. Offsetting does not seek to set euro cash "values" for wildlife, but instead determines the necessity and costs of conserving that wildlife and provides a mechanism by which societies can pay for the conservation.

Our entire GDP relies ultimately on the goods and services provided by nature. Conventional methods give these zero value, and as long as this remains the case, developments will continue to erode nature at alarming rates. We need to act fast to counter this trend, and offsetting is one way to do so. Any risks of offsetting can be managed such that it delivers on its substantial ecological and economic promise. Ecosystem services science has a key role to play in this regard by providing the scientific evidence required to underpin the metrics, to monitor compliance, and to evaluate ecological outcomes.

Chapter 27

SKB, SNOWMAN,
and Ecosystem Services

Simon W. Moolenaar[1] and Jos Brils[2]

[1]*Programme Manager SKB, Chair of the SNOWMAN Network, Strategic consultant Environment &
Sustainability at Royal Haskoning DHV,* [2]*Senior adviser with Deltares in the area of sustainable
management of natural resources; Co-Coordinator of the Dutch Community of Practice (CoP) on
Ecosystem Services*

Dr. Simon Moolenaar is currently involved with the SKB and SNOWMAN programmes on sustainable land use and sustainable soil and groundwater management. Ecosystem services of the soil-water-system play a central role in the projects with consortia that relate science-policy-practice. Scale is national/regional/EU. At Royal HaskoningDHV Simon is working on the implementation of Biodiversity and Ecosystem Services (BES) within companies (Corporate Responsibility; Corporate Ecosystem Services Review).

Ecosystem Services. http://dx.doi.org/10.1016/B978-0-12-419964-4.00027-5

Jos Brils, originally trained as biologist is working as a senior adviser with Deltares (www.deltares.nl) in the area of sustainable management of natural resources. His specific areas of interest and expertise are: river basin management, ecosystem services (ES), science-policy interfacing (SPI) and international networking.

SKB AND SNOWMAN

SKB[1] is the independent Dutch foundation for knowledge development and knowledge dissemination in the field of soil and groundwater. Through its program Sustainable Sub-surface Development (SSD), SKB aims to identify and share knowledge for the sustainable use of soil and subsurface. The SKB SSD program has several objectives, such as integrating natural sciences and social sciences and exploring the relevance of the "soil-water" system for addressing societal challenges such as climate change, declining natural capital, water safety and water quality, sustainable cities, and sustainable agriculture. Furthermore, SKB was the founding member and right from the start was a participant in the SNOWMAN network.[2] SNOWMAN is a transnational group of research funding organizations and administrations in the field of soil and groundwater in Europe. SNOWMAN, like SKB, aims to develop and share relevant knowledge for the sustainable use of soil and groundwater. The concept of ES is very useful to the work of both SKB and SNOWMAN as it very well contributes to the above described aims and objectives. This issue will be further elucidated in this chapter.

ES PERSPECTIVE

SKB and SNOWMAN regard ES as an integrative and very promising concept for: (i) bridging different fields of science and expertise, (ii) discovering win-win ES combinations in support of multifunctional land use, and (iii) exploring and defining the role of the soil–water system while addressing these challenges.

---

1. See www.skbodem.nl
2. See www.snowmannetwork.com

The purely scientific challenges are significant due to the inherent complexity of soil–water systems. What is often underestimated is that the practical applicability of the ES concept too is challenged by great complexity and is in demand of specific experience. Thus the SKB and SNOWMAN programs aim to obtain more experience on how to apply the ES concept in real-world practice—that is, in sustainable, regional development trajectories in which soil and ground water (should) play an important role. The objective is to increase that experience through an increase in the amount of practical experiments. As a result, within the SKB program about 20 innovative, ES-projects and initiatives have started.[3] To ensure the practicality of these projects and initiatives, SKB aims at involving stakeholders (landowners, land managers, NGOs, authorities, policy makers, and businesses). They, as end-users, will have to use the results in their day-to-day practices.

Furthermore, we—the authors of this section—think that:

- An ES approach helps to identify and quantify the ecological and socioeconomic trade-offs and synergies on which soil-sediment-water related decision making should be based;
- ES concepts can help to effectively link science, policy, and practice by making the trade-offs more transparent;
- Networking of people with a shared interest—that is, an interest in the soil-sediment-water system—is a key to knowledge-building and capacity-building: locally, nationally, and internationally.

ES NETWORKING

One specific SKB ES initiative is the Dutch Community of Practice (CoP) on ES.[4] In this CoP, Dutch-speaking individuals who share a passion for the ES concept, or who are at least intrigued by it, through regular meetings share practical experiences in applying the ES concept and thus learn how to better implement ES in practice [1, 2, 4]. The CoP was set up by applying guidance as developed by the European Commission-funded project PSI-connect [3]. In this CoP, science, policy, and practice are connected by addressing conceptual themes, regional projects and running initiatives such as the TEEB-Nl studies. In this context, an interesting exchange with the Belgian BEES community has developed as well. It is now understood how to actively stimulate mutual exchange of ES experiences, practices, problems, and solutions and how to further build on the expertise of policy-relevant ES research, collaborative research, and knowledge interfacing. This could work by combining several running ideas and initiatives in the Netherlands and Belgium, thus raising critical mass and sharing generic and specific

---

3. See http://www.skbodem.nl/project?thema=ecosysteemdiensten
4. See http://www.skbodem.nl/project/43 and see also: http://www.linkedin.com/groups? gid=4272045&trk=myg_ugrp_ovr

lessons. Furthermore, SKB has become a member of the global Ecosystem Services Partnership (ESP[5]).

Although the NL CoP ES, BEES, ESP, and the like are already very helpful, we think that even more active capacity building is needed. The expertise of science–policy interfacing and of practical application can gain a lot from the further exchange and sharing of experiences and knowledge. This capacity building transcends the context specificity of practical policy contexts in many respects: the challenges posed to both experts, policy makers, and stakeholders can benefit a great deal from generic lessons from a diversity of practical cases. With regard to practical "science–policy" and "science–society" interface knowledge, we sense that there still remains quite a large *knowledge gap* to fill, especially focusing on the practical applicability of ES-concepts.

Hence, what we need—in our opinion—is even more networking, that is:

- Networking **People**: exchange of knowledge and experience (learning together)
- Networking **Cases**: collaboration on projects involving end-users (putting theory into practice)
- Networking **Events** and creating social network facilities
- Interfacing with **other (international) initiatives** regarding practical ES applications such as BEES, the EEA-TEEB initiative, and the ESP.

WHAT NEXT?

It would be worthwhile first to start a deeper investigation of the added value and effectiveness of ES networks such as the Dutch CoP-ES, BEES, ESP, and other (inter)national ES networks in bridging the gap between the scientific explorations on ES and the practical applications of ES in policy, management, and business. How does capacity building with respect to ES work in these networks? What are the lessons learned? What are the challenges and which are the ways forward?

REFERENCES

1. Brils, J., and van der Meulen, S. 2010. Delen van ervaringen met ecosysteemdiensten. SKB, Gouda 2010 (in Dutch). Available via: http://www.skbodem.nl/project/43
2. Brils, J, and Moolenaar, S. 2011. Op weg naar een ecosysteemdiensten praktijkgemeenschap. Jos Brils & Simon Moolenaar. SKB, Gouda (in Dutch). Available via: http://www.skbodem.nl/project/43
3. Magnuszewski, P., et al. 2010. Report on the prototypes of knowledge brokering instruments. PSI-connect, Delft. Available via: www.psiconnect.eu
4. Moolenaar, S., and Penning de Vries, L. 201. Verslag Startbijeenkomst: Community of Practice Ecosysteemdiensten, September 28, 2011. SKB, Gouda (in Dutch). Available via: http://www.skbodem.nl/project/43

5. See www.es-partnership.org

Contribution of DG Environment of Federal Public Service Health, Food Chain Safety and Environment

Lucette Flandroy, Sabine Wallens, Kelly Hertenweg and Saskia Van Gaever

DG Environment of Federal Public Service Health, Food Chain Safety and Environment

Lucette Flandroy - Federal public Service Health, Food Chain Safety and Environment, DG Environment, Service of Multilateral and Strategic Affairs. Presently involved in Biodiversity and Biosafety issues, at national, European and international level. Concerning Biodiversity, mainly involved in biodiversity and health links, as a part of ecosystem services and their valuation, in the national Community of Practice related to this issue. Concerning Biosafety, mainly involved in application and evolution of legislations and international agreements concerning risk evaluation and management of GMOs and concerning the socio-economic considerations related to them. National focal point of the Cartagena Protocol.

Ecosystem Services. http://dx.doi.org/10.1016/B978-0-12-419964-4.00028-7

Sabine Wallens is an expert in Biodiversity at the Federal Public Service Health, Food Chain Safety and Environment. She is working on strategic issues and on the sectorial integration of biodiversity and ecosystem services, in particular to key market players such as businesses, consumers, NGOs, trade unions... She is engaged in the specific aspect of the sustainable use of biodiversity and ecosystem services.

Kelly Hertenweg is an expert in international environmental relations at the Federal Public Service Health, Food Chain Safety and Environment. She is engaged in the protection, conservation and sustainable management of forests, their biodiversity and their role in the fight against and adaptation to climate change. As a member of the Belgian delegation she follows REDD+ and other forest related negotiations within the UN Climate and the Biodiversity Conventions.

Saskia Van Gaever completed her doctoral research on the ecology of meiobenthos associated with reduced deep-sea environments at Ghent University in 2009. Since 2009 she is working as marine expert for the Belgian Federal Government. She is appointed as national coordinator for the implementation of the EU's ambitious Marine Strategy Framework Directive. At the international level she is involved in the marine-related UN processes and is a member of the Group of Experts of the first World Ocean Assessment.

We are members of the federal Ministry of Environment of Belgium, working toward the sustainable use of biodiversity, one of the pillars of the Convention on Biological Diversity (CBD). The fundamental relationship between biodiversity and preservation of ecosystems has been recognized by the CBD. Biodiversity and ecosystems functions offer a wide variety of services to humans and the rest of nature.

Among the worthwhile services of biodiversity are those related to health, notably: the production of healthy and sufficient food in a sustainable way; product standards with less potential impact on the environment (including biodiversity) and health; medical inspiration for research in physiology and the production of new drugs; prevention of epidemics of old or new infectious diseases. All necessitate the preservation of biodiversity and of resilient, equilibrated ecosystems that stabilize each other at some scale (cf. e.g. Overview of the BE Contact group on Biodiversity and Public Health).

It is essential to develop the best possible indicators and methodologies to assess their status and to evaluate potential and acceptable impacts on biodiversity and ecosystems (and linked ecosystem functions and services), in order to insure their interrelated resilience. (It is hard to imagine that the loss of any element of biodiversity is not a potential loss of a treasure for new drugs, even if the ecosystems seem to survive with the disappearance of some species). Limits of adaptation, no-return points, have to be defined. This is necessary in order to argue for their preservation facing various pressures. It is also important to recognize that current scientific knowledge is still limited and filled with uncertainties; it cannot offer all the information to always take purely science-based decisions; the precaution principle is thus to properly take into account in such decisions.

Currently, we are tackling this problem, and at the same time that we are facing climate change in our environment, through our involvement in:

- environmental risk assessment and socioeconomic considerations relative to GMOs (e.g., GM plants) and the agroecosystems in which they are used.
- the integration of biodiversity in several federal public sectors by defining specific actions (see Federal Plan (2009–2013) for the Integration of Biodiversity in four key federal sectors), by organizing specific interactive training on biodiversity and ecosystem services and on the interrelationship with the work/professional activities.
- the development of scientific and legal instruments relative to invasive species.
- collaborative projects (including informative seminars) with market actors (producers, companies, retailers, and consumers), people from the civil society (NGOs, trade unions), on products and services issued from biodiversity and ecosystem services, with the aim of sustainable use.

- the negotiations for and implementation of the international Access and Benefit Sharing instrument under the CBD, which should help to preserve the biodiversity and ecosystems of the regions' providers of products.
- the REDD+ negotiations in order to ensure that REDD+ for forest biodiversity and its ecosystem services is being recognized and exploited to the fullest extent possible.
- forest ecosystem services by creating synergies between our climate and biodiversity policies.
- the preservation of bees, which pollinate many cultivated plants and are excellent indicators of ecosystem health.
- the development of the so-called *bioeconomy*, which develops various new products/services (including ecosystem services) through diverse new technologies.
- the recent initiatives of the European Commission, such as the transition (ecological, social, and economic) toward EU 2050, including caring for resource efficiency and achieving a low-carbon economy.

When developing methods to evaluate–in particular, economically–the services offered by ecosystem functions, the everlasting nature of these services (providing ecosystems are preserved) should be taken into account. (By contrast, similar human-made services necessitate costly maintenance and reinvestment.). This is worth wile in the perspective of the transition towards a more sustainable society.

In the framework of the EU GMO legislation and under the Cartagena Protocol, it is important to evaluate impacts on ecosystem services.

The evaluation of ecosystem services is essential in the context of socioeconomic considerations that may be taken in decisions relative to GMOs at the international (Cartagena Protocol) and EU levels. In this view, it is crucial to define indicators for ecosystem functions and methods to evaluate their services (economic as well as social, cultural, aesthetic, and spiritual) in order to assess potential impacts ex ante and/or ex post. Particular attention should be paid to the sustainability of ecosystem services, specifically in relation to the wording of Art. 26.1 of the Cartagena Protocol that focuses on "socio-economic considerations arising from the impact of living modified organisms on the conservation and sustainable use of biological diversity, especially with regard of the value of biological diversity to indigenous and local communities."

The relationships between impacts on biodiversity elements and ecosystem functioning should be clarified in order to more accurately assess the impact of GMOs on specific biodiversity elements. (In this regard, in the absence of a sufficiently clear position in the CBD, interpretation of the EU GMO regulation seems somehow different from this in the EU pesticide regulation, as if each element of biodiversity would be more important to preserve in the pesticide regulation, whereas ecosystem functions are represented by functional groups in the GMO regulation.)

THE MARINE ENVIRONMENT

In 2008, the EU made a fundamental change in its traditional approach to the regulation and management of the marine environment, marine natural resources, and marine ecological services. The Marine Strategy Framework Directive requires the member states of the European Union to put in place measures to achieve and maintain the good environmental status of marine waters by 2020. Apart from redressing a longstanding gap in EU law, the Directive also marks the EU's first concerted attempt to apply an ecosystems-based approach to the regulation and management of the marine environment, marine natural resources, and marine ecological services. This Directive may serve as the principal source of marine environmental management measures in the EU for many decades to come (Long R., 2011).

It is evident that pressure on natural marine resources and the demand for marine ecological services are often too great and that the EU member states need to reduce their impacts on marine waters regardless of where their effects occur.

By applying an ecosystems-based approach to the management of human activities while enabling a sustainable use of marine goods and services, priority should be given to achieving or maintaining good environmental status in the marine environment, to continuing its protection and preservation, and to preventing subsequent deterioration.

Marine strategies shall apply an ecosystems-based approach to the management of human activities, ensuring that the collective pressure of such activities is kept within levels compatible with achieving good environmental status and that the capacity of marine ecosystems to respond to human-induced changes is not compromised, while enabling the sustainable use of marine goods and services by present and future generations.

Generally speaking, if the notion of ecosystems services is helping us to raise awareness among different private and public sectors concerning the usefulness and necessity of preserving biodiversity, it also brings challenges and threats. Valuation of ecosystem services can indeed be a hard and risky task. First, because some functions of biodiversity, like those linked to culture and to mental health, can hardly be quantified; quality indicators should thus also be developed and accepted, but these are difficult to use to assess the impacts of pressures and to make comparisons. Second, because even quantitative valuations of some ecosystem services cannot always be translated in economic terms, even if it could be easy to have one single monetary parameter for comparisons and decisions. Third, because especially economic quantification of ecosystems services, which may be a useful information, when feasible, for the preservation of ecosystems, may also raise the threat - like happens for climate and Green House Gases (GHG) emissions - of trade-offs on biodiversity, of buying credits, of compensatory projects that could not always be equitable neither could they replace the concrete local loss of biodiversity and ecosystems services Fourth and finally, because the status of biodiversity and services is not static even in the absence of any man-made pressure; this underlines the difficulty to define adequate baselines for comparisons.

Finally, good knowledge and understanding of biodiversity, ecosystems, and their functioning, as well as good insights into the impacts and dependence on biodiversity and ecosystem services from various human activities, are crucial to facilitate policy decision making. An active BEES community offers the means to discuss and exchange ideas and allows us to extend our network on this ES concern.

Chapter 29

Relevance of an Ecosystem Services Approach in Southern Belgium

Marc Dufrêne

ULG-GxABT

Since 2013, **Prof. Marc Dufrêne** is developing a new research team on Ecosystem Services evaluation, mapping and modelling at ULg (Gembloux Agro-Bio Tech). He is in charge of developing interdisciplinary researches and innovations within GxABT particularly associated with the development of agroecology in rural landscapes.

The awareness and appropriation of the ecosystem services (ES) concept are in their early stages in Wallonia. At the political level, no mention of ES is made in the Déclaration Politique Régionale 2009–2014, while this text calls for a "regional plan for biodiversity" that indeed covers part of the problem. Within the administration, awareness of the concept is also only latent, with only two studies evaluating the return on investment for the Water Framework Directive. This is why the Département d'Etude du Milieu Naturel et Agricole(SPW-Wallonie) has launched a research convention with the University of Namur to provide a synthesis and a regional assessment of ES, with the first economic assessment of forestry finalized at the end 2012.

Ecosystem Services. http://dx.doi.org/10.1016/B978-0-12-419964-4.00029-9

The ES approach is considered a real opportunity to give meaning to biodiversity conservation and management in the perspective of both the outstanding and more common components of our natural capital.

The status of biodiversity is far from being positive: At least 40% of species are extinct or declining, even though the restoration of some populations has occurred in birds, dragonflies, and butterflies, which are related with wild game regulation, ageing forest dynamics, and more locally by large recent restoration actions (LIFE projects). A large majority of habitats are in an unfavorable conservation status. Some problems may be quite simple to manage, as is the case for deadwood and old trees in forests, or they may be more difficult to solve as in the case of reducing eutrophication or controlling vegetation colonization in open habitats.

A major limitation for the outstanding biodiversity continues to be the limited investment to protect biodiversity hotspots and the absence of specialized teams to manage them. At present less than 12,000 ha[1] of areas are protected by of the Law on Nature Conservation, and around 4500 ha are integral forest areas, resulting in less than 1% of the territory being devoted to natural structures and processes. The difference with Flanders is quite impressive considering that the same regulation framework was in force before regionalization. For more common biodiversity, constraints needed to allow sustainable management in rural landscapes are also still too limited or not well oriented to maximize the efficiency of actions.

So, what should be proposed as a framework for an integrated strategy? In Southern Belgium, the rule is to base biodiversity management on the ecological network concept. This concept allows only two complementary ways for action: (1) some space must be set aside for nature, and (2) more space needs to be given to natural processes in production processes. On the field, this is transposed through the well-known landscape partition in core areas, sustainable use areas, and the landscape matrix. The network design should be devoted to persistence of species population by maintaining colonization rates that are always higher than extinction rates.

Historically, two different strategies have been promoted: first, a segmentation approach aiming to focus only on outstanding biodiversity in core areas, and second, an integrative approach to increase the stakeholder's awareness in other parts of the landscape.

We still need a combination of both the segmentation and integration approaches, especially in Wallonia; a region that has a great need to increase protected areas and consequent ES devoted areas in almost every component of the landscape.

The segmentation approach is still necessary because a large part of the outstanding biodiversity occurs where ecological conditions are more or less extreme. This can be achieved through natural processes or specific management

---

1. With one site reaching more than 5500 ha (Réserve naturelle des Hautes-Fagnes).

organized by specialized technicians with the help of local managers. A realistic aim would be to reach about 40,000 to 50,000 ha of such core areas, that is, 2.5 to 3% of the territory.

We also need an integration approach to develop sustainable use of the resources in more common land-use classes over all landscapes. Such an integration should be mainly developed:

- by promoting management adaptations within the parcel such as precision agriculture and agroecological practices to develop production ES in a more sustainable way;
- by developing interfaces for other ES such as regulation and cultural services, using core area optimization but also hedgerow, woodland, and grassland strip design;
- by maximizing joint ES instead of focusing on one ES in field management measure as in set-aside measures, and nitrate and erosion management.

We also need greater cohesion between the different landscape compartments.

Is it still possible to restore such areas without facing important problems with managers and landowners? Field experiences with large *LIFE* restoration projects give us some effective insurance. If the evolution of expected income is plotted on a graph with land-use intensity, the curve generally shows an asymptotic shape (Figure 29.1).

The land exploitation mainly occurs in easily accessible areas, while it is limited in isolated ones. For biodiversity, the evolution is supposed to be favorable at the beginning with the landscape diversification, but a progressive accelerated decrease is expected with time. On the field, the economic curve could be different from what was expected, with some capital losses in cases where the ecological

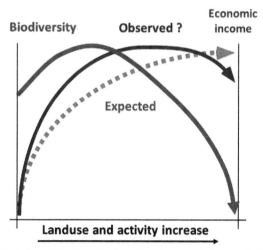

FIGURE 29-1   **Evolution of biodiversity (in red) and economic income (in brown and green) in function of land-use and activity increase.**

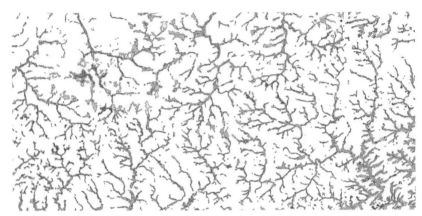

FIGURE 29-2   **Example of soils that are marginal for production as peat (brown), wet soils (blue gradient), alluvial soils (green), huge slopes (gray), and superficial soils (orange). The network design is quite clear.**

conditions are too restrictive. In such a case, there is a real opportunity to restore biodiversity because we can also restore ES that show a similar global evolution.

In Southern Belgium, almost 18% (300,000 ha) of the territory is covered with soils that are marginal for intense primary production (Figure 29.2). It concerns mainly wet soils but also some dry ones and high slopes in the Ardennes. Moreover, these marginal soils show a real natural and continuous network structure. These soils therefore have very high biological potential for restoration with limited impact on production, which is hardly profitable. For example, for spruce plantations, foresters have reported that one tree out of four (35– 40.000 ha) is planted in ecological conditions that do not allow benefit. On peaty and wet soils, such plantations result in important disturbance of ES. A similar pattern is observed for 40,000 hectares cultivated on alluvial soils, while the 80,000 hectares of grasslands occurring near rivers and rivulets positively contribute to regulating and supporting services.

It is thus evident that there is a high potential for restoring ES linked to carbon stocks, water regulation or purification, and sediment regulation on a significant part of marginal production conditions, with a more profitable production of traditional production ES on convenient soils.

How and where to work forward? The first step consists in establishing regional conservation aims and action plans resulting from consistent analysis of the status of biodiversity and ES. This step includes the release of budgets and financing tools, the reevaluation of legal constraints, and the design of a common green infrastructure. The green infrastructure is not an easy task as the green infrastructure network is a structure nested in different ecological networks acting at different geographical levels, with bottom-up and top-down processes.

At the regional level, we first need a general frame with the definition and delimitation of several main ecological networks such as wet lowlands, chalky grasslands, peatbogs, wet highland valleys, and other ecological networks for forest habitats or some peculiar species. More locally, we need to organize in homogeneous zones the network of sites around existing core areas. We should concentrate management and restoration actions in order to reach critical and efficient size and efficiency. This is the only way to develop at the same time significant species populations (that can act as sources for colonization), effective ecosystem services, infrastructures for a large public, and a specialized team for monitoring and management.

For example, in 2002 an ambitious wetland restoration program was launched for peatbog restoration. Six different LIFE projects were approved to work on all the high plateaus in the Ardennes, restoring around 2500 ha of peatbogs as well as a large diversity of wetlands (wet heathlands, fens, mires, alluvial forests). These projects were initially designed to restore biodiversity hotspots, but they will also have locally high impacts on ES for carbon stocks, water regulation, and cultural services. Similar actions should be designed for alluvial habitats in other parts of Southern Belgium.

At the site level, the focus should be placed on optimizing ES and biodiversity by designing an optimal network of hedgerows and other punctual landscape elements. The most important opportunities for actions related to ES are those concerning agri-environmental payments from rural development plan and actions related to the control of erosion and flooding in cultivated areas. The LIFE program allows a new option for financing the restoration of efficient green infrastructure in intensive agriculture areas. A new financing option for agro-forestry should also be further considered. But such initiatives need to be optimized, considering outstanding biodiversity improvement.

Working for conserving and developing outstanding biodiversity on one hand and ecosystem services on another might be seen as similar, but this is not the case. It is necessary to develop a Geographical Information System (GIS) model that at the same time provides information about biodiversity hotspots and the result of a clear assessment of ecosystem services, in addition to the real impacts on socioeconomic agro-food systems, wood, water, and energy production at different scales. On the field, the multiscaled green infrastructure should result from optimization of the two conservation strategies.

Such a decision support system will stimulate a lot of interesting debates. It should increase awareness about the significance of biodiversity in landscape planning with a cautious attitude on the biodiversity monetization process and reality.

Chapter 30

A Participatory Approach to Wildlife Management in Walloon Farmlands

Layla Saad

Fauna and Biotopes asbl

Layla Saad - is an ecologist. She is the coordinator of Fauna & Biotopes, an environmental association which aims at studying and managing wildlife and habitats. The actions of the association take place at the regional scale (Wallonia) and include the conception and implementation of integrated action plans in favor of wildlife, territorial mediation, applied research studies, public sensitization and training.

Agricultural intensification and specialization have caused the loss of wildlife habitats. At the same time, agriculture provides positive environmental benefits, including preservation of public goods such as farmland biodiversity and agricultural landscapes.

In the urgent context of the erosion of farmland biodiversity, our association, Fauna and Biotopes, seeks to match competencies and stakeholders to improve ecosystem management by:

1. focusing on wildlife conservation (fauna and flora) and habitat management;
2. searching for a consensus among stakeholders, leading to sustainable actions and integrated management.

Ecosystem Services. http://dx.doi.org/10.1016/B978-0-12-419964-4.00030-5

One of our missions is to supply knowledge and technical help to land managers, improve management tools, and apply sound management techniques to aid in wildlife and habitat efforts.

While we do not specifically use the terminology of ecosystem services, our daily actions, which are clearly aimed at enhancing the positive noncommodity outputs of agricultural production, such as habitats for species, are part of supportive ecosystem services.

AN ES APPROACH TO WILDLIFE MANAGEMENT?

With the support of the Walloon public service, our association carries out several projects aimed at developing a model of biologically sound management of wildlife on a large scale (with a minimum of 2000 ha). This aim involves a multi-actor approach (including farmers, hunters, local authorities, foresters, and conservationists), directed at enhancing the habitat for both game and non-game wildlife species.

The approach embraces fieldwork, including an inventory of the existing ecological network (trees, hedges, agro-environmental schemes), with the objective of conserving it while focusing on improving its quality and promoting the implementation of new network elements.

If taking conservation decisions to benefit species and habitats is quite straightforward, the current challenge lies in interlinking ecological and social systems. In our daily work, we have the knowledge to directly act on the biological attributes that provide ecosystem services (e.g., enhancing the partridge population through habitat design), but the attributes of the people whose well-being benefits from such a service also need to be considered.

The multi-actor participatory approach dimension is therefore a key step in our projects, for it aims at reaching a sustainable consensus. Indeed, considering each stakeholder's private "optimum" for land use will always lead to an imbalance (cropping profits, recreational hunting, renewable energies development, biodiversity conservation), while collective action can result in a more sustainable land-use pattern. In this process, identifying the stakeholders' values and their relationship to wildlife is essential. How do people value wildlife? How do they affect or are affected by wildlife? How do they want it to be managed?

Considering wildlife management, the drivers of stakeholders' engagement include broad social ideals (good governance, biodiversity conservation) as well as the specific traits of individuals (motivation to hunt, photograph, and observe wildlife). In order to reconcile all these interests, wildlife management needs to achieve benefits at individual, group, and societal levels.

A participatory approach that links landowners/farmers to beneficiaries of ecosystem services appears to be an effective use of ecosystem services. Using the concept to discuss land-use planning could facilitate effective consensus building on critical issues.

FROM SCIENCE TO PRACTICE

As practitioners, the implications of embracing the ecosystem services concept are still not crystal clear, but two aspects of the approach seem especially important to us:

1. The need to think globally about how any given project, proposal, or plan would impact on service provision and subsequent human well-being.
2. The need to manage ecosystem services in relation to wider stakeholder values, needs, and priorities.

Using the science of ecosystem services can help support more integrated and sustainable land-use decisions. Thereupon, even more than in any other field, interactions between researchers and practitioners are essential (for example, through a BEES community in Belgium) and would definitely bridge the gap between science and public demand.

THE VALUATION OF ES—OPPORTUNITY OR THREAT?

Valuation of ecosystem services is potentially a useful tool, but it needs to be applied with care and only when sufficient scientific understanding exists. Although the overall ecosystem services approach is useful when applied in a user-friendly and participatory way, we believe that as presently constituted quantitative/monetary valuation is not practical. At our level, the ecosystem services approach can be practically implemented through building good relationships among services providers (farmers, foresters, land managers) and beneficiaries (hunters, naturalists, citizens).

On a broader level, government intervention to internalize biodiversity benefits in the economic decision making of farms would increase the positive externalities of agricultural production as well as global social welfare. Current policies tend toward pro-environmental change, with important incentive tools such as agri-environmental schemes or support of organic production. However, additional measures are needed to reward environmental benefits and to be consistent with wildlife conservation in agricultural areas. With the help of researchers, the challenge for policy makers today is to comprehend the multifunctional character of agriculture in relation to social-ecological thinking. Attention should be paid to the fact that any change in agricultural support might significantly affect the provision of ecosystem services from agriculture.

Chapter 31

Ecosystem Services for Wallonia

Cédric Chevalier

Political advisor, Cabinet of the Walloon and the Wallonia-Brussels Federation Minister of Sustainable Development and Research

Cédric Chevalier - Cabinet of the Walloon and the Wallonia-Brussels Federation Minister of Sustainable development and Research; political advisor for regional economics and sustainable development public policies

The concept of ES[1] is of crucial importance for the Walloon government. Generally speaking, the concept is very important for virtually every public executive body at each governing scale (local, regional, national, continental, and global). Indeed, the concept could help to embody the very idea of "nature" in the political sphere and could put this idea on the table of the deciders. The scientific consensus is that we are living a multilevel and multiscale environmental crisis never experienced by humanity (climate change, biodiversity extinction, ecosystems erosion and collapse, pollutions of water, air, and soils). If current trends continue,

---

1. We use the term "sacred" in a non-religious meaning, as a value or a threshold considered as non-negotiable by a human group. Examples of such absolute limits to human activity are the interdiction of esclavagism or of children work. In these domains, no "compensation mechanism" is tolerated; only a strict ban enforced by law applies. A lot of these examples concern the human person in itself. Nevertheless it could be argued that some parts of nature are so important for Humanity that destroying them can be considered as a crime against a "sacred" value.

Ecosystem Services. http://dx.doi.org/10.1016/B978-0-12-419964-4.00031-7

catastrophic consequences are not only possible, but certain. Already, some places in the world are measuring significant environmental problems.

Political scientists have also shown that current governance systems fail to take into account "nature" in their decisions.

These two facts—global environmental crisis and systematic governance failure—are not mutually exclusive. Being blind to the effects of our current socioeconomical system on nature, governments are unable to measure the problems and to reverse the trends. It could also be asked if this blindness is not responsible for the crisis in the first place.

To explain this governance failure, following are some possible causes:

- **Cognitive failure**: Expertise, and awareness on the socioeconomic, but above all biological, dependence of humanity on nature is painfully lacking at higher levels of governance. Knowledge of the reality of environmental crisis is more common, but without an understanding of our vital dependency, this crisis can be disregarded by governments as a collateral preoccupation. The expertise on the environmental crisis emanates principally from the academic sphere. So we think that politicians are not sufficiently aware of the importance of this environmental crisis, that they are not sufficiently environmentally literate to understand the stakes, that scientists do not have sufficient influence on political thinking and knowledge, and we think that policy/science interface is too weak.

- **Political/Philosophical/Moral failure**: Even when there is sufficient knowledge of the environmental crisis and of our dependency on nature, governance systems seem to lack the elementary will to put these aspects on the table and to take them into account. This failure can be explained by a lack of altruism, a lack of moral character, a fear of popular unrest with unpopular decisions taking nature into account, a political partisan weakness to impose this subject to political partners, and the like. Even when benevolent, a political leader often lacks strategic information on the environmental consequences of his or her decisions. When confronted by adverse particular interests, this lack of strong data tends to weaken the politician's ability to take the best option.

- **Information/Technical/Scientific failure**: When consciousness of the environmental crisis, our vital dependency, and a will and capacity to act politically are gathered, governance systems still lack the necessary tools. Although socioeconomic statistics have been available for a long time with the help of statistical public and private institutions, environmental statistics are still young in comparison. Data on local pollution are now common, and data on global problems such as climate change, ozone layer, and overfishing have also become common. But data on ecosystems and biodiversity, a fortiori their systemic importance at local, regional and national levels, is quasi-invisible to deciders. Last but not least, when some tools already exist, they remain a recent and an active field of research, not totally mature. So more research is still needed on the concept, its definition and scope, its tools, its conditions of use, its effects and its practice at political levels. But because ES are a blind spot in the political radar, it has not been given enough priority. More could be done in research and academic circles, for example, with the help of government financing.

- **Societal acceptability**: These problems are increased by the complex relationship between politicians and the citizens of their constituencies. If citizens can have the same problems listed above as politicians, politicians must take into account mainstream opinion and the will of the population to change. Even with a strong ES research sector, a nice interface between science and politics, optimal and robust tools available at higher levels, even with a strong awareness and political will, if politicians hesitate to take decisions and are afraid of popular reactions, and if citizens oppose reforms, nothing will move.

So because of these problems, the ES concept is of paramount potential importance for the Walloon government. As a would-be tool for governance, it has the potential to address each of these problems:

- Mainstreaming ES awareness can increase population awareness of the environmental crisis and our vital dependency on nature (ecosystems, biodiversity) and give citizens necessary information to increase pressure on their political leaders.
- Increasing ES research can help scientists to conceptualize and define useful concepts, create new and robust data and decision tools, empirically test ES tools use at all governance levels, and check the results of their use.
- Increasing policy/science interface could enhance political awareness of the environmental crisis and human vital dependency on nature and help channel data and tools for governance use.
- Structurally incorporating ES data into public statistics, via statistical organizations, and presenting it in official reports to the government could give strong objective arguments regarding environmental impact to the politicians at the decision table.
- If ES data comes on the governmental level, it could be hoped that willing politicians would be able to strengthen their case for action for nature, against political partners and opposition, with the help of scientists, aware population, and strong scientific data.
- Finally, if ES becomes mainstream, the cognitive failure of some deciders could disappear or become a minority point of view, as has been the case for climate change in recent decades with the help of the IPCC.

In this respect, change can come from global initiatives like IPBES, and regional ones, like the BEES community, working in conjunction in all these areas.

With regard to future trends, the major opportunity lies mainly in helping to regulate human activities at all decision levels with respect to biospheric boundaries, implemented at the regional level. Some other opportunities could also be pointed:

- Continue mapping all regional Walloon ES.
- Prioritize financing and intervention for the most precious ecosystems and species.

- Stop biodiversity and ecosystem degradation.
- Improve the status of biodiversity and ecosystems.

The most important advantage of using ES in the population, any government, and in particular in the Walloon government would be to put a name, a concept, and data on a biophysical reality that is not accounted for at the moment in our societies. That is: Allow the mental revolution first. ES could be a powerful tool for awareness. Other advantage, depending on the efficiency and robustness of ES developed tools, could be as follows:

- Quantify, qualify, and monitor ES status and trends in Wallonia: geographical areas, fluxes, biophysical parameters, economical importance, governance pertinent perimeter, species, ES categorization.
- Raise politician and citizen awareness and increase systemic ecological thinking in governance.
- Help political arbitrage between conflicting objectives.
- Give advice and sound the alarm when necessary; avoid environmental irreversibilities.
- Give data needed to define other human activity regulation tools such as taxes, subsidies, and regulations.
- Give data for every actor of society: citizens, NGO, firms, public administration, governments, etc.

But ES also present disadvantages and limitations:

- ES complicates further the political process with more constraints.
- ES is only a tool and cannot replace a philosophical/political stance for preserving and improving nature. Some choices will not be easier with ES, like some cruel choice between areas to preserve where no significant ES data could be available.
- ES is only "one" tool in a "toolbox" that must be widened. Other significant and necessary tools are rules, laws, taxes, penalties, subsidies, education, research, social and moral norms. ES must be used as a partial solution together with others tools and in the context of a wider political strategy.
- Ecosystem and biodiversity degradations are the results of complex socioeconomic forces. These fundamental causes must be politically addressed in order to reverse the trends. ES can only be an information and regulation tool in this process and cannot replace politics, law, and social norms targeted at these fundamental causes.
- ES could be of important use for mainstreaming environmental awareness and help arbitrage between different options, in human-made and semi-natural ecosystems especially. In purely natural ecosystems, "biodiversity hotspots" (like Amazonia) and "vital global ecosystems," no tool ever will be able to compensate for an absence of political will to enforce preservation

and protection. For these cases, we think that only the rule of law and social norms, under a common recognition that these ecosystems have a "sacred" character for Humanity, can be effective. We use the term "sacred" in a non-religious meaning, as a value or a threshold considered as non-negotiable by a human group. Examples of such absolute limits to human activity are the interdiction of esclavagism or of children work. In these domains, no "compensation mechanism" is tolerated; only a strict ban enforced by law applies. A lot of these examples concern the human person in itself. Nevertheless it could be argued that some parts of nature are so important for Humanity that destroying them can be considered as a crime against a "sacred" value.

- A tool cannot replace the engineer. For example, Belgian environmental laws and regulations are extremely severe. However, even with these legal tools, rules are not enforced because of several other "transmission of rule" problems: notably, lack of awareness in local stakeholders, lack of finance for regulation, lack of formation for administrative police, lack of will from judiciary powers.

And last but not least, ES must be studied and used with caution. Here we list some potential threats:

- Deviant use of the ES concept in combination with monetization can destroy the advantage of the concept and, worse, be a way to merchandize nature. In particular against monetization, the inherent instability of prices in market economies has fatal consequences for people who believe dogmatically in monetary valuation of nature. Nature has shown particularly complex characteristics that are incompatible with the "continuous and linear" aspects of markets, including strong inertia, path dependency, irreversibilities, complexity, and threshold effects. So the ES community must make perfectly clear that ES concepts, tools, and practices must help to adapt human activity, stop ecosystem and biodiversity degradation, and improve their status as a result.
- Use of ES could inadvertently increase the utilitarian point of view on nature and weaken the case for preserving "zero utility nature," with only moral and ethical grounds remaining, for example, by finding no "data" justifying the protection of species being of relatively null utility for human societies, especially vulnerable species. So ES as a way to "collectivize" nature's importance could also produce the mistaken assumption that "zero utility nature" is only a private preference whose protection must be arbitrated by governments. ES can only serve nature's cause inside a global moral and ethical doctrine arguing for preservation and enhancement (strong sustainability ethics)
- ES could be perfectly instituted at highest levels and well crafted but stay a "beautiful toy" never taken seriously by political leaders, as is already the case for some political institutions.

- ES will not be able to solve the permanent arbitrage problems of humanity. The existence of humanity implies a nonzero environmental footprint. If we want to minimize this footprint by all our means, this implies that arbitrage between options, all negatively impacting environment, must be made. Even if we "sanctuarize" some areas for exclusive use by "Nature," the question will remain for human-made ecosystems (where human is the primary user of the ecosystem) and interface areas (where humans share the use of ecosystems with nature). So the temptation will always be to unify all available options under a common currency unit to make them comparable and commensurable, and monetization will often seem the best solution in order to do prepare that arbitrage. But as said elsewhere, we think that monetization has very strong defects.
- For example, if ES conare used to evaluate the impact of the implantation of a new airport (in a case like the Aéroport du Grand Ouest Project in France for example) and to calculate environmental compensation if this airport is built, ES concepts won't say if the "decision to build a new airport" or the "decision to develop popular low-cost regional flying" is good or not in the first place.

The main challenge is to deal with the cognitive failure of the population and the politicians, which impedes the mainstreaming of nature awareness and ES-like tools. If a sufficiently strong minority can make the case for the ES concept and tools, research financing, population awareness, statistical implementation, and political backup could help to make ES a solution for the crisis in which we live.

But reasoning along a transition path, the first challenge is to secure a success story linked to ES concept research. For example, one should be able to show at a local specific level that ES is pertinent, useful, and usable and that it brings significant positive results during the decision process and for the population in general and stakeholders in particular. So it is urgent that we build partnerships between researchers that go on the ground to test their tools with local administration and municipalities, population and stakeholders. In this respect, the BEES Community offers strong hopes.

Other challenges are more technical and will be linked to the articulation of the political decision process with the incorporation of ES data at relevant moments.

As stated earlier, ES science has tremendous potential relevance for policy making. The main scientific challenges are technical. After the theoretical ideas, details will make the difference between a neglected gadget and a real and serious tool used by politicians and the wider society. Some other challenges could be listed:

- Incorporate at the local level biospheric boundaries found at higher levels.
- Prove the capacity of ES tools to help decision making in real situations.
- Build robust ES tools.
- Propose practical advice on ES use.
- Always link technical research with ethical and philosophical considerations on nature.

In order to advance ES research and possible use, an active BEES community seems to be of utmost importance. Indeed, the BEES community could do the following:

- Increase networking between ES stakeholders: scientists, politicians, public officers, businessmen, NGOs, citizens.
- Form a unique central reference for ES research and communication.
- Increase governments' awareness of ES.
- Exchange practice.
- Feed research with ground data and knowledge.
- Help build and mainstream ES concepts, tools, and practice.

We have seen that there is an absolute need for strong policy/science interfaces that channel scientific data and offer robust tools to political leaders. The BEES community, based on the model of transdisciplinary science, offers a strong hope in this respect.

Chapter 32

Relevance of the Concept of Ecosystem Services in the Practice of Brussels Environment (BE)

Machteld Gryseels

Brussels Environment, Direction Quality of the Environment and Nature Management Gulledelle

Machteld Gryseels - Director of the "Direction Quality of the Environment and Nature Management" of Brussels Environment. At the scale of the urban Brussels Capital Region, she is responsible for the strategic and operational management of the forests and semi-natural sites (10 % of Brussels surface), the collection and analysis of environmental data and indicators, the follow up of air quality (in-and outdoor), and the knowledge development on the relation human health and environment.

Since 1989, Brussels Environment (BE)—Brussels Institute for Management of the Environment (IBGE- BIM)—has been the administrative body responsible for the environment and energy in the Brussels Capital Region.

IBGE-BIM acts, from the regulatory standpoint, as a research, planning, advisory, and information body, as well as an issuer of permits and a survey and control agency. It has authority in the areas of green spaces, nature, waste, air quality, noise, water, soil, and energy.

Ecosystem Services. http://dx.doi.org/10.1016/B978-0-12-419964-4.00032-9

The concept of ecosystem services is especially useful to the Institute regarding its responsibility to "nature," in particular:

- protection and conservation of biodiversity and the monitoring of flora, fauna, and natural resources;
- management of green spaces (in particular regional parks);
- management of natural and seminatural sites (forests and nature reserves) and the relation with the other responsibilities, in particular those regarding to the monitoring of air quality (indoor and outdoor); noise problems; management of water; and the major consequences to human health.

Although formal use of the term *ecosystem services* is a rather new issue in the working of the Institute, BE has long based its nature and green space policy on the development of "green infrastructures" in the urban area, essential for ecosystem services. In 1994, a vision was developed, and guidelines were presented to "promote" the "biological patrimony" of the Brussels Capital Region. The Green and Blue Network program was laid out in 1995 and taken up in the Brussels Regional Development Plan. New green spaces were created, with "new" objectives for the urban area: preservation of natural vegetation in the urban area, restoration of pond banks, recuperation of water, and improvement of soil quality. Simultaneously, efforts have been made to reconnect green spaces, for example, by reopening formerly covered rivers. A good example is the restoration of the Woluwe Valley in southwest Brussels. It is a model combination of improvements in biodiversity, landscape, microclimatic conditions, noise problems, and soft mobility along a green and blue ribbon in the urban context.

However, the ecosystem services concept has been used formally since 2010 to promote and to strengthen the "Nature Plan" also in the urban area. The Report on the State of Nature in the Brussels Capital Region, published in 2012, explains the contribution of nature to the quality of life, through the advantages of the ecological and social functions, the "ecosystem services." The Nature Plan, based on the recommendations of the Nature Report, has been elaborated with all stakeholders in a very participative way and should be proposed to the Brussels government in 2013.

It is an encouraging development that the concept has now been accepted throughout the Institute, as well as outside the sector of green spaces and nature conservation, as an opportunity to help improve the quality of life in the city.

Indeed, the concept was used for the first time in an official external advice in 2012 to the Brussels administration of town planning, regarding a new regional Soil Destination Plan. This new soil plan is being developed within a demographic framework in Brussels and is concentrates on increasing urbanization (houses, schools, etc.), in answer to the expected population growth during the next decades. In this advice, the importance of green spaces and nature has been emphasized in the context of the life quality in an urban area:

The green character as well as soil permeability are absolutely necessary to maintain ecosystem services, which maintain the quality of life in the city and thus the well-being of the population.

Emphasis has been placed on the numerous ecological functions of green spaces and their contribution to a healthy urban environment, such as:

- they contribute to a better air quality (capture of PM, oxygen production) and the effects on the urban heath island;
- they contribute to a better noise environment as green spaces are "quiet" zones;
- they assure regulation of the carbon cycle and attenuating climate change;
- they constitute infiltration zones for water (and thus help prevent flooding and soil erosion) and alimentation of groundwater;
- they contribute to a better water quality;
- they are an essential support for biodiversity.

With regard to social functions, which are so important in urban areas, green spaces open to the public:

- are places to meet and exchange in the local neighborhood;
- offer opportunities for all types of recreation and relaxation for different ages;
- invite physical open air activity;
- offer diverse opportunities for artistic expression and pedagogical support;
- have a positive effect on "urban stress."

…

This positive relation between equitable access to green spaces, contact with nature for all categories of population, which also includes the most vulnerable segment of the population (children and the poor), and human health has been described in numerous scientific publications.

Thus, BE strongly recommends that the town planning administration guarantee those ecosystem services. BE asks for general and forcing legal prescriptions in the future soil and town plan to ensure the "green" quality of the habitat environment. General aim is the development of a "Brussels ecological network" (by the preservation of existing green spaces and green/blue connections, creation of new ones, promotion of green roofs, walls, ...).

The main challenge for BE is to convince all stakeholders involved in the management of the region that green spaces and nature create invaluable ecosystem services, in both rural and urban areas—services that are essential for human health and quality of life.

The concept of ecosystem services can help to find arguments for fulfilling the Brussels Capital Region's goal to have Brussels become a real "green, ecological capital."

To develop those arguments, an active BEES community to discuss and exchange science and practice would certainly add value to the BE's work.

Chapter 33

Contribution of the Agency for Nature and Forests

Jeroen A.E. Panis

Agency for Nature and Forests, Government of Flanders

Jeroen Panis - Agency for Nature and Forests, Government of Flanders. Advisor on area based nature policy with the emphasis on the exploration of new concepts. Current topics are the ecosystem services concept, temporary nature and nature conservation objectives for Natura 2000.

SENSE AND SENSIBILITY

The Agency for Nature and Forests (*Agentschap voor Natuur en Bos*, ANB) is part of the Flemish government and as such is responsible for nature and wildlife policy and the management of the Flemish public nature reserves, forests, and green areas. The agency focuses on the conservation, restoration, and development of nature, forests, and green areas. It promotes sustainable management of nature and wildlife and seeks to improve the quality of the natural environment and fulfil the social needs of both present and future generations. ANB places nature in the center of society in an effort to enhance the interaction between society and nature, to inspire the society to engage itself in preservation, and also to emphasize the benefits of nature to society.

The ecosystem services (ES) concept proves the importance of nature for the well-being and prosperity of humankind and society. The concept therefore

Ecosystem Services. http://dx.doi.org/10.1016/B978-0-12-419964-4.00033-0

backs our approach and goals. It is important to remember four points about ES. First, it is a powerful argument for a strong nature policy. Second, it helps integrate nature conservation targets in other policy domains. Third, it can be used as a conceptual framework for optimal development and management of an area ("green infrastructure"). And lastly, it can provide a holistic framework bridging the gap between different sectors and enabling win-win situations.

ECOSYSTEM SERVICES IN PRESENT DAILY OPERATIONS

The ES concept surfaced policy-wise only in 2005 with publication of the Millennium Ecosystem Assessment (enhanced by the publication of the TEEB reports in 2010–2011); similar ideas and concepts had popped up earlier. The idea of multifunctionality in the Flemish forest policy, for instance, dates from 1990. The main idea is that a sustainable forest management needs to balance the economic function (mainly wood production) with ecologic, environmental, social, and scientific functions. This balance is described in criteria and indicators. A system of management planning and grants supports the approach.

ES in policy is still in its infancy. Of course, nature and environmental policy have ties to the concept, such as grants for public access to forests and nature reserves, but by and large the concept as such has not yet come into use. One reason is policy makers' lack of familiarity with the concept or indifference toward it, over rather recent existing legislation and different political priorities (e.g., focus on growth in GDP) to a difficult access of ecosystem services proponents to policy makers. A large part of the nature and environmental legislation and policy is a transposition of European directives that do not refer to ecosystem services either. Fortunately, this will change with adoption of the new European biodiversity strategy, the roadmap to a resource efficient Europe, and so on. As a consequence, integration in other policy domains is growing, albeit slowly, especially regarding trade-offs. So, the challenge is to introduce the concept in our policy and its uptake in other policy domains. The advent of the nature value explorer was a big step forward on that front.

Therefore, at present, the concept itself is used mainly in concrete projects by ANB and partners—for instance, in projects aimed at protecting against flooding in the Scheldt Basin, the Grote Nete Valley, and the Dijle valley and also in projects concerning tourism. Nature is used as the foundation of tourism in the Hoge Kempen National Park, the Bosland project, and the Zwin area. ANB has a project called KOBE (*KennisOndersteuning bij Beheer en Economie van natuur-, groen- en bosdomeinen*), which deals with optimization of the benefits of nature and forest management without impact on biodiversity in ecosystems. At this time it focuses on wood and other types of biomass (e.g., grass clippings).

Ecosystem Service Challenges

The loss of biodiversity and natural ecosystems may be caused by not taking into account the worth of it when making decisions. Lack of awareness or acknowledgment of these values causes a warped decision-making process with suboptimal results. The scientific challenge is to provide accurate, workable, and credible values and information to correct this situation. The policy challenge lies in acceptance of the fact that biodiversity and ecosystems have value and that not taking these values into account leads to suboptimal decisions with negative ecological, social, and, with emphasis, economic consequences.

Another policy challenge is to sort out the legal status of ecosystems and ecosystem services. Can someone own an ecosystem or parts of it? What about biophysical structures and processes, ecosystem functions, ecosystem services, benefits, and values? A stable and transparent approach is needed for adequate protection, compensation, and payment schemes.

The challenge for ANB and others who manage land or ecosystems or do spatial planning is to incorporate the ES approach as a guideline and condition to develop and manage (public) space.

The final and probably most important challenge remain the mainstreaming of the worth of biodiversity, ecosystems, and their services to society, and acknowledgment of biodiversity and ecosystems as the foundation of the Flemish and the global economy.

DARK CLOUDS ON THE HORIZON

ES can be very beneficial for nature conservation policy, but one should be wary of certain things. A focus on ecosystem services over ecosystems and biodiversity can lead to a predomination of the financial valuation of nature and the reduction of ecosystem services, ecosystems, and biodiversity to economic goods that can be traded and replaced at will. Price-tagging nature elements can create a perception of a "license to pollute" (or to destroy).

TEEB advocates capturing the value of biodiversity and ecosystems. This is clearly necessary to maintain, manage, and develop the essential ecosystems that deliver indispensable services. Investment should be made in systems that divert a part of the benefits for society to concerned landowners and other stakeholders as a way to compensate for efforts and income forgone. But a lot of questions remain: Who will pay? Who will be paid? What will be paid for? How much will be paid? These questions have to do with efficiency and effectiveness, but also with ethics and transparency. Perverse effects are not unlikely and need to be averted.

BELGIAN ECOSYSTEM SERVICES COMMUNITY

To identify the issues and to assess the ecosystem service, to demonstrate the values of nature and biodiversity, and to capture these values and find solutions for the issues that threaten them, collaboration and dialogue are needed on various different levels: local and regional; scientific, policy, business and practitioners. This means providing a forum to exchange ideas, knowledge, and experience. Success requires the uptake and broad acceptance of the concept in every layer of society: policy, science, business, and society as a whole.

The scientific field of ecosystem services is developing rapidly. It is difficult for a single institution to keep up with everything. A community that can keep track of the evolutions and provide a regular state of the art assessment will be useful, without losing track of the main goal: implementing the concept and protecting our ecosystems.

Finally, a lot of activity is taking place on international and European scenes, policy or science oriented, or both, in working groups and forums. An active BEES community can play a role not only in disseminating their results in Flanders, by extension Belgium, but also in forming common Belgian viewpoints and introducing these insights at various levels. These roles will be mostly informal or facilitating. In addition, lessons learned in Flanders, Wallonia, and Brussels should be disseminated in Europe and beyond.

Chapter 34

Integrating Ecosystem Services in Rural Development Projects in Flanders

Jan Verboven and Paula Ulenaers

Vlaamse Landmaatschappij regio West

Jan Verboven - Flemish Land Agency. As a biologist and project manager, Jan Verboven is involved with the planning and implementation of nature and rural development projects in East-Flanders and coordinates land banking for farmers in port expansion related nature development projects (port of Antwerp) and flood plain development (Sigma). He a.o. represents VLM within the Belgian Ecosystem Services Community (BEES) and the co-leading of a EU agro-ecosystems pilot study as part of the EU objective on mapping and assessment of ecosystems and their services (MAES, EU Biodiversity Strategy).

Ecosystem Services. http://dx.doi.org/10.1016/B978-0-12-419964-4.00034-2

Dr. Paula Ulenaers - Flemish Land Agency. As a biologist and project manager, Paula Ulenaers is involved with the implementation of nature development projects in the province of Limburg. With her practical experience of working with ESD's in rural development projects, she participates in a pilot study case in ECOPLAN.

The aim of the Flemish Land Agency (Vlaamse Landmaatschappij, VLM) is to preserve open space in an increasingly populated Flanders and to improve "quality of place" by means of projects and processes in rural and peri-urban areas. Although ecosystem services are not yet regarded as an integrated part of VLM's operation, several development projects and processes of VLM incorporate different ecosystem services. This chapter presents an overview of VLM's experience with ecosystem services in land and nature development projects, land consolidation projects, and environmental stewardships with farmers, and also assesses the strengths and weaknesses of ecosystem services.

ECOSYSTEM SERVICES VERSUS OTHER ENVIRONMENTAL AIMS OF VLM

Ecosystem services (ES) focus on the benefits of ecosystems to human society. A similar orientation is seen in VLM's core concept to improve "quality of place" in rural and peri-urban areas [3]. The concept of quality of place originates from urban development in the USA, later adopted in the UK. The objective of quality of place is to ensure that living, working, and relaxing in these areas remains enjoyable, and that residents feel comfortable and are willing to identify themselves with the areas where they live. Although some environmental indicators of quality of place may include services provided by local ecosystems (e.g., clean water and air, visual screen, access to nature areas), other indicators are more closely related to social aspects such as employment, culture, tolerance, and safety as part of the general human pursuit to improve one's quality of life.

Other VLM projects and processes have the specific aims of improving *natural biodiversity*, as intended by international and Flemish nature conservancy policies (e.g., Convention on Biodiversity, EU Bird- and Habitat Directives, Natuurdecreet). Measures to improve natural habitat vary according to species-dependent

Measure	Ecosystem Services	Other Environmental Goals
Flood basin construction	Controlled flooding, sediment capture	Improved biodiversity: migratory aquatic birds Adaptation to climate change
Water course enlargement	Controlled flooding (increase storage capacity)	
Bank vegetation restoration and/or enlargement	Improved water quality (self-cleaning capacity), improved fisheries	Improved biodiversity: aquatic macroinvertebrates, fish spawning and nursery, aquatic mammals
Reeds and marsh construction	Improved water quality (self-cleaning capacity), wastewater treatment	Improved biodiversity: wetland habitats, reed birds, amphibians, fishes,
Pond construction	Drinking water for cattle	Improved biodiversity: wetland habitat, amphibians, aquatic mammals Nature education and awareness
Wet (submersible) grassland restoration	Controlled flooding, improved water quality, improved fisheries	Improved biodiversity: wetland habitats, meadow birds, fish spawning, migratory birds
Management of riparian strips to adjacent farm land	Buffer against fertilizer and pesticide intrusion, improved water quality	Improved biodiversity (limited): wildlife passages
Extensively used (flower) strips in farm land	Soil erosion prevention, improved gaming, natural pest control	Improved biodiversity: farmland birds, mammals, predatory insects
Hedgerow planting and management	Fencing, wood/biomass production, shade for cattle	Improved biodiversity: hedgerow birds, mammals, amphibians
Construction of woods and (urban) forests	Wood/biomass production, carbon storage, climate conditioning; recreational amenities, visual screen (quality of place)	Improved biodiversity: woodland birds, mammals, terrestrial invertebrates
Construction of pathways for walking, cycling, bridling	Recreational amenities (quality of place)	Nature education and awareness
Construction of wild-life crossings (tunnels, overpasses, fish ladders)	Improved road traffic safety	Improved biodiversity: migratory amphibians, mammals, fish

ecological requirements, as determined by specific nature conservancy objectives. Although some of these natural habitats may deliver ecosystem services and improve quality of place [4], this is not the main intention, or is scientifically yet unknown as the improvement of natural biodiversity is multispecies oriented (ecocentric) and not solely for the benefits of human society. Economic valuation of ecosystem services has proven inadequate in these cases as biodiversity values are not included at all or just poorly so [2].

The conservation and restoration of biodiversity and ecosystem services can play a key role in helping human societies to *adapt to the adverse effects of climate change* [6]. Some VLM projects address this aspect, albeit on a relatively local level. More robust adaptation measures of climate change might require a geographically more extensive river basin approach, for example.

OVERVIEW OF ECOSYSTEM SERVICES IN VLM PROJECTS AND PROCESSES

The accompanying table presents an overview of measures executed in VLM projects and processes, such as land and nature development projects, land consolidation projects, and environmental stewardships with farmers, with an indication of ecosystem services delivered, as well as other environmental goals intended.

ECOSYSTEM SERVICES-ORIENTED APPROACH IN NEW DEVELOPMENT PROJECTS

Although the ES concept is still in its experimental stage, it has been recently used as a starting point for elaborating plans and measures in some new development projects by VLM in Flanders.

Within a rather large area of about 20,000 ha, 13 important stakeholders, including the Province of Limburg and seven municipalities, asked for territorial development in De Wijers." Several previous attempts to start up this process had not succeeded. Therefore the focus was put on the potential of the area rather than on any existing bottlenecks. After discussing several useful concepts such as "cradle-to-cradle" and "participatory futuring," "ecosystems goods and services" was used as an approach to organize stakeholder involvement in order to develop a common vision in the De Wijers project. The use of this methodology was new for VLM and was supported throughout the process by the Flemish Institute of Nature Conservation (INBO).

Several workshops were carried out at neutral locations in De Wijers and facilitated by unbiased facilitators. In addition to the initial 13 stakeholders, other public and private partners were invited.

A first set of workshops focused on cultural services, and partners managing touristic attractions in De Wijers were invited. Both the level of participation and the outcome of the workshops were satisfactory. Participants agreed that the innovative approach stimulated the "win-win" thinking and the search for synergies. The results were used to carry out two specific studies emphasizing the possibilities for recreation and tourism in De Wijers.

The next sequence of workshops focused on provisioning and regulating services. In a first step, all such services for De Wijers were identified and clustered in land-use systems such as ponds and river valleys, the canal and peri-urban systems, forest, and heathland. These workshops revealed a greater awareness of and insight into the importance of the water system and the need for integrated water management in De Wijers. Opportunities for restoring the relationship between the physical system and human activities were also identified.

A final workshop focused on social and economic opportunities. The qualities of the landscape and the services they provide were taken as a starting point for the social and economic development of the area. Unfortunately, participation in this workshop was significantly lower than that in the previous ones. Nonetheless, the workshop was quite productive.

Use of the concept of ecosystems goods and services stimulated stakeholders to think on a more regional level and in a more integrated and sustainable way about the De Wijers project area. The final report, "Challenges for De Wijers," was the result of the ideas generated by the working groups. Now there is a need to translate the challenges presented in this report into strategic and operational objectives within a Master Plan for the project area.

The EU Interreg IVA Flanders-Netherlands Solabio-project ("Species and landscapes as carriers of biodiversity") focused on so-called functional agrobiodiversity (FAB). FAB on the scale of agricultural fields or landscapes provides ecosystem services that are inherent to sustainable agricultural production on a regional level, improving regional and global environment as well [5].

Soil life is strongly influenced by soil management by farmers such as plowing, application of fertilizers and pesticides, crop choice, and rotation. The Demeter project includes elaboration of a decision support tool for soil management by the University of Ghent, the assisted application of this tool on 50 farms in Flanders, and information exchange through a website and at demonstration days and agricultural fairs.

MAIN CHALLENGES FOR AN ECOSYSTEM SERVICES APPROACH IN DEVELOPMENT PROJECTS

With most of its development projects in Flanders, VLM focuses on a multidisciplinary, integrated, and area-oriented approach, involving many social sectors and stakeholders, and negotiating as much as possible for sustainable win-win solutions to local problems. Such an approach has led to several measures delivering different ecosystem services. Undoubtedly, there are clear parallels between integrated and ES approaches, and from this point of view the ES approach is not completely novel.

A more explicit ES approach within VLM projects is still in the experimental stage and is struggling with:

- a need for information on ecosystem services to project leaders, thematic experts, and agricultural advisers;
- questions on the importance and added value of an ES-oriented approach versus the actual integrated approach with development projects;
- the conversion of general ideas on ES into concrete planning and measures; and
- the lack of an indicator for monitoring ES outcomes.

The future development of VLM-projects also partly depends on Flemish policy directives and regulations on an ES approach.

REFERENCES

1. Temmerman, F. 2012. Onderzoek naar het effect van akkerranden op functionele biodiversiteit en natuurlijke plaagbeheersing. Final report commissioned by VLM. (in Dutch; English abstract in ELS-FAB Newsletter no. 6, May 2012).
2. VITO. 2012. Natuurwaardeverkenner. Nature value calculator version 1.1.0., internet application (in Dutch).
3. Vlaamse Landmaatschappij. 2011. Collaborating on quality of place. Green Works! Conference.
4. Meiresonne, L., and Turkelboom, F. 2012. Biodiversiteit als basis voor ecosysteemdiensten in regio Vlaanderen. Mededelingen van het Instituut voor Natuur- en Bosonderzoek 2012(1). INBO, Brussel (in Dutch).
5. Vlaamse Landmaatschappij. 2012. Uitdagingen voor De Wijers (in Dutch).
6. UNEP. 2011. Ecosystem-based approaches to adaptation and mitigation—good practice examples and lessons learned in Europe (http://www.bfn.de/0502_skripten.html).

Chapter 35

Reflection on the Relevance and Use of Ecosystem Services to the LNE Department

Tanya Cerulus

Environment, Nature and Energy Department, Environmental, Nature and Energy Policy Division, Government of Flanders

Tanya Cerulus - Environment, Nature and Energy Department, Environmental, Nature and Energy Policy Division, Government of Flanders. Economically underpinning the Flemish environmental policy, mainly on topics such as the economic valuation of ecosystem services and a.o. the integration of them in social cost-benefit analysis. Coordinating the Nature Value Explorer to enable policy makers and others to do so in an accessible way.

A USEFUL CONCEPT

The concept of ecosystem services (ES) is useful to the Environment, Nature, and Energy Department (LNE) in several different ways:

The Policy Paper 2009-2014 of Environment and Nature states that the concept offers a fresh perspective on the interaction between society and nature in which nature and forest are seen as necessary allies in promoting a sustainable society and economy. The ES approach is a promising tool for identifying desired social win-win or trade-off situations for a multifunctional open space

Ecosystem Services. http://dx.doi.org/10.1016/B978-0-12-419964-4.00035-4

in Flanders, and it provides a valuable framework for bringing those with different interests together.

It may also provides a broader base for nature conservation in Flanders. In conflicts between nature conservation and economic development, more informed policy choices for Flanders' sustainable development can be made by describing, quantifying, and monetizing the changes in ES of planned projects or policies. Doing so makes it possible to account for the effects on ecosystem services in assessment frameworks such as societal cost-benefit analysis (SCBA).

LNE could, for instance, use the ES concept to:

- better coordinate SCBA and environmental impact assessment (EIA) and the information both decision frameworks need from each other;
- better support other assessment frameworks such as the Environmental Costing Model that LNE uses for cost-effectiveness analysis to achieve environmental objectives at the lowest possible cost;
- develop compensation mechanisms for taking measures to maintain or achieve ecosystem services (payments for ecosystem services);
- determine the amount of administrative fines to be imposed on offenders who have violated the environmental law by quantifying and monetizing the damage to the environment or to impose equivalent recovery measures to the agent that causes significant damage to soil, water, habitats and protected species; and
- contribute to the greening of the economy.

WITH A LOT OF WORK AHEAD

Although the ES concept itself is rather new in Flanders, many existing policies developed and implemented many years ago can be placed under this heading.

During the "introduction of the ES concept in Flanders," which is one of the measures included in the Environmental Policy Plan 2011-2015, LNE gained its first practical experience with this concept by outsourcing the study Economic Valuation of Ecosystem Services for SCBA [1]. This study resulted in several quantification and valuation functions and monetary values that can be used to quantify and value changes in ecosystem goods and services of (semi)-natural land use. In order to reduce the labor intensity and risks of miscalculation and misinterpretation by doing so, LNE, VITO, and UA recently provided both a manual and supporting online calculator called Nature Value Explorer that policy makers, nature and environmental organizations, research agencies, and others can freely use via www.natuurwaardeverkenner.be. The goal is to constantly update this toolkit with new scientific knowledge of ecosystem services as well as with exchanges of experiences and examples of the users' good practices.

Thanks to this tool and even more to the ever-growing expertise network on ecosystem services (e.g., the BEES community), LNE gained new insights into the qualitative, quantitative, and monetary valuation of ecosystem services and illustration of case studies. Thus, almost all parts of the introduction phase (expertise network, inventory, demonstration projects, communication, and ES-based policy instruments) started off well. Only the development of ES-based financing instruments in LNE remains less explored. However, by finishing the Legal Study of Ecosystem Services, LNE can hopefully facilitate this difficult process in which adverse effects need to be averted and attention must be paid to ethics and transparency.

It is important to continue ongoing research (like the ECOPLAN project) to fill the remaining knowledge gaps in the provision and valuing of ecosystem services, the variables of their supply and demand, the way to make them more spatially explicit, and so on. In this regard, the right balance between what is scientifically correct and what can be pragmatically applied in practice needs to be pursued. Only then will the relevant target groups be able to effectively apply the ES concept in their choices and decisions (policy, management, production, consumption, etc.). Once enough concrete examples have been gathered that demonstrate how to do so, others will follow and the ES concept might become broadly integrated.

Despite all the opportunities offered by the ES concept, its role in LNE's daily practice is still very minor. Up till now, it played a supportive role in determining the fine to be imposed for polluting the River Zenne. In addition, the Nature Value Explorer has been used by others in different scenarios, including estimation of the economic value of a small forest in Antwerp in order to preserve this natural area. The ES concept has already been used in several SCBAs as well, such as the one LNE did on ecologically engineered banks along navigable waterways in Flanders. Now that the Mobility and Public Works department is about to publish its recently extended Standard Methodology for SCBA of transport infrastructure in which they integrally refer to the Nature Value Explorer as the appropriate tool to take effects on nature and landscape into account, application of the ES concept in SCBA will only increase.

The challenges and threats that confront the ES concept might be one reason why it still is not used more extensively. For example, constantly improving scientific results might give the false impression that all values are taken into account, but nature's intrinsic value cannot be captured in the valuation of ecosystem services. We need to be careful that the focus on ecosystem services does not reduce ecosystems and biodiversity to economic goods that can be traded and replaced at will. Although LNE supports the economic valuation of ecosystem services, its use is not recommended in all cases; it must always be carried out in a controlled and transparent way and ideally must be part of a broader assessment like multicriteria analysis. LNE shares the view of the Agency for Nature Forests(ANB) that price-tagging nature elements can create a perception of a "license to pollute" (or to destroy), an attitude we need to prevent.

Other challenges when making policy decisions based on the ES concept are the need to have an overall vision with regard to multiple ecosystem services; increase awareness and acknowledgment of the overall values of eco-systems and biodiversity for society; and to involve stakeholders in participa-tory processes from the start in order to create more support. For this purpose, the complex ES concept and the entailed terminology must be presented in a more comprehensive language.

A supportive BEES community can help overcome the challenges and threats ES faced.

SUPPORTED BY THE BELGIAN ECOSYSTEM SERVICES COMMUNITY

Networks and forums for ES bring scientists, experts, policy makers, and stake-holders into contact with each other, so that they can exchange best practices and find a medium for dialogue and collaboration. When scientific research is followed by an overarching, and therefore neutral and interdisciplinary network, it ensures fulfillment of the practitioners' actual needs.

The BEES community can also facilitate access to all kinds of data, even internationally, and keeps track of the scientific and demonstrational devel-opments and lessons learned in Belgian, European, as well as international settings.

REFERENCES

1. Liekens I., Schaafsma M., Staes J., De Nocker L., Brouwer R., Meire P., 2009. Econo-mische waarderingsstudie van ecosysteemdiensten voor MKBA. Studie in opdracht van LNE, afdeling milieu-, natuur- en energiebeleid, VITO, 2009/RMA/R308. http://www.lne.be/themas/beleid/milieueconomie/waardering-van-baten-en-schaden/economische-waardering-van-ecosysteemdiensten-voor-mkba

Chapter 36

Obstacles to Use an Ecosystem Services Concept in Agriculture

Sylvie Danckaert and Dirk Van Gijseghem

Flemish Ministry of Agriculture, Division for Agricultural Policy Analysis

Sylvie Danckaert - Policy-advisor for the ministry of agriculture and fisheries of the Flemish government (Regional government) in the field of agro-environmental issues, urban and rural planning, greening of the CAP, etc.

Dirk Van Gijseghem Policy-advisor for the ministry of agriculture and fisheries of the Flemish government (Regional government) in the field of agro-environmental issues, urban and rural planning, greening of the CAP, etc.

Ecosystem Services. http://dx.doi.org/10.1016/B978-0-12-419964-4.00036-6

Sylvie Danckaert and Dirk Van Gijseghem

The concept of ecosystem services (ES) is being used increasingly within the Flemish Ministry of Agriculture and Fisheries. Different studies introduced the concept within the Ministry:

- Green and Blue Services in Flemish agriculture and horticulture [1], where blue services are regarded as an operationalization of the ES concept.
- Agro-biodiversity. "The Pillar of the Third Generation of Agri-environmental Measures" [2], where ES was used to evaluate current and possibly new agri-environmental measures.

The Ministry of Agriculture and Fisheries uses the ES concept in agricultural education for farmers. The agri-environmental measures and other policy measures (e.g., support for organic agriculture, demonstration projects, short supply chains) are regarded as an operationalization of the concept within the agricultural policy domain.

The ES concept shows that services play an important role in agriculture. Agriculture needs regulation (like pollination, natural pest control, water and atmospheric regulation) and supporting services (like soil structure, nutrient cycling, and genetic biodiversity) to produce the provisioning services (food, feed, fiber, fuel) and other cultural services such as an esthetic landscape. Using those regulating and supporting services correctly—thereby minimizing disservices such as habitat loss, nutrient runoff, pesticide use—comes close to the concept of sustainable agriculture. The ES concept makes nonproductive services more visible.

The concept might play an important role in developing new agri-environmental measures and in advising activities. Ecosystem services can provide a more holistic and integral approach. When evaluating a bundle of ecosystem services, it might be possible to develop new, more multifunctional measures that have a more positive effect on the environment. It might also play an important role in advising activities.

There are still obstacles to use an ES approach in agriculture. First, there are application barriers. There is need for education of the farmers and for pioneering. Other implication barriers are legal certainty and the possibility of integration into farming practices.

Second, there are some gaps in knowledge: What are win-win situations? What is the advantage of multifunctional land use? What is the value of functional agro-biodiversity? Which services are provided by which agricultural sector? Who gets the benefits for the services provided by agriculture (e.g., blossoms in the fruit region—benefits go to the local hotels, restaurants, and cafes)? What is the value of cultural and regulating ecosystem services of different types of agriculture for the society? Which type of agricultural landscape is more valued than others?

Third, there is the problem of trade-offs between different ecosystem services. The relation between biodiversity and ecosystem services is not one on one. In case of any trade-off, how does society choose?

Fourth, there is the problem of the reference level. Is there a clear line between legal obligation for a farmer, what society expects from farmers, and what farmers can do voluntarily? Where is the line between remedying agricultural disservices and delivering a service? How do the costs incurred by farmers to deliver services remedy disservices relate to the benefits for society and the benefits for the farmer? To what extent is the delivery of ecosystem services linked to the viability of the farming sector?

All of these questions lead to the conclusion that the ecosystem services approach to agriculture also entails some risks and threats. The concept of ecosystem services is a difficult one to communicate to farmers and policy makers. There is also fear of an increasing reference level. It is our hope that an active BEES society can help answer some of the questions posed in this article and help us to communicate the concept clearly to a nonscientific public.

REFERENCES

1. Danckaert S., Van Gijseghem D. & Bas L. (2011) *Groene en blauwe diensten in Vlaanderen. Praktijkervaringen,* Departement Landbouw en Visserij, afdeling Monitoring en Studie, Brussel.
2. D'Haene K., Laurijssens G., Van Gils B., De Blust G. & Turkelboom F. (2010) *Agrobiodiversiteit. Een steunpilaar voor de 3de generatie agromilieumaatregelen?* Rapport van het Instituut voor Natuur- en Bosonderzoek (INBO) i.s.m. het Instituut voor Landbouw- en Visserijonderzoek (ILVO). In opdracht van het Departement Landbouw en Visserij, afdeling Monitoring en Studie. INBO.R.2010.38

Chapter 37

The Concept of Ecosystem Services

Leen Franchois

Research Department, Boerenbond

Leen Franchois - Advisor in Nature and Water policies at the research department of Boerenbond. She follows the implementation of various directives, policies and regulations related to water and biodiversity. Boerenbond is the largest Flemish agricultural organization with about 16 000 members from all agricultural sectors. Boerenbond defends the rights of its members, gives advice at farmers and organizes training sessions and information meetings. Boerenbond participates in various projects, from research over technical innovations to sustainable initiatives.

Ecosystem services (ES) play an important role in agriculture. Agricultural systems produce provisioning services. They also provide and consume a range of other ecosystem services, including regulating and cultural services. Examples are carbon sequestration, water retention, and agricultural landscapes. Despite the important role of ecosystem services, the concept is rather theoretical for Boerenbond and his farmers. On the field, ecosystem services are produced or used without knowledge of the concept.

THE ROLE OF ECOSYSTEM SERVICES

Several Boerenbond projects investigate the possibilities of delivering ecosystem services to farmers. ECO², the most well-known project, is a collaborative approach in which farmer groups implement agri-nature preservation

Ecosystem Services. http://dx.doi.org/10.1016/B978-0-12-419964-4.00037-8

and management in order to create economic benefits and environmental sustainability. We create farmer groups participating in site-specific landscape, nature, and environmental management on an economic basis by taking advantage of system innovation. Farmers receive support in knowledge sharing and capacity building from agri-environmental experts from an agro-management center. One of the most important aspects of this program is the fact that the farmers themselves take the initiative.

ECO² projects are diverse, including the following:

- water conservation: by placing little dams, water levels can be controlled in way that creates advantages for agriculture and environment;
- water quality: creation of buffer strips by a group of farmers;
- pollination and natural pest control: creation of flower-rich field margins;
- maintenance of landscape elements such as hedges and sunken roads.

BOTTLENECKS AND CHALLENGES

The Flemish Situation

The Flemish situation is very specific: 34% of Flanders is urbanized, compared to 11% for the European average. This fact adds pressure on land, leading to very high prices for Flemish agricultural land. It is mainly for this reason that Flemish agriculture is intensive and that it is more difficult than in other regions to find win-win situations that combine economic efficiency and ecological value. Agri-environmental measures are relatively successful (10% of agricultural land is under one or more measures), but the success is limited to those measures that easily fit in the usual agricultural business and do not have a big impact on land use (because of the pressure on land). One of the principles behind the proposed greening of the Common Agriculture Policy (CAP) is "public money for public goods." Spending public money should be efficient and effective; that is, good results must be achieved with a minimum of investment. One criticism of the greening proposals is that they are so general for every European farmer that it cannot easily be efficient and effective. Is there a way the ecosystem concept can make a good trade-off between ES food production, which generates concrete economic value for the farmer, and environmental-related services, whose value for the individual entrepreneur is not always clear and in that way make the most efficient and effective choice?

Economic Valuation of Ecosystem Services

As mentioned earlier, farmers provide not only productive ecosystem services but also ecological ecosystem services and some cultural ecosystem services as well. Some of the services are the consequence of agricultural exploitation, whereas others ask for more effort. In our project ECO² seeks to stimulate groups of farmers to take the initiative related to biodiversity, landscape, and other environmental initiatives. In practice, it is not always easy to translate this initiative into economic value for the farmers because society is not always prepared to pay for the services even though they generate an economic loss for the farmer. Can the ES concept provide a solution to this problem?

Legal Certainty

Farmers are skeptical about environmental measures, especially those related to biodiversity. If society/the government wants to stimulate farmers to take environmental measures, they should give them the necessary legal certainty that their farming activities are not being threatened. Until now that has not been the case. The "new" biodiversity in the agricultural area has a great impact on further agricultural operations in the area, mainly in the context of licensing and spatial planning. The applicable European and sectorial regulations, such as the Flemish Nature Conservation Decree and Habitat and Birds Directives, have a disproportionate effect on the realization of the main destination of the area.

Field Research

The literature already contains a lot of information about ecosystem services related to agriculture. However, there is a need for translation to the field. Here are some examples:

- Are flower-rich field margins really effective as a natural pest control?
- How can a farmer combine an economically viable agricultural activity with the protection measures for Natura 2000?

THE BEES SOCIETY

We see a role for the BEES Society to solve the problems and questions posed in this chapter. The challenge is to translate the rather scientific approach into practical applications.

Chapter 38

Ecosystem Services in Natuurpunt

Wim Van Gils

Policy Director, Natuurpunt

Wim Van Gils - Head of policy department of Natuurpunt. Natuurpunt is the largest Belgian NGO working on the protection of nature. The policy department focuses on spatial planning, water management, agriculture and of course biodiversity policy. Lobby and campaign work is done mainly on the Flemish level. At the EU level we collaborate with the Birdlife Network.

Natuurpunt is the largest Belgian NGO dedicated to the protection of nature. Its main goal is the long-term protection of important habitats, species, and landscapes. To achieve this, the organization protects nature in Flanders by buying and managing land, studying species and habitats, raising awareness, introducing educational programs for a general as well as a more specific public, and lobbying local and regional governments.

ECOSYSTEM SERVICES IN PRACTICE AT NATUURPUNT

The slogan "Nature for everybody" reflects the value Natuurpunt places in preserving the joys of nature in the scenery. Through birth forests, playing forests, barefoot routes, silent walks, etc., the organisation brings nature closer to the people (as part of its cultural services). That is the beating heart

Ecosystem Services. http://dx.doi.org/10.1016/B978-0-12-419964-4.00038-X

of Natuurpunt: members, volunteers and sympathizers attach great importance to these cultural services and are committed to experience and improve them. These services are hardly if not at all "valued" with a price: Natuurpunt does not charge users for its services; at the most it asks for a small contribution for its guides and teachers. Natuurpunt doesn't commercialise these services: that would go against the organisation's basic values. "Provisioning services (wood from our forests, hay from our fields, meat from our cattle) are not a priority for Natuurpunt. However, for financial reasons, these services received more and more attention over the past years. Nevertheless, the management of its nature reserves is never planned or executed only to increase provisioning services, but they do search actively to find ways to reach their goals and maximize these services financially. This tendency is expected to continue, given the effect on nature policy of the financial challenges that the government faces.

The "regulating services" have gained importance only the last few decades. Because of the increasing attention to flood problems and space scarcity in Flanders, water managers mainly take the lead. Theydetermine the added value of nature reserves as for water infiltration and storage by mounting multifunctional projects together with Natuurpunt. Other regulating services (pollination, air quality regulation, pest control) are less well known and do not lead to actions on the field: the beneficiaries (e.g., agriculture) do not recognize/acknowledge the value, and/or there are no fixed budgets that can be used. Regulating services also play an important role in policy decisions. (see below)

The "supporting services" are mostly unexplored terrain.

EVERY MEDAL HAS TWO SIDES: INTRINSIC AND MONEY VALUES

One of the main causes of biodiversity loss is the systematic underestimation of the value of ecosystem services (ES). That is why the link between nature and the services it delivers to society increases the (political) support for nature conservation. Everyone understands the language of money, and therefore it presents an interesting perspective for policy discussions on a larger scale. With regard to the policy, pricing the services that nature delivers "for free" increases their visibility and, as a consequence, their protection value.

Monetizing ecosystem services in Environmental Impact Assessments or Societal Cost Benefit Analysis give a better view of the pros and cons of certain policy intentions, ideally followed by choices leading to win-win situations. The best known example in our region is the Sigma plan, in which ecosystem services were decisive in the choice of a flood protection scenario. The challenge is to increase, improve, and consolidate such actions.

However, the monetary valuation of ecosystem services comes with some dangers. Ecosystem services are part of a bigger whole, characterized by irreversibility and uncertainty. In other words, an economic valuation language is too restricted. A second challenge for ecosystem services is to bridge the gap between a functional approach to nature and biodiversity that follows from the ecosystem approach, and the social reality where value is easily ignored if it is not inserted in the doing and thinking of people.

Knowing that biodiversity is economically and/or intrinsically valuable is not a guarantee for the needed support. The ES concept does not appeal to the people: The often very personal ties between groups of people and species or landscapes are much stronger than the often difficult-to-grasp ecosystem services (e.g., carbon sequestration). This fact, combined with methodological problems, makes the concept of ecosystem services on a microscale of little use to make progress for species, habitats, or landscapes.

Chapter 39

Ecosystem Services in Nature Education in the Province of West Flanders

Kris Struyf and Leo Declercq

Zwin Nature Centre, KNOKKE-HEIST

Struyf Kristiaan - Currently I'm responsable for the follow up generally of the 'content' in the New Zwin Nature Centre, which will be finished in 2015, e.g. the realisation of the permanent exhibition, educational displays inside and outside, the development and organisation of the educational departement, the cooperation with scientific institutions, etc.

Declercq Leo - At the moment providing content for the new Zwin visitor center, especially on bird migration and the links between exposition, nature guides and the general public. Organising courses in the Zwin area in order to train guides in nature, heritage, history … of the Zwin region. Working on EE for the new visitor center, also introducing issues like ecological history, ecosytem services and shifting environmental baselines.

Ecosystem Services. http://dx.doi.org/10.1016/B978-0-12-419964-4.00039-1

In 2007, the concept of ecosystem services (ES) was introduced by the Millennium Ecosystem Assessment. Anyone who cares about environmental and nature education is looking for arguments to clarify the importance of nature conservation. A question often heard is "why all this money for some animals or plants?" From practice, we know that not only the larger public, but also informed people (active in nature conservation, for example, nature guides who have an important educational role) struggle with these questions. In the trade-offs between the ecological, the social, and the economic (sustainable development), nature seems to have a difficult position in many discourses.

During the last decade, a whole range of concepts related to sustainable life were conveyed to the sector of nature and environmental education. People active in this field have the interesting, but often very difficult, task of inspiring all kinds of people to start living in a nature and environmentally friendly and sustainable manner. Newly introduced concepts such as *education for sustainable development, biodiversity, climate change,* and *ecosystem services* deal with complex issues, which makes public understanding of these issues not easy.

Concerning ecosystem Services (ES), the arguments for its use and profits, also economic, can be very powerful. This is why the nature and environmental education service of the province of West Flanders initiated a process to draw attention to ES in order to provide concrete arguments for nature guides, civil servants, and the like and to show that "the birds and the bees" are essential for the functioning of society.

Although the term *ecosystem services* can be very informative for insiders, for outsiders it is not appealing. On the contrary, *Eco* probably reminds them first of all about taxes, recycling, or, in the best case, subsidies. The word *system* sounds rather abstract, and *services* sounds like government institutions. Even if the content of the concept is interesting, still it needs an attractive name to appeal to people. Not everyone is a priori open to the concept, but this attitude often changes when ecosystem services are presented in terms of hard currency.

In the Zwin Nature Center of the Province of West Flanders, we introduced the concept of ecosystem services for the first time during the 2010 year of biodiversity. This was part of an insect exhibition in which the concept was launched in a very understandable way:

Biodiversity is more than listing species. Also the mutual relations between species and their living environment (water, air, soil) are part of it. We call an area with all of its plants and animals living an ecosystem. For example insects provide society goods and services that are invaluable. Ecosystem services may include:

- Provision of food and water,
- Regulation services such as flood control and disease control,

- Cultural services, spiritual, recreational, and cultural,
- And supportive services, such as oxygen, nitrogen, and carbon cycles, and other vulnerable conditions to sustain life on Earth.

A recent research project regarding the economics of ecosystems and biodiversity, *showed that losses in nature have a direct impact on the economy. This is often underestimated. When species disappear, it is final. This not only is disastrous for these species, but also endangers human prosperity and well-being. Many human beings depend on biodiversity, such as in agriculture, fisheries and tourism. Biodiversity and ecosystem services are the natural capital and will play a major role in future economic* strategies to assure growth and prosperity on our Earth.

(K.Struyf, 2010) .

Most concrete examples, however, were not local and could be regarded as irrelevant to a Flemish context. This is why the Province of West Flanders commissioned an ES study for the Zwin area as part of an EU project. The main question was to explore the locally relevant ecosystem services and provide the first accessible translation toward education. The project's presentation was initially very extensive and too scientific with respect to language, statistics, and presentation style. Therefore, to introduce ecosystem services to nature guides of the Zwin area, an accessible presentation was worked out, stressing locally relevant examples. One of these examples, the life story of the flatfish sole, illustrates how this species throughout its different life stages is dependent on several specific habitats within the coastal natural areas and very specific prey species (all having no significant "economic" value), before it finally ends up as a well-known local dish. The concrete and local approach proved to be successful not only for nature guides but later on also for the general public.

During a workshop of the European project *Natura People* in 2012, in which the province cooperates with UK and Dutch partners, the importance of the protected nature areas was explained to an audience of bed and breakfast and camping ground owners, museum and visitor center representatives, and the like. After this workshop, it was very clear that the salient examples raised awareness much more efficiently than impressive calculations and tables with economic values. Concrete examples relating to the daily life of policy representatives, entrepreneurs, and other stakeholders have an essential awareness-raising function, and only when their awareness is raised, will they open up to scientific data, if presented in a comprehensible and transparent manner.

In summary, valuation of nature and its services and capture of the benefits has become a very prominent issue. This new prominence has become important as the economic benefits of nature, environment, and biodiversity are hardly taken into account in public decision making, and also because their value has remained unknown. Awareness of this value will rise as more studies provide illustrations and proof.

Up till now, however, bridging the gap between theory and practice has been difficult because the ES concept (regardless of some evident examples such as pollination) is hard to convey to the larger public. Examples from their own contexts are often lacking, and calculations of the economic values are complex and not transparent, even for some policy makers and certainly for the general public. Therefore, much work is still needed to provide this translation. In the case of the Zwin area,"locally embedded," "close to daily life," and "concrete" made the difference.

Chapter 40

Integrating the Concept of Ecosystem Services in the Province of Antwerp: The Inland Dunes Project

Lieve Janssens

Province of Antwerp, Department of Environment, Nature & Landscape Team

Lieve Janssens - Advisor, Province of Antwerp - Department of Environment - Nature & Landscape Team. Project Manager Inland Dunes project, (regional) landscape management. Project Manager Interreg IVB-project Green Infrastructure For Tomorrow-Together! Project Manager (regional) Landscape Management Plans. Advisor landscape policy in general and more specific contributing to processes on (regional) landscape development (Ecoduct - Ecoducten, Antwerp fortress belt - Antwerpse Fortengordel, Lake District - Kempense Meren, ...)

INTRODUCTION

Until recently, the Province of Antwerp did not explicitly apply the concept of ecosystem services (ES) in its practice. With an ecosystem services-based process on landscape planning in the area of the inland dunes, this will be tried for a first time. We will first introduce the main sources of inspiration for applying an ES-based approach. Next we will introduce the project and the context in which it came about. Finally, we will briefly focus on the expected added

Ecosystem Services. http://dx.doi.org/10.1016/B978-0-12-419964-4.00040-8

value of the concept for the Province of Antwerp and the Belgium Ecosystem Services community.

THE INLAND DUNES PROJECT AND ITS INSPIRATION

Three projects can be mentioned as inspiration to the ES-based process on landscape planning in the area of the inland dunes. (1) The Interreg IVB project Green Infrastructure for Tomorrow-Together! (2) The Wijers project of the Flemish Land Agency. (3) The Belgium Ecosystem Services (BEES) project that formed the basis of this book and of the BEES community. For more information on the Wijers project, see Chapter 34. For more information on the BEES-community, we refer the reader to the introductory chapter of this book. Here we will focus on the Interreg IVB project Green Infrastructure for Tomorrow-Together! which was an important source of inspiration for the Province of Antwerp.

GREEN INFRASTRUCTURE FOR TOMORROW-TOGETHER!

The Green Infrastructure for Tomorrow-Together! project is a three-year GIFT-T! project (www.gift-t.eu) involving seven partners from three countries: Great Britain, The Netherlands, and Belgium. In five case study areas, a prototype method for planning green infrastructure based on the ecosystem services approach is developed. The inland dunes project is a case study within this project.

The project aims to implement the concept of ecosystem services into everyday policy making (bringing it from the study shelf to the planning table). Key features of the methodology developed in this project are:

1. Goal finding and long-term ambition: collecting stakeholders' demands for ecosystem services—building up a shared vision and a partnership of "shareholders"
2. Diagnosis and design of spatial alternatives: green infrastructure and ecosystem services analysis and mapping—cost-benefit analysis
3. GI business plans and business cases—search for (new) funding

THE INLAND DUNES AND ITS CHALLENGES

The inland dunes project is situated on the territory of the municipalities of Balen, Mol, Meerhout, and Geel. The identity of the area is closely linked with the inland dunes and the watercourses of Molse Nete and Grote Nete (special protected areas of European importance) embedded in a dense landscape with lots of small-scape landscape elements such as tree rows, and hedges and cultural heritage elements such as windmills, watermills, chapels, and homesteads.

Currently, the inland dunes are mostly covered with 50- to 70-year-old pine trees. Choices are to be made for the future land use and management of the inland dunes. Eighty % of the woods are in private ownership. The forest owner's organization, Bosgroep Zuiderkempen, requested a policy framework on forest management.

FIGURE 40-1

This bottom-up demand anticipated the Province of Antwerp's policy on landscape and green infrastructure, inspired by the European Landscape Convention. The Province of Antwerp's targets are the enhancement of the environmental quality and the community-based development of integrated visions on a multifunctional use of green infrastructure on a regional scale.

Project Partners

The project was initiated and conducted by a partnership within the government of the Province of Antwerp; Department Environment, Bosgroep Zuiderkempen vzw (Forestry Group), Kempens Landschap vzw (Landscape conservation organization), Regionaal Landschap Kleine en Grote Nete vzw (Regional Landscape), Rurant vzw (Countryside organization), and Toerisme Provincie Antwerpen vzw (Tourism Organisation).[1]

This partnership will be supported by policy makers, companies, stakeholder organizations, and citizens. In a process of co-creation, they will be developing a shared vision on the inland dunes' ecosystem services, responding to the needs of all involved in the process. Simultaneously, a strong partnership will be built up.

The project will be developed in collaboration with and supported by experts in ecosystem services, ecology, participatory processes, business planning, and economics.

The Project Approach

The challenge for the inland dunes area is the development and implementation of a future scenario for the inland dunes and surroundings. A scenario for the preservation and quality enhancement of the inland dunes and sustainable

---

1. "Province of Antwerp" further in the document refers to this partnership.

multifunctional use of the inland dunes. A scenario optimizing the ecosystem services needed and creating value out of the ecosystems delivered by the inland dunes.

The scenario building will be based on an analytical, deliberative process: a combination of top-down expert assessment and bottom-up stakeholder deliberation. All relevant stakeholders will be involved. In this process the Province of Antwerp will be supported by the Research Institute for Nature and Forest (INBO) and the University of Antwerp.

Spatial alternatives will be designed, based on an analysis of the landscape (ecology, connectivity for species, archeology), and on the input coming from the stakeholder deliberation. The design of spatial alternatives will be underpinned by analysis and mapping on the optimization of provided ecosystem services. Inspired by the GIFT-T! project (see above), the process will include the search for funding and investment in green infrastructure. New opportunities will be searched for in new alliances between green infrastructure and local economy, cultural heritage, tourism and recreation, and win-win situations between green infrastructure and businesses. The valuation of ecosystem services will underpin these assessments.

EXPECTATIONS REGARDING AN ECOSYSTEM SERVICES-BASED APPROACH AND THE BELGIUM ECOSYSTEM SERVICES (BEES) COMMUNITY

It is still too early to present a concrete experience-based evaluation of the relevance and use of ecosystem services. Belief in the usefulness of ecosystem services for landscape planning is based largely on the inspiration from and the experiences within the GIFT-T! project; partly on contacts in this project with experts from the Flemish Land Agency involved in the Wijers project; and partly on acquaintance with several experts of the BEES community. The ES concept is believed to be useful in indicating cross-cutting themes and stimulating responsibility from all sectors. It is also believed to help disclose connections possibly leading to funding and investment in green infrastructure. Within the BEES community the Province of Antwerp hopes to further share experiences and assessment of the usefulness of the ecosystem services concept in landscape planning.

Chapter 41

Bosland: Application of the Ecosystem Services Concept in a New Style of Forest Management

Pieter Vangansbeke[1,2], Leen Gorissen[1] and Kris Verheyen[2]

[1]*Unit Transition Energy and Environment, The Flemish Institute for Technological Research (VITO)*, [2]*Forest and Nature Lab, Ghent University*

Pieter Vangansbeke is a PhD candidate at Ghent University in collaboration with and funded by the Flemish Institute for Technological Research (VITO). His research focusses on system and process innovation of forest managent in Bosland. System innovation is elaborated by introducing the ecosystem services concept in a sustainability analysis of biomass harvesting. The final research goal is to develop an operational framework for forest management supported by multiple stakeholders that simultaneously optimizes biomass harvest and other ecosystem services.

Ecosystem Services. http://dx.doi.org/10.1016/B978-0-12-419964-4.00041-X

Leen Gorissen is a transition researcher at the Flemish Institute for Technological Research (VITO) in the "Transition Energy and Environment" Unit. She graduated as biologist and holds a PhD in Ecology, Evolution and Behaviour. Her current activities at VITO are concentrated on the integrated (systemic) sustainability approach interlinking cleantech, land use, ecosystems, biodiversity, climate change and transition management, with a focus toward what society can learn from nature in respect to adaptability, redundancy and resilience (biomimicry).

Kris Verheyen is professor in Forest Ecology and Management at Ghent University. His PhD research focused on the recovery of herb layer vegetation in secondary forest (2002; KU Leuven). In October 2004, he became head of the Forest & Nature Lab. Since then, his research interests developed around the study of the impact of multiple global change drivers on (plant) communities, the research on the importance of biodiversity for ecosystem functioning and applied research aimed at the optimization of ecosystem management for ecosystem service delivery.

BOSLAND

Bosland (literally forestland) is a statutory partnership clustering several forests with a previously fragmented management in Flanders (Belgium). The project covers the area of two municipalities (Hechtel-Eksel and Overpelt) and a town (Lommel) in the northwest part of Limburg Province, all together about 22,000 ha (Figure 41-1). Currently, the project is managed by six partners: the four different owners (the two municipalities, the town, and the Agency for Forest and Nature Management of the Flemish region—Agentschap voor Natuur en Bos, ANB) and two main stakeholders (Regionaal Landschap Lage Kempen (RLLK), a local organization for landscape conservation and Tourisme Limburg (TL), a provincial organization promoting tourism).

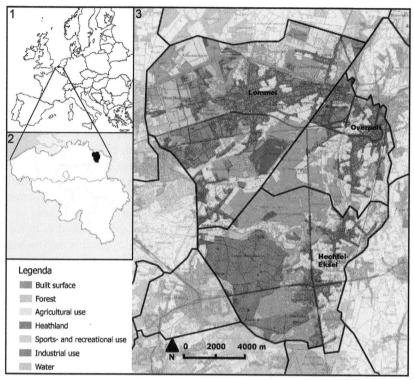

FIGURE 41-1   Situation of the project in Belgium and in Europe (1 and 2) and a land-use map of Bosland (3). The core area of Bosland consists of forests that used to be managed by the different owners and are now managed together.

Bosland lies on the border of the Campine Plateau, and most soils are characteristically sandy and nutrient-poor. Until the middle of the 19th century, Bosland was mainly covered by an extensive heathland. Gradually, afforestation with conifers took place. Approximately 17,000 ha of the Bosland area consists of rural space, containing 10,000 ha of nature and forest area. Public forests cover more than 4500 ha, with Scots pine (*Pinus sylvestris*) and Corsican pine (*Pinus nigra var. corsicana*) as the main tree species [1]. Bosland is an important area for wood production, with an annual harvest of about 25,000m³ wood. Tourism also plays a major role: no less than 1.12 million overnight stays of 279,708 different visitors were booked in the Bosland-area in 2010. Besides its role in biomass production and recreation, Bosland being one of the few large forest areas in a highly urbanized region plays a vital role in maintaining and supporting biodiversity. Different habitats and species that occur in Bosland are rare on a Flemish and even on a European scale. Protected species include European nightjar (*Caprimulgus europaeus*), smooth snake (*Coronella austriaca*), and ladybird spider (*Eresus sandaliatus*). In addition, Bosland provides several other ecosystem services that are not yet quantified and marketed. Typical examples are regulating services such as climate and local air quality regulation, pollination, and biological control.

A NEW WAY OF FOREST MANAGEMENT

Traditional land-use planning and management strategies are not well-equiped to deal with the systemic nature of the sustainability problems facing society today. In the past decade however, new research and policy fields have emerged that are explicitly based on systems thinking, complexity, and uncertainty in which the intrinsic features of the societal system such as dynamics, interdependence, and actor-networks are taken into account. In this interface between substance (management of ecosystem services) and process (shaping/guiding the transition process; also referred to as transition management), the innovation potential can be harvested fully: by means of shaping a new discourse with a higher ambition level that is co-created by the involved stakeholders to fuel legitimacy, empowerment, and local identity. It is in this light that the Bosland case is an interesting and promising frontrunner in finding new approaches to forest managenent.

To analyze the Bosland project, we used a learning history-like approach [2, 3]. We executed a literature study on all available policy documents, mainly forest management plans, reports of participatory events, envisioning studies, and the resulting master plan. Based on this material, we reconstructed the development of the project. This history was discussed with several key stakeholders in the project development in semistructured interviews. Both sources of information were finaly combined to perform a transition analysis on Bosland. Although the project was not explicitly set up as a transition experiment as such, our analysis identified many typical characteristics of transition experiments as defined in transition literature. Transition experiments typically have a socially broad context, a long-term timeframe and goal, a systemic approach, and a learning process that questions the underlying rules of the regime [4]. Moreover, our analysis showed that the Bosland case differs from the traditional forest management regime in three specific features, as follows.

1. Shifting the incumbent paradigm towards coherence
   Bosland reflects a distinctive paradigm shift from management of isolated forest patches by different owners to collaborative, more systemic management across various owners on a landscape scale (management for coherence). The triggering device for this shift was the legislation on extensive forest management plans that stimulated new forms of collaboration to lighten the workload, with a strong focus on co-creation. The traditional top-down approach was abandonded, and the involved parties are now collaborating as equals, resulting in what might be called an interactive democracy [5].
2. Connecting long-term visions to short-term actions

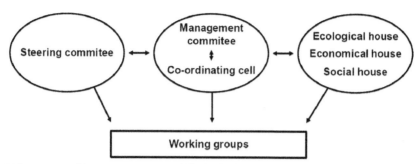

FIGURE 41-2   Management structure of the Bosland project, with the Bosland parliament as a participatory sounding board, on the right.

Bosland started off with the development of an overarching long term vision for the forest land; assembling all forest fragments of the entire region and underpinned by an extensive inventory of both the forest fragments and the accompanying social needs of the involved stakeholders. Because of the evolution of the Bosland project and the entry of two non-profit organizations in the partnership, the long term vision has been updated and widened to a master plan for the whole region. Based on this updated vision, supported by all partners, operational and strategic goals were determined. These strategic goals form the basis of further concrete planning and are elaborated in the master plan. In general, the short-term action agenda is based on the long-term vision.

3. Focusing on participation

The initiators of Bosland recognized that building legitimacy and empowering societal actors is essential in the transition to more sustainable forest management. During the two envisioning processes, several events took place to involve a wide diversity of stakeholders; this involvement was later anchored in "the Bosland parliament," consisting of three equal "participative houses," each of them oriented to a specific pillar of sustainable forest management: an ecological, a social, and an economical house (Figure 41-2). This innovative participatory approach allows stakeholders to be more actively involved in forest management.

APPLICATION OF THE ECOSYSTEM SERVICES CONCEPT IN BOSLAND

4. Ecosystem services in the long-term vision

One of the strategic goals for Bosland, written down in the master plan, is the sustainable management of ecosystems to deliver goods and services to society. This explication of the ecosystem services concept in the overarching long-term vision is a first step in implementation. It implies an ambition to maximize the provisioning (wood and biomass provision, but also other forest-related products) and regulating services. A point of particular interest in the vision is quantification of regulating services and communication of these values in educational infrastructure. The ecosystem services concept will therefore also be used here as a communication tool to increase local support.

Strengthening biodiversity on a species, habitat, and landscape level as a supporting service is a separate strategic goal. This goal is underpinned by an earlier elaboration of a long-term ecological vision for the separate forests and for the total area in relation to the adjacent forest and nature areas. Another separate pillar of the long-term vision for Bosland is recreation as the main cultural service. The project wants to attract local visitors and tourists by developing recreational infrastructure, specific recreational products, and a univocal touristic communication.

5. Striving for equilibrum between ecosystem services in real-life management

Translation of the long-term vision to actual management plans is one of the spearheads of the Bosland project. Forest managers use the long-term vision as a means of giving direction to short- and medium-term action. They develop trajectories leading from the current situation to the desired state (vision) and have set targets along the way to implement these targets in management plans on a shorter term. The management plan of each larger part of Bosland aims for a balanced four-leaf clover of supporting, regulating, provisioning and cultural services.

Applying this multifunctional forest management sometimes leads to tensions between the different stakeholders. To unite, for example, a wood harvest operation with the several forms of recreation (hiking, mountain biking, horseback riding, among others) with the supporting function for biodiversity sometimes seems an impossible mission. However, with a smart prioritization and zonation within the larger Bosland parts, forest managers do succeed in combining services at a landscape level. To track down and to resolve the existing frustrations of certain stakeholders, meetings of the Bosland parliament are organized and stakeholders are brought together in discussion walks whenever a larger management choice is taken. This increases mutual understanding between stakeholders, builds reflective capacity, and increases the appreciation for the different tasks of the forest managers who often operate as mediators.

6. Research on ecosystem services in Bosland

As mentioned earlier in this section, the quantification and valuation of ecosystem services present a fundamental challenge, and so several collaborations with scientific institutes have been set up.

A long-term projection for wood provision has been developed in collaboration with K. U. Leuven. This unique project combines actual data of standing stock with growth data based on a field survey and a soil map to predict future harvest volumes and standing stocks under different management scenarios for the next 60 years. These scenarios were analyzed with a multicriteria analysis and brought back to a stakeholder group, uniting forest managers and wood purchasers who decided on the final scenario that is currently being used as a basis for the forest management plans [6].

In 2013, the FORBIO-site was set up in the heart of Bosland. On this experimental site of eight ha, the main tree species for wood production of Bosland have been planted in different plots. These plots have four different grades of tree species diversity, varying from monocultures up to four species mixtures. The differences between ecosystem service delivery of monocultures and mixed plots are monitored throughout the stand development [7]. The results can be used in future forest management in and outside Bosland. The FORBIO experiment has two other sites in Belgium and is part of TreeDivNet, the largest biodiversity experiment in the world, which groups similar experiments around the globe.

A rising demand for biomass, mainly for renewable energy production purposes, could lead to higher harvesting rates in Bosland. This could impact other ecosystem services, such as maintenance of soil fertility, carbon sequestration and storage, habitats for species, recreation, and tourism among others. These different ecosystem services are currently quantified, and possible trade-offs and synergies are determined in a study by Ghent University and the Flemish Institute for Technological Research (VITO) [8].

Bosland also hosts top-level biodiversity research. Tracing Nature vzw tags European nightjars with a radio transmitter to determine habitat preferences and social patterns. Use of this device has led to unique insights into conservation of the species and the related habitat. Based on this research, forest managers are already applying measures to increase the habitat of the nightjars [9].

CONCLUSION

To evolve toward ES-based and actor-supported forest management, a new style of forest management needs to be adopted. The Bosland project has introduced several necessary innovations in this respect. A strong focus on (i) coherence on a landscape scale, (ii) long-term envisioning, and (iii) stakeholder participation makes the introduction of the ecosystem services concept in forest management possible. Today the concept is in use on different scales in Bosland: (i) as a leitmotiv to express the long-term vision and as a communication tool to increase local support; (ii) as a target for setting up multifunctional forest management plans; and (iii) as a guideline for the development of targeted research programs. Bosland can thus be considered a frontrunner that is exploring a new style of forest management and new roles: based on co-creation with and co-ownership of stakeholders and a focus on a more systemic, ecosystem services-based approach.

REFERENCES

1. Coördinatiecel Bosland, Master plan Bosland—challenges for the future (in Dutch). 1–53. 2012. Hechtel-Eksel, Drukkerij Grafico. Available online at http://www.bosland.be/over-bosland/publicaties/19/masterplan-bosland-uitdagingen-voor-de-toekomst.
2. Kleiner, A. and G. Roth. 1996. *Field manual for a learning historian*. MIT, Center for Organizational Learning and Reflection Learning Associates, Boston.
3. Vangansbeke, P., L. Gorissen, F. Nevens, and K. Verheyen. (no date). Transition towards co-ownership in forest management: Bosland (Flanders, Belgium) as a frontrunner. Technological Forecasting and Social Change. Unpublished work.
4. Raven, R., S. van den Bosch, G. Fonk, J. Andringa, and R. Weterings. 2008. Competency kit for transition experiments (in Dutch). Competentiecentrum Transities. 1–145. Utrecht, AgentschapNL.
5. Edelenbos, J., and R. Monnikenhof. 2001. Local interactive policymaking (in Dutch). Uitgeverij Lemma bv, Utrecht.
6. Moonen, P., V. Kint, G. Deckmyn, and B. Muys. 2011. Scientific support of a long term planning for wood production in Bosland (in Dutch), 1–79. K.U.Leuven, Leuven.
7. Verheyen, K., K. Ceunen, E. Ampoorter, L. Baeten, B. Bosman, E. Branquart, M. Carnol, H. De Wandeler, J.C. Grégoire, P. Lhoir, B. Muys, N.N. Setiawan, M. Vanhellemont, and Q. Ponette. 2013. Assessment of the functional role of tree diversity: The multi-site FORBIO experiment. *Plant Ecology and Evolution* 146(1), 26–35.
8. Vangansbeke, P., L. Gorissen, and K. Verheyen. (no date). Impact of increased wood and biomass harvest on soil fertility and carbon stocks in Bosland. Unpublished work.
9. Evens, R., E. Ulenaers, and D. Gorissen. 2013. Scientific report on the European nightjar (*Caprimulgus europaeus*).

This book is an outcome of the *BElgium Ecosystem Services* (BEES)-project (funded by the Belgian federal science policy BELSPO) and the *BEES-community* that resulted from it *(www.beescommunity.be)*. The BEES-community is supported by a secretariat facilitated by the *Belgian Biodiversity Platform*. In 2011 the BEES project promoters decided to start with this book project as to reap the fruits of discussions throughout a series of workshops, in which a large group of Belgian and invited international experts took part. Gradually this project got a more international flavour, as international peers became involved in reviewing chapters, joined as co-authors, or were invited to contribute with short reflections on ecosystem services science and practice.

BEES project promoters: Patrick Meire (University of Antwerp), Leo De Nocker (The Flemish Institute for Technological Research VITO), Francis Turkelboom (Research Institute for Nature and Forest INBO), Luc De Meester (KULeuven), Edwin Zaccai (Université Libre de Bruxelles ULB), Alain Peeters (Natural Resources Human Environment and Agronomy RHEA), Hendrik Segers (Royal Belgian Institute of Natural Sciences RBINS).

BEES community coordinators & editors: Sander Jacobs, Nicolas Dendoncker & Hans Keune

Authors: Aertsens Joris, Bastiaensen Johan, Bauler Tom, Baveye Philippe C., Beauchard Olivier, Braat Leon, Brils Jos, Broekx Steven, Brosens Dimitri, Caesar Jim, Cerulus Tanya, Chevalier Cédric, Cornell Sarah, Danckaert Sylvie, De Bie Tom, De Nocker Leo, De Vreese Rik, Declercq Leo, Deliège Glen, Dendoncker Nicolas, Dhondt Rob, Dufrêne Marc, Duke Guy, Flandroy Lucette, Fontaine Corentin, Franchois Leen, Ghys Frédéric, Goethals Peter, Gomez-Baggethun Erik, Gorissen Leen, Gowdy John, Gryseels Machteld, Haest Birgen, Hermy Martin, Hertenweg Kelly, Heyrman Hilde, Huybrechs Frederic, Jacobs Sander, Janssens Lieve, Keune Hans, Kretsch Conor, Landuyt Dries, Liekens Inge, Markandya Anil, Martens Pim, Meire Patrick, Meiresonne Linda, Moolenaar Simon, Panis Jeroen, Peeters Alain, Pipart Nathalie, Prieur-Richard Anne-Hélène, Raes Leander, Raquez Perrine, Saad Layla, Schaafsma Marije, Schneiders Anik, Segers Hendrik, Sharman Martin, Simoens Ilse, Smeets Nele, Spangenberg Joachim, Staes Jan, Stevens Maarten, Struyf Kris, Sukdhev Pavan, Teller Anne, Thoonen Marijke, Turkelboom Francis, Ulenaers Paula, Van der Biest Katrien, Van Gaever Saskia, Van Gijseghem Dirk, van Gils Wim, Van Reeth Wouter, Vangansbeke Pieter, VanHecken Gert, Verboven Jan, Verheyen Kris, Villa Ferdinando, Vrebos Dirk, Wallens Sabine, Wittmer Heidi.

Reviewers: Corbera Esteve (chapter 21), Giampietro Mario (chapter 15), Gomez-Baggethun Eric (chapter 1), Hauck Jennifer (chapter 12), Maes Joachim (chapter 3), Markandya Anil (chapter 2), Morse-jones Sian (chapter 11), Potschin Marion (chapter 18), Prieur-Richard Anne-Hélène (chapter 16), Van Huylenbroeck Guido (chapter 22), Villa Ferdinando (chapter 14), Wittmer Heidi (chapter 5, 13), Zaccai Edwin (chapter 17)

Note: Page numbers followed by "f" denote figures; "t" tables; "b" boxes.

Printed and bound by CPI Group (UK) Ltd, Croydon, CR0 4YY

03/10/2024

01040420-0007